Principles and Practice in Mining Engineering

Principles and Practice in Mining Engineering is an up-to-date introduction to the scientific principles and technological practices of mining engineering. This book introduces the processes involved in surface and underground mining, and covers many topical issues common to mining engineering practices, including mining and quarrying methods, environmental protection measures, finance and investment, policy and mining education. Recent technology and innovations (technovations) in the mining and mineral industry, including digital mines, IoT/IIoT, AI, and machine learning, are also discussed.

Seven case studies of mines and mining operation from different parts of the globe are included to demonstrate how various minerals, including lithium, potash, copper, gold, uranium, and coal, are extracted. These case studies are written by experienced industry professionals working for reputable companies. Suggested readings, references, websites, and conversion tables for mining engineering applications are included at the end of the book for the reader's reference.

Principles and Practice in Mining Engineering gives practical, real-world knowledge to the mining workforce engaged in the mining and minerals industry globally. This book is also aimed at students, scientists, academics, NGOs, and professionals just entering the mining industry.

Principles and Practice in Mining Engineering

Abhay Kumar Soni, Ishwardas L. Muthreja, and Rajendra R. Yerpude

CRC Press
Taylor & Francis Group
Boca Raton   London   New York   Leiden

CRC Press is an imprint of the
Taylor & Francis Group, an **informa** business

A BALKEMA BOOK

Front cover: Abhay Kumar Soni, Back cover: shutterstock

First published 2024
by CRC Press/Balkema
4 Park Square, Milton Park, Abingdon, Oxon, OX14 4RN

and by CRC Press/Balkema
2385 NW Executive Center Drive, Suite 320, Boca Raton FL 33431

CRC Press/Balkema is an imprint of the Taylor & Francis Group, an informa business

© 2024 Abhay Kumar Soni, Ishwardas L. Muthreja and Rajendra R. Yerpude

British Library Cataloguing-in-Publication Data
A catalogue record for this book is available from the British Library

ISBN: 9781032228266 (hbk)
ISBN: 9781032228198 (pbk)
ISBN: 9781003274346 (ebk)

DOI: 10.1201/9781003274346

Typeset in Times New Roman
by codeMantra

Contents

Preface

An engineer is a person who applies the basic knowledge of science and engineering for the benefit of society and industry. Most mining engineers realize that their field is dominated by practicals. Listening and learning several things about the industry to which they are professionally engaged and answerable require field experience quite often to understand the technicalities involved. In this regard, practices and principles, applicable to mine (s), as depicted in this book, are written to leave concise information to the readers who need to know about the overall mining, from a theoretical and practical perspective(s).

The mineral sector, regardless of whether it is engaged in exploration or exploitation, mineral processing, or metal marketing, is going through a paradigm shift from back-bencher to front-runner in this digital era of information and communication. The production/excavation of coal or metallic ore (mining) especially in mineral-rich countries is significant because it contributes to the national economy. This book, a text cum reference book, is an attempt to illustrate some aspects being witnessed in this growing mineral industry. With pride, we can say that our contribution as this book's author is to help and explain to the young budding engineers and concerned industry professionals the role and linkages in key areas of mining such as production, system, safety, and technical innovations. Within limited pages, it was not possible to cover all the topics of each subject in their entirety, but our effort is to make things clear so that further steps will be taken by the engineers involved.

Chapter 1 deals with introductory mining engineering issues required for an engineer. Rudiments of mining, finances, and taxes concerning the mines and the mining operation have been professionally covered in this chapter. *Rock engineering* (rock mechanics) and the *Mine environment* along with *Mine climate engineering* are those subjects that deal with most of the technical issues about the mines ranging from support to the impacts of mining on various activities and are very widely covered in the literature by several researchers, academician, and authors. Hence, we have only touched them very superficially. Mineral policy, organizational policy, industry–institution interaction, mineral inventory, and the role of industry and academia have been described.

Chapter 2 narrates production and productivity with a focus on mines and ancillary mining operations, which involve so many engineering aspects. More production and better productivity are essential parts to achieve the mine targets. The name, fame, and overall environmental health of a mine are linked to the production and productivity quite intimately. To make headway on the path of progress, production must be achieved with a minimum industrial disturbance which could be a mine or a processing plant for metal extraction.

Chapter 3 deals with *Mine System Engineering*, a comparatively new and emerging subject the professionals dealing with mine areas. This chapter is equally applicable to a beginner in mining engineering and international readers of the mining subject. Indeed, industrial growth is necessary, which is possible with new and updated system engineering knowledge.

Chapter 4 deals with safety engineering and its different essential aspects concerning mines and mining engineering. Accidents, risks, fundamental knowledge of safety performance, accident prevention methods, safety education and training, employee participation in safety, human factors, and related issues are all the parts of safety which are required to be understood by an engineer who is devoting his/her career to the mining industry. Implications of safety at the workplace or in various working environments are significant as well as important. Hence, awareness of various methods of safety management is critical to managing productivity and operational effectiveness in almost all mining organizations, including captive mines of a plant producing a commodity for public use. Those who are concerned about managing safety by initiating appropriate measures will benefit from this chapter.

All over the world, the mineral industry is a shining industry with technovations (Technology + Innovations). For both, the economy as well as industrial growth, mines, mining operation, and mineral processing, are equally important for the growing industry. Chapter 5 covers vital aspects that include the economy as well as industrial growth for the entire mining industry in very limited pages concisely.

Chapter 6 is one of the most important aspects of mining engineering describing the case records from different parts of the globe to know how minerals are extracted all over the world. Some of these case studies are contributed by experienced field professionals who are directly engaged in respective mines in a responsible capacity. We have tried to cover: How potash is extracted in Canada (Solution mining in Canada); How lithium is extracted in Bolivia (A Latin America case study of salt flats from Bolivia); How mining is practised in Poland (Underground Longwall Mining of Coal); Opencast coal mining from India by mechanized means; How copper ore is extracted in Brazil (Salobo Copper Mine, Brazil); How mining of strategic mineral (uranium ore) is mined/extracted; and what are the methods of gold extraction in America. Our contributors are as follows:

  i. Zbigniew Burtan and Piotr Małkowski (for Longwall coal mining in Poland)
 ii. Ajay Ghade and K.K. Rao, UCIL, India (for the case study of strategic mineral – uranium)
iii. Arun Kumar Rai, Utah, USA (gold mining in the USA)
 iv. T.N. Suryavanshi (for case record of coal mine closure in India)
  v. Neil Burns, Wheaton Precious Metals Corp., Vancouver, BC, Canada (for review and technical assistance of copper mining in Brazil)

Chapter 7 describes the training and skill development necessary for mining engineering. Virtual Mine for field training, Skill development, Fire and rescue training, First aid for mines, and Gas testing certification are some practical aspects and tools needed at the mines. This add-on is very useful in mining engineering. Concerned websites related to the mining sector, MDOs, i.e., the list of the mining companies and involved organizations, mining software for technical solutions, and some conversion factors to change from one unit to another, useful for mining engineering applications, have been given in this chapter. Such knowledge is required for handling the mining equipment, machinery, and instruments. Suggested readings and references given in this chapter are helpful for gaining additional knowledge.

In this book, we have not demarcated any line between coal and non-coal minerals, though both are mined differently. A coal mining face is considered best when produces more, whereas a stope (an actual site in an orebody for metallic ore production) requires more safety to be productive. Industry studies clearly show that the cost of production, worker safety, minor/major accidents, and technology adoption all affect the mineral production from mines, e.g., Vertical Crater Retreat (VCR) method is one of the latest methods of mining operation in metal mines and the Longwall/Room and Pillar methods of coal mining are widely practised methods. To mandate low- and lean-grade ore exploration and its beneficiation, one should ensure the utilization of the resources as well. Depending on the encountered field conditions in mines and the mining leaseholder's capabilities, the principles and practices may be applied.

Before the readers start reading the main text, I would also like to share the book's origin story. While I was engaged in the research, I realized that the things I learned should be shared and not go as a waste of knowledge in due course of time when I step down or not remain active. Both of my co-authors, who are polished academicians with long experiences, also have a similar realization. With this thought in mind, we all collaborated and contributed to the book's conception, research, and writing. Another problem that surfaced while writing is the practical utility of the book and the vast amount of already available literature on various subjects and topics. The kind of material that should and shouldn't be included in the book was a big problem. Researching in more detail, we came up with the content as given in this book in its present form. Our contributors were greatly helpful in our book's creation, and I acknowledge all of them in a separate acknowledgements section. We are sure and confident that the young budding engineers and professionals of the mining, geology, and mineral engineering disciplines will be benefited by this book irrespective of their nationality.

Abhay Kumar Soni
Ishwardas L. Muthreja
Rajendra R. Yerpude

About the authors

Abhay Kumar Soni graduated in Mining Engineering from Ravishankar University, Raipur, Chhattisgarh in 1983. He completed his postgraduate degree from Birla Institute of Technology and Science (BITS), Pilani and his Ph.D. in Environmental Science and Engineering from Centre of Mining Environment, Indian School of Mines (ISM), Dhanbad in 1998. He was awarded the William Greenwood Scholarship of Association of Geo-Scientists for International Development (AGID), University of Sao Paulo, Brazil. Dr. A.K. Soni is Chief Scientist, retired as Scientist-in-Charge, and Head of the CSIR – Central Institute of Mining and Fuel Research (CSIR-CIMFR) Nagpur office and engaged in research in mine environment and allied areas. His area of interest for research is 'Geo-hydrological problems related to mines'. In the past, he was associated with the development of eco-friendly techniques of mineral extraction and has worked for R&D in the area of 'Environmental management in fragile/sensitive areas with particular reference to mining operations and hill type areas (Himalaya)'. He developed an Environmental Degradation Index (EDI) for applications in mining in ecologically fragile areas and has a special interest in policy issues on mining and the environment.

He had served Coal India Limited (at SECL subsidiary) and Department of Mining Engineering, Government Engineering College, Raipur (Chhattisgarh) before becoming scientist in October 1987 at CSIR-CIMFR. Dr. Soni has more than 30 years of experience in research, academics, and the mining field in various areas of environment and subsurface sciences. Dr. Soni has more than 12 years of administrative experience too in science and R&D administration, as Head, of CSIR-CIMFR, Nagpur Research Centre. To his credit, Dr. Soni has authored two books, namely, *Mining in the Himalayas* (published by CRC Press, Taylor & Francis, 2017) and *Limestone Mining in India* (published by Springer). He is Academic Editor for two open access books titled *Mining Techniques: Past, Present and Future* (2021) and *Sedimentary Rocks and Aquifer: New Insight* (2023) that are published by Intech Open, London. He had visited the USA and UK in connection with research work and widely travelled across the Indian peninsula.

He is actively associated with technical and professional organizations and societies, e.g., Institution of Engineers (India), Mining Engineers Association of India (MEAI), International Mine Water Association (IMWA), and Indian Society for Rock Mechanics and Tunnelling Technology (ISRMTT). He is associated with a large number of mining and tunnelling projects in India and is invited by

Indian universities to deliver lectures and conduct examinations for post-graduate students. He is also associated with the Bureau of Indian Standards (BIS) and
Nagpur University in the capacity of members in various technical committees.
To his credit, he has more than 130 technical publications on mining- and envi-
ronment-related topics in national and international journals, conference proceedings, workshops monographs, book chapters, etc. in Hindi and English and has
handled more than 100 R&D projects in the capacity of principal coordinator and
principal investigator. He has received the best R&D paper award from a nationally accredited technical society, viz. Indian Geotechnical Society (IGS), in 2003
and the best citizen award in 2006. He had delivered lectures in reputed institutes
like IIT-Roorkee, Government Engineering College (Now NIT-Raipur), and Indian
School of Mines-Dhanbad and conducted training courses, seminar, and workshop
of both international and national importance. Several important events/functions,
e.g., World Environment Day and Engineers Day, are organized by him in various
capacities.

Ishwardas L. Muthreja, Professor of Mining Engineering at Visvesvaraya National Institute of Technology (VNIT), Nagpur, is a graduate of Mining Engineering from
the Ravishankar University, Raipur (1979). He obtained his M.Tech. in Mining Engineering at the Indian School of Mines, Dhanbad and his Ph.D. from Visvesvaraya
National Institute of Technology, Nagpur. He has worked in the Department of
Mining Engineering at VNIT Nagpur since its inception in 1983. Presently he is
holding the charge of Head of the Mining Engineering Department. It is his fifth
tenure as Head of the Department. His areas of interest are mine environment,
mine planning, and slope stability. He has published more than 70 papers in various
national and international journals and has presented at national and international
conferences. Prof. Muthreja has vast experience of more than 40 years in the teaching, research, and mining industry in India. He has completed two international
collaborative research projects in the area of 'Mine Environment' with the University of Exeter, UK and Camborne School of Mines funded by The British Council.
In addition, he has coordinated several research projects funded by the Ministry of
Human Resource Development, Govt. of India; Western Coalfields Ltd., Nagpur;
MOIL, Nagpur; Northern Coalfields Ltd.; etc.

Rajendra R. Yerpude is Professor of Mining Engineering at Visvesvaraya National Institute of Technology (VNIT), Nagpur. He joined the institute in 1990 and was elevated as a professor. He holds a Ph.D. in Mining Engineering from IIT-ISM (Indian
School of Mines, Dhanbad, 2000) and has a rich teaching and research experience of
31 years. His research areas and interests include system engineering, mathematical
and statistical applications to mining engineering, modelling applications, computer
applications to mining engineering and management, simulation, computer graph-
ics, virtual reality, mine safety engineering, economics of mining engineering, mine
environment slope stability, and rock mechanics applications. He has supervised six
Ph.D. scholars awarded with the degree and two Ph.D. scholars are working. He has
one patent published and completed various research projects in mining areas.

List of contributors

Zbigniew Burtan
Department of Mining Engineering and Occupational Safety
Faculty of Civil Engineering and Resource Management
AGH University of Science and Technology
Krakow, Poland

Ajay Ghade
Former Executive Director
Uranium Corporation of India Limited (UCIL)
Jaduguda, India

Piotr Małkowski
Department of Mining Engineering and Occupational Safety
Faculty of Civil Engineering and Resource Management
AGH University of Science and Technology
Krakow, Poland

Arun Kumar Rai
Experienced Mining Professional
Utah, USA

K.K. Rao
Deputy General Manger
Uranium Corporation of India Limited (UCIL)
Jaduguda, India

T.N. Suryavanshi
Western Coalfields Limited (WCL)
Nagpur, India

Acknowledgements

The herculean task of book writing is that kind of work which takes considerable time and involvement of many. We have carried out this for more than one and a half years. In this sojourn, technical impediments were encountered and obstacles faced which were overcome with the help of all our colleagues at our respective offices, college, and mine sites in the field. We thankfully acknowledge all of them without naming all who helped us directly and indirectly and were involved in technical discussions.

We as book authors are extremely thankful for the authorities namely the Director, CSIR-CIMFR, Dhanbad and the Director, VNIT, Nagpur with whom we were associated during the manuscript preparation of this book. Indeed, organizational help was taken from time to time in some of the research work reported in this book. We feel that writing a case study by those directly involved at mine either for production or for associated ancillary activities is best because their experience matters. With our contributors roped into sharing for this book, the weightage of case studies is increased manifold, and we would like to put on record and duly acknowledge all our contributors, namely: Zbigniew Burtan and Piotr Małkowski, Poland – for Longwall coal mining in Poland; Sri Ajay Ghade and Sri K.K. Rao, UCIL, India – for the case study of uranium, a mineral of strategic importance; Sri Arun Kumar Rai, Utah, USA – for gold mining in the USA; Sri T.N. Suryavanshi, India – for case record of coal mine closure in Western Coalfield Limited (WCL), India; and Mr. Neil Burns, Wheaton Precious Metals Corp. Vancouver, BC, Canada – for review and technical assistance of copper mining in Brazil. To frame some tutorials, we appreciate and acknowledge the contribution of scientists of CSIR-CIMFR, Nagpur, namely Dr. A.K. Raina, Dr. John Loui P., and Dr. (Mrs) C.P. Verma. The artwork improvement required for the book figures is done by Ranjit K. Mandal and is thankfully acknowledged.

Thanks are also due to all our colleagues, i.e., faculty members, scientists, and all support staff, who encouraged, enlightened, and supported us in this endeavour. We would like to put on record and duly acknowledge our publication partner (T&F) and its team – Leon, Gagandeep, Lukas, Kirsty, Jahnavi – and the production department team for executing their jobs very aptly in a professional manner. At the last, we are also thankful to Mr Ganesh Pawan Kumar Agoor, Project Manager, Code Mantra for careful editing and handling just before the printing / the book production stage, thereby taking the present shape of the final book visible to all of us before our eyes.

Acknowledgements

List of abbreviations

AITUC	All India Trade Union Congress
BCCL	Bharat Coking Coal Limited (A subsidiary of CIL)
BEML	Bharat Earth Movers Limited
BGL/AGL	Below ground level/above ground level
BGML	Bharat Gold Mines Limited
BIS	Bureau of Indian Standards (The National Standards Body of India)
CCL	Central Coalfields Limited (A subsidiary of CIL)
CIL	Coal India Limited (A state-owned major coal mining corporate in India with 10 subsidiaries: A *Maharatna* coal production organization/mining company)
CIMFR	Central Institute of Mining and Fuel Research *(Formerly, Central Institute of Mining Research or CMRI)*
CMPDIL	Central Mine Planning and Design Institute Limited (A mine planning and design subsidiary of CIL)
CMRI	Central Mining Research Institute
CPCB	Central Pollution Control Board, Govt. of India
CSIR	Council of Scientific and Industrial Research, Govt. of India
DGMS	Director General of Mines Safety
ECL	Eastern Coalfields Limited (A subsidiary of CIL)
ESSEL	A Group of Companies engaged in the mining of minerals
GDP	Gross domestic product
GOI	Government of India/Govt. of India
GSI	Geological Survey of India
HCL	Hindustan Copper Limited
HEMM	Heavy earth-moving machines
HMS	Hind Mazdoor Sabha
HZL	Hindustan Zinc Limited
IBM	Indian Bureau of Mines (Ministry of Mines, Govt. of India)
IIT	Indian Institutes of Technology
INTUC	Indian National Trade Union Congress
MCL	Mahanadi Coalfields Limited (A subsidiary of CIL)
MDOs	Mine development operators
MMTC	Minerals and Metals Trading Corporation Limited
MOC	Ministry of Coal (Govt. of India)

MOEFCC	Ministry of Environment, Forest Wildlife and Climate Change (Govt. of India)
MOIL	Manganese Ore India Limited
MOM	Ministry of Mines (Govt. of India)
MT	Million tonnes
NCL	Northern Coalfields Limited (A subsidiary of CIL)
NEC	North Eastern Coalfields Limited (A subsidiary of CIL)
NGO	Nongovernmental organization
NIT	National Institutes of Technology
NMDC	National Mineral Development Corporation
RL	Reduced level (elevation concerning a datum)
PIG	Państwowy Instytut Geologiczny (*In Polish*) - The State Geological Institute of Poland
ROM	Run of mine
SAIL	Steel Authority of India Limited
SCCL	Singareni Collieries Company Limited
SECL	South Eastern Coalfields Limited (A subsidiary of CIL)
SPCB	State Pollution Control Board (A statuary govt. organisation at the state level (under CPCB))
SSM	Small-scale mines
STC	State Trading Corporation
TISCO	Tata Iron and Steel Company Limited
TM	Trademark
TPA/Tpa	Tonnes per annum
TPD/tpd	Tonnes per day
TPY/tpy	Tonnes per year
UCIL	Uranium Corporation of India Limited
VNIT	Visvesvaraya National Institute of Technology (A govt. engg. college in India)
WCL	Western Coalfields Limited (A subsidiary of CIL)
WHO	World Health Organization

Chapter 1

Mining engineering

An introduction

> Wrong thinking is the only problem in life and right knowledge is the ultimate solution
> to all our problems.
>
> (Bhagwad Gita)

The excavation of minerals from the Earth forms the core of mining engineering. It is largely focused on 'mineral', 'engineering', and 'environment', which is everybody's concern. The mining and mineral industry is large and labour-intensive. Hence, a concerted effort to explain the mineral industry and mining enterprises, mineral-rich countries, and industry globally, including the trends, has been made in the introduction part of the book. The first chapter of the *Principles and Practices in Mining Engineering* gives a concise description and abridged explanation of the whole mining industry irrespective of the country and continent. The *rudiments of mining* explain the basics of the life cycle of a mine and the impacts of mining from an environmental angle to know what are the theoretical and practical aspects of mining engineering. For those who want to practice ideas in future in the field, for real application, the 'history of mining' is essential as it enables them to know how minerals were/are excavated in the past and present.

Mining finances, investment in mining, and mineral policy are those key areas that are equally important to understand the whole industry as a commercial entity. Each stakeholder in the mining and mineral industry has their defiled role for the best results to be obtained through interaction of industry and participants, be it on the organizational matter, policy matter, or in other areas. Therefore, this first chapter is a must-read for students, beginners, and new entrants to the mining and mineral industry. However, it may be referred to by all as the ideas apply to the entire industry. Four sub-sections of the chapter, Parts I, II, III, and IV, explain all these aspects in detail.

PART I

1.1 Introduction

Minerals, petroleum (rock oil), and their products, such as metals, are the gift of nature on which human survival depends. Mines are the place where the art of digging minerals and mineral processing were available for centuries. Thus, mines, minerals, and metals are not new to human civilization. Over time and since older days, the 'art

DOI: 10.1201/9781003274346-1

of mining' is slowly refined and developed as a full-fledged 'science'. Thus, it is obvious that the mining and mineral sector is a big and grown-up industry that can satisfy human needs and greed with modern-day inputs and engineering skills. This sector, as a whole, involves a large number of operational mines of all sizes and types supported by other core and ancillary enterprises from several sectors.

Mining/excavation of different mineral types, namely, fuel (coal), metallic, non-metallic, and atomic minerals from the surface and underground mines, is being done in the industry all over the world. In yet another classification and characterization of minerals, they are differently divided as follows: coal minerals and non-coal minerals as well as major minerals, minor minerals, and industrial minerals, according to their use and application in the industry.

1.2 Mineral industry and mining enterprises

The mega mineral industry has more than one hundred solid minerals to excavate, excluding natural gas, which is also a natural gaseous mineral. Some minerals are also mined/excavated in the liquid state from the Earth, e.g. salt and potash.

The geographical distribution of minerals across the globe is not even. All countries are not endowed with mineral resources. Some countries possess a sufficient quantity of minerals of one type, but they are deficient in respect of other mineral types. Thus, the distribution and availability of minerals for economical exploitation, either for domestic consumption or for export promotion, or both, become an important factor for the mineral and mining industry. The size of the industry and enterprise drives the sector, specifically when it is mining because its contribution to the economy becomes significant and its impact on people remains noticeable.

The top 40 global mining companies engaged in mineral production by market capitalization are listed in Table 1.1. As of 2021, these companies, largely in private sectors, barring a few which are public sector enterprises like Coal India Limited (India), have their share and contribution so vast that they impact the regional economy of the area and drive the country, government, and people.

Table 1.1 Top 40 Global Mining Companies by Market Capitalization (as of 31 December 2019)

2020 rank	2019 rank	Change from 2019 rank	Company	Country	Year-end	Commodity focus
1	1	–	BHP Group Limited	Australia/UK	30 June	Diversified
2	2	–	Rio Tinto Limited	Australia/UK	31 Dec.	Diversified
3	3	–	Vale S.A.	Brazil	31 Dec.	Diversified
4	5	1▲	China Shenhua Energy Company Limited	China/Hong Kong	31 Dec.	Coal
5	6	1▲	MMC Norilsk Nickel	Russia	31 Dec.	Nickel
6	4	2▼	Glencore Plc	Switzerland	31 Dec.	Diversified
7	9	2▲	Newmont Corporation	United States	31 Dec.	Gold
8	7	1▼	Anglo American plc	UK/S. Africa	31 Dec.	Diversified
9	11	2▲	Barrick Gold Corporation	Canada	31 Dec.	Gold

(Continued)

Table 1.1 (Continued) Top 40 Global Mining Companies by Market Capitalization (as of 31 December 2019)

2020 rank	2019 rank	Change from 2019 rank	Company	Country	Year-end	Commodity focus
10	25	15▲	Fortescue Metals Group Limited	Australia	30 June	Iron Ore
11	10	1▼	Grupo México S.A.B. de C.V.	Mexico	31 Dec.	Diversified
12	13	1▲	Freeport-McMoRan Inc.	United States	31 Dec.	Copper
13	8	5▼	Coal India Limited	India	31 Mar.	Coal
14	16	2▲	Newcrest Mining Limited	Australia	30 June	Gold
15	20	5▲	Zijin Mining Group Company Limited	China/Hong Kong	31 Dec.	Diversified
16	21	5▲	Public Joint Stock Company (Polyus)	Russia	31 Dec.	Gold
17	24	7▲	Agnico Eagle Mines Limited	Canada	31 Dec.	Gold
18	12	6▼	Saudi Arabian Mining Company (Ma'aden)	Saudi Arabia	31 Dec.	Diversified
19	26	7▲	Shandong Gold Mining Co. Ltd.	China/Hong Kong	31 Dec.	Gold
20	18	2▼	China Molybdenum Co. Ltd.	China/Hong Kong	31 Dec.	Diversified
21	19	2▼	Shaanxi Coal Industry	China/Hong Kong	31 Dec.	Coal
22	New	–	Hindustan Zinc Limited	India	31 Mar.	Zinc
23	23	–	Antofagasta plc	United Kingdom	31 Dec.	Copper
24	22	2▼	ALROSA	Russia	31 Dec.	Diamond
25	14	11▼	Teck Resources Limited	Canada	31 Dec.	Diversified
26	35	9▲	AngloGold Ashanti Limited	South Africa	31 Dec.	Gold
27	33	6▲	Kirkland Lake Gold Ltd.	Canada	31 Dec.	Gold
28	15	13▼	South 32 Limited	Australia	30 June	Diversified
29	30	1▲	Sumitomo Metal Mining Company	Japan	31 Mar.	Diversified
30	29	1▼	China Coal Energy Company Limited	China/Hong Kong	31 Dec.	Coal
31	New	–	Impala Platinum Holdings Limited	South Africa	30 June	Platinum Group Metals
32	17	15▼	Mosaic Company	United States	31 Dec.	Potash
33	36	3▲	Polymetal International plc	Russia/UK	31 Dec.	Gold
34	32	2▼	First Quantum Minerals Ltd.	Canada	31 Dec.	Copper
35	31	4▼	Jiangxi Copper Company Limited	China/Hong Kong	31 Dec.	Copper
36	New	–	Sibanye Stillwater Limited	South Africa	31 Dec.	Platinum Group Metals
37	37	–	Tianqi Lithium Industries Inc.	China	31 Dec.	Lithium
38	28	10▼	Fresnillo PLC	Mexico	31 Dec.	Diversified
39	34	5▼	Yanzhou Coal Mining Company Limited	China/Hong Kong	31 Dec.	Coal
40	New	–	Kinross Gold Corporation	Canada	31 Dec.	Gold

Source: Price Waterhouse Cooper (PwC) (2020).

Mining enterprises, across the globe, are mostly capitalistic, involving very large cash flow as their input. The current trend of these enterprises these days is towards diversification of business because of social and environmental pressure and climate change issues, e.g. MOIL Ltd. (earlier Manganese Ore India Limited) – a mining company, producing manganese ore principally for steel and ferro-alloys industry – has diversified from core mining to wind power generation. Similarly, CIL – the giant coal mining company in India – is exploring avenues for import substitution of coal through planned investment and joint ventures for alternative energy sources and the fertilizer sector. We are observing the closure of brown coal mines in Germany, Poland, and other European countries. Great Britain (UK) is phasing out the power generation from coal and moving to generate power without burning coal. Slowly and steadily, the exploration companies of precious metals, base metals, speciality metals, and industrial mineral compounds are adopting cost-effective, technology-driven techniques. To develop mineral properties, the mega mineral industry and mining enterprises are seeking mine development, mining, and production of coal/ore through new digital initiatives and remote operations in new contractual practices. As per the current trend, the preferred choices have been altered significantly all over the globe in mining practices.

1.3   Mineral-rich countries and industry

A country with ample *mineral resources* at its disposal, either for present-day excavation or for future use, is categorized as a 'mineral-rich country'. Important mineral-rich countries around the globe are Canada, Australia, the USA, the UK, India, Russia, Mexico, Peru, Chile, South Africa, China, Brazil, Germany, and Poland. India is one of the developing nations that is mineral rich and also has a vibrating mining and mineral processing industry, posing several unforeseen and site-specific challenges. Thus, unique problems faced by the industry all over the world are so vivid that they are to be tackled with engineering ingenuity. In this aspect, mining engineering plays an important role.

India, South Africa, and Australia, all of which were once a part of Gondwana land (about 180 million years ago), share similar geological settings and mineral potential. Australia and South Africa have rich gold deposits. India has a thriving mineral industry (Annexure 1.1) with ample iron ore, bauxite, limestone, and coal reserves. Many other mineral types of metallic, non-metallic, and industrial categories form the mineral inventory of many mineral-rich countries. Many of the core sector industries, namely, cement, iron and steel, power and electricity, fertilizer, metallurgical (ferro-alloys), paint, and chemical industry, are dependent on mining because coal and mineral form their raw material feed. Thus, the mining sector is part and parcel of the whole industry and industrial scenario.

It is beyond doubt that the mineral sector, when rich, drives trade practices across the various parts of the globe and contributes significantly to the national economy and gross domestic product. New mineral searches (mineral exploration), governance, the productivity of mines, and compliance with regulations together with the new and changed adoption in extraction methodology at different mine stages are some of the notable challenges faced by the mining and mineral industry. It is quite apparent that the growth of industry and infrastructure build-up can gain momentum only when

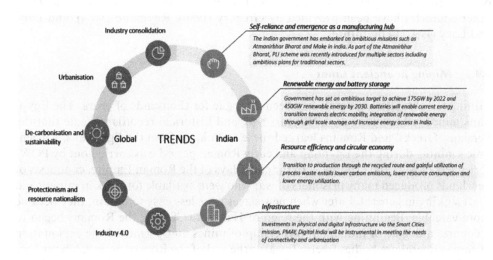

Figure 1.1 Global trends: the World and India.

these broad area challenges are met amicably without affecting resource use, environment, health, and safety.

Some of the megatrends that will shape the industrial scenario of the mining world over the next decade are depicted in Figure 1.1. All these broader areas will have a significant impact on the mineral sector globally including in India.

1.4 History of mining

The history of mining is fascinating and parallels the history of civilization in different parts of the world. A practical mining engineer should know how mining systems and societies were built in the past. How they operated and changed with time. The mining cultures, minerals uses, and technologies of the past help us to paint a detailed picture of how mining was and is today.

It is quite obvious that the *history of the mining world* is very vast and rich. Its description in the limited pages of this book is rather difficult; if done, it will defocus the central theme of the book and make the reading monotonous. Every mineral-rich country has its history according to the cultural background, ideologies, society, and governments that develop and built into the nation it is today. A brief of mining history (ancient mining and metallurgy) in general is as follows:

The history of mining has travelled through many civilizations and cultural eras, viz. stone age, dark age, iron age, bronze age, etc., and developed as a modern applied science in its present form from the art and craft of earlier periods. From prehistoric times to the present, mining has played an important role in human existence. Historical records indicate that in the earlier period, 'mining' and 'processing of minerals' were not considered separate. Bare minimum tools and infrastructure resources were used and no machinery was available. Our ancestors were digging minerals from the Earth's womb with quite an ingenuity. The mining practices in India were according to the cultural diversity of the region but different from other parts of the world. Some interesting aspects of mining and metallurgical history of other countries have been provided by Gregory (1980), Raymond (1984), and Lacy and Lacy (1992) in literature.

1.4.1 Mining in ancient times

Mining for obtaining minerals has been in vogue for thousands of years. The Egyptians mined gold as long as 4,000 years ago, and historical records indicate that the Persians, Greeks, and Romans learned these techniques from the Egyptians. Ancient time's mining during the Egyptian and early Roman period was carried out by POWs (prisoners of war) and criminals. In the early days of the Roman Empire, conquests of new lands produced many prisoners of war who were available for work in mining and shaft sinking in general. Later when new slaves were less easy to obtain, they became more valuable. Beginning with the reign of Hadrian (138 CE), the Romans began to recognize a degree of individual ownership of mines and permitted the exploitation of some mineral/ore bodies freely. Towards the end of the Roman period, labour laws were passed that mandated improved working conditions for the workers in the mines, namely, sleeping & bathing, accommodation, food, and specific hours of work. In the early days, underground mines, shafts, and tunnels were supported solely by timbers to prevent ground collapse. These techniques were grossly inadequate from a safety viewpoint endangering the lives of miners.

The fall of the Roman Empire during the latter part of the 5th century was followed by widespread political and economic chaos that persisted in Europe for more than four centuries (the dark ages). The social chaos, incessant warfare, diseases like the plague, and general economic instability from the 5th to the 11th century resulted in a marked reduction in mining. From 1100 CE to 1500 CE, the status of the miner did not change much from Roman times. The trade of mining, which included shaft sinking, became a respected profession later. During the medieval period, miners in places like Frieburg, Goslar, and Joachimstal were exempted from military service and taxation. English tin miners had the right to prospect anywhere except in churchyards, or on highways, orchards, or gardens. Such privileges and freedoms are a marked contrast to those of the miners under Egyptian or early Roman rule. *De Re Metallica* by Georgius Agricola, 1556 CE provides a fairly comprehensive knowledge of mining and shaft sinking techniques that our ancestors had.

In 1168, silver was first discovered near the town of Freiberg in Saxony. A silver rush spread across Europe in the late 12th and early 13th centuries, with strikes in Bohemia, Moravia, Hungary, the Alps, and Sardinia. To develop these mines, Saxon workers were normally brought in. One of the greatest silver mines of all time was discovered at Joachimsthal in Bohemiain in 1516. It was in this town that Agricola resided and his book on 'mining and shaft sinking' is based on techniques employed in this area. With growing stability in the 12th to the 16th centuries, shaft sinking and mining activity increased. In central Europe, the Avars, Czechs, and Saxons mined gold in Bohemia, Transylvania, and the Carpathians. This particular mining revival was led mainly by the Saxons and other German people. The period from antiquity to 1600 CE covered a huge period with many changes in civilization (Allchin and Alchin, 1982). However, from the early mining by the Egyptians, Romans, etc., through the dark ages/times and then the medieval period, the techniques for mining and sinking shafts changed a little.

Mining in its simplest form began with Palaeolithic humans some 450,000 years ago, evidenced by the flint implements that have been found with the bones of early humans from the *Old Stone Age* (Lewis and Clark, 1964). Our ancestors extracted pieces from loose masses of flint or easily accessed outcrops, and using crude methods of chipping the flint, shaped them into tools and weapons. By the *New Stone Age,* humans had progressed to underground mining in systematic openings 2–3 ft (0.6–0.9 m) in height and more than 30 ft (9 m) in depth (Stoces, 1954). However, the oldest known underground mine, a hematite mine (iron ore) at *Bomvu* Ridge, Swaziland (Gregory, 1980), is reported from the *Old Stone Age* and is believed to be about 40,000 years old. Early miners employed crude methods of ground control, ventilation, haulage, hoisting, lighting, and rock breakage. Nonetheless, mines attained depths of 800 ft (250 m) by early Egyptian times. Metallic minerals also attracted the attention of prehistoric humans. Initially, metals were used in their native form, probably obtained by washing river gravel in placer deposits. With the advent of the *bronze and iron age*, however, humans discovered smelting and learned to reduce ores into pure metals or alloys, which greatly improved their ability to use these metals. The first challenge for early miners was to break the ore and loosen it from the surrounding rock mass. Often, their crude tools made of bone, wood, and stone were no match for the harder rocks, unless the rock contained crevices or cracks that could be opened by wedging or prying. As a result, they soon devised a revolutionary technique called *fire setting*, whereby they first heated the rock to expand it and then doused it with cold water to contract and break it. This was one of the first great advances in the science of rock breakage, and it had a greater impact than any other discovery until dynamite was invented by Alfred Nobel in 1867. Mining, like all industries, languished during the *dark ages*. Several political upheavals and social developments changed the standing of mining and the status of miners from poor to improved.

1.4.2   Tools and techniques of old mining

The earliest miners sought flint[1] for tools and weapons. In ancient times, a form of the shallow shaft was commonly sunk as deep as 30 m (Stuart, 1879) in the chalk beds of northern France and southern England in the neolithic period (8000 BCE to 2000 BCE). Some Roman shafts were quite deep, e.g. the shafts at El Centenillo in Spain went down to (198 m/650 ft). Advance rates at the end of this period were probably in the range of 1–2 m per month (Stuart, 1879).

In the ancient period, only primitive tools of excavations were wedges and pick made from deer antlers and shovels made from the shoulder blades of oxen in the flint mines; in the metal mines, stone hammers, antler tools, and wooden shovels were used. The waste from the shaft sinking was hauled to the surface in leather bags or wicker baskets by one or two men. Fire setting was practised for assistance in fracturing the rock and making it easier to remove. Czechs and the Saxons used fire quenching as a method of breaking the rock. Ventilation methods were also primitive, often limited to waving a canvas at the mouth of the shaft.

The valleys of the Tigris, Euphrates, and Nile were home to the first metal-using cultures. Copper and gold were the first metals gathered in notable quantity, with copper being particularly important. A civilization using considerable amounts of copper was established in Mesopotamia by about 3500 BCE and in Egypt by about 3000 BCE.

Copper was used to make tools and weapons. From Egypt and Mesopotamia, the knowledge of metals spread across Europe. The copper-based cultures of the world were replaced by cultures using bronze by about 1500 BCE. This development led to significant improvement in the quality of weapons and tools. Iron was not successfully smelted until about 1400 BCE.

Underground mining by the Egyptians was carried out over a wide area with two places, in particular, being well known – the *Nubian Desert* in northern Sudan and the *Timna Valley* (now Israel). The Greek historian Agatharcides, in his work from around 200 BCE, gives a vivid description of mining under the Egyptians. Agatharcides wrote of fire setting and breaking of the rock with chisels, miners who wore candles on their foreheads, and 'overseers who never cease with blows'. The Egyptian miners who worked both in the mines of *Nubia* and *Timna Valley* used metal chisels and hoes and excavated very regular, circular shafts with footholds in the walls for moving up and down. Some of these shafts were over 30 m deep. Mining operations in the *Timna Valley* peaked in the 14th to 12th centuries BCE; the main period at *Timna* is now dated to the Early Iron Age, ~1,000 BCE. Egyptian miners of those days wore loincloths, perhaps headbands, and if they were a prisoner, ankle manacles. An oil lamp was used for lighting. Fire quenching was the rock-breaking method of those days, not at *Timna Valley*.

The Romans followed the Greeks as leaders of the then-excavation world. Rome explored all around the Mediterranean for mineral wealth to support its rising empire. Digging, deep vein mining, boring, and shaft sinking were recorded in Roman literature. Shafts during Roman times were square-shaped, small, and braced with wood to prevent collapse. Inclined or vertical shafts were necessary to provide access, ventilation, and a means of removal of the minerals, and these shafts were not very deep because of occasional collapses. As a whole, shaft sinking techniques under the Romans were not very different from those employed by the Egyptians earlier. Generally, hardwood or metallic tools were used to enlarge the fractures in the rock and assist in breaking it away from the rock face. Single- and double-headed hammers were used in combination with pointed bars and wedges. Besides metallic/iron tools, the Romans used fire to fracture the rock for removal.

1.4.3 Mining's contribution to civilization

The evolution of mining has paralleled human evolution and developed with the advancement of civilization and certainly one of the first industries (Gregory, 1980) and still getting matured with time. Many milestones in human history – Marco-Polo's journey to China, Vasco-de-Gama's voyages to Africa and India, Columbus's discovery of the new world, and the 'gold rushes' that led to the settlement of California, Alaska, South Africa, Australia, and the Canadian Klondike (name of a region in N-W Canada) – were achieved with minerals providing a major incentive (Rickard, 1932). The chronological development of mining through the ages with important world events is outlined in Table 1.2.

To get a useful insight into mining history either full or partial, concerning any country, readers should refer to the literature of an individual country, e.g. a suggested reading has been given for India (Soni, 2020). In brief, the history of mining is valuable and beneficial for the readers directly or indirectly, whether connected with the mining/mineral industry or not.

Table 1.2 Ancient Mining History: Chronological Development

Date	Event
450,000 BCE.	First mining (at the surface), by Palaeolithic humans for stone implements
40,000	Surface mining progresses underground, in Swaziland, Africa
30,000	Fired clay pots used in Czechoslovakia
18,000	Possible use of gold and copper in native form
5,000	Fire setting, used by Egyptians to break the rock
4,000	Early use of fabricated metals; start of Bronze Age
3,400	The first recorded mining of turquoise by Egyptians in Sinai
3,000	First use of iron implements by Egyptians; Probable first smelting of copper ore with coal by the Chinese
2,000	Earliest known gold artefacts in New World, In Peru
1,000	Steel was used by the Greeks
100 CE	Thriving Roman mining industry
122	Coal used by Romans in the present-day United Kingdom
1185	Edict by the bishop of Trent gives rights to miners
1524	The first recorded mining in New World was by Spaniards in Cuba
1550	First use of the lift pump at Joachimstal, Czechoslovakia
1556	First mining technical work, De Re Metallica, published by Georgius Agricola, Germany
1585	Discovery of iron ore in North America in North Carolina
1600	Mining commences in the eastern United States (iron, coal, lead, gold)
1627	First use of explosives in European mines in Hungary (Possible prior use in China)
1646	The first blast furnace was installed in North America, in Massachusetts
1716	First school of mines established, at Joachimstal, Czechoslovakia
1780	Beginning of Industrial Revolution; pumps were used in mines as the first modern machines
1800	Mining progresses in the United States; Gold rushes help open the west
1815	Sir Humphrey Davy invents the miner's safety lamp in England
1855	The Bessemer steel process first used, in England
1867	Dynamite invented by Nobel applied to mining
1903	First low-grade copper porphyry in Utah; Era of mechanization and mass production opened in US mining with the development of the first modern mine (an open pit), subsequent operations were underground as well
1940	The first continuous miner introduced which is an era of mining without explosives
1945	Tungsten carbide bits developed by McKenna Metals Company (now Kennametal)

Source: Hartman and Mutmnasky (2002).

1.5 Global scenario

Mines and minerals are the source of wealth that provides raw material security for infrastructure growth all over the world. Once a back-bencher industry is now taking rapid strides in terms of modernization and remote operation. The global scenario of the mining industry is fast changing on nearly every front.

All the stakeholders of the industry, right from miners to managers, mine lessee to private entrepreneurs, industrialists to financers, policymakers to educationists, including the government either at the centre or at the state, have an impeccable and integral role as well as contribution to the industry. Largely, being capital intensive, the mining sector is either in the hands of public sector companies or with the major private players on the investment fronts. The mining sector has very little to do with the public as it is not a public dealing sector. Irrespective of any country, many important economic activities of the poor or rich countries are dependent on mining. Contrary to the mining sector's potential, struggle on the environmental front, mechanization, and automation (through digitization), the global scenario and trend of the mining industry are heading from a conventional to a smart industry in the 21st century.

1.5.1 Changing trends

Modern science, technology, and engineering have pole-vaulted humans to unthinkable levels. The mining industry is not untouched by it. The global mining industry is witnessing great leaps in coal and minerals production, along with the retrenchment on account of environmental hullabaloo (e.g. Carmichael coal mine project of Adani Ltd. in Australia, Kudremukh Iron ore mining closure, Goa iron ore mining, and greenfield steel making and mining project of Korean Giant Posco in India in 2017). In terms of socio-economic development, quality of life and skills, the mining sector has more awareness to reach elevated levels on par with other industries.

To match and augment the global industrial trend, the major identifiers that remain decisive to the sound industry are *revenue, tax, and tax incentives* (Part III, Section 1.18). Therefore, the key task for the policymakers is to design fiscal regimes for the mining industry that raises sufficient revenue while providing an adequate inducement to invest. Mineral resource-rich countries are better equipped to identify it in cost potential terms. The mineral & mining sector for metal sales, pricing, and trading through the *London stock exchange* is an industry orientation globally. For many developing countries, since the major source of revenues is often none other than receipts from mining, the exchange of know-how and international industry participation from different countries keeps the changing industrial trend rejuvenated and healthy.

1.5.2 Technological advances in mining

As mining operators augment profitability by reducing operating costs, raising asset utilization, and improving safety performance, there has been a greater need for digitization within the mining sector (Sheo, 2020). With mines growing deeper, underground mining becoming safe, and new mineral reserves being discovered in remote-forest areas, it is becoming clear that we are not left with any other choices except to opt for technology that has a huge role to play. By 2030, in a short span of 10 years from now, the global mining industry will take advantage of systems that are simple, risk-free, and cost-effective, and that will upgrade and improve performance within the mine while offering easy adaptability and expansion options to meet future requirements (Paul et al., 2021).

With industry 4.0 beckoning an increasing number of next-generation technology such as sensors, automated or self-controlled equipment, and big data automation, solutions are being adopted to allow for smooth and efficient mine operation. All of this, however, requires reliable, high-bandwidth connectivity (information & communication technology in mining). In this ongoing 21st century, up-gradation of technology, advances, and innovations for better productivity through automation routes has been observed at many mine sites. MDOs have focused on many unit operations of mining (drilling, blasting, loading, transportation, etc.) for advancement. Notable among them are:

a. **Larger, more durable, and more efficient dumpers and haul trucks with autonomous operation.**
    One person alone can remotely operate a small fleet of autonomous trucks. Improvements in software (algorithm-driven computer programs) are likely to allow this to be performed even more efficiently. Driverless technology, less fuel consumption, and low maintenance costs are some special features of autonomous haul trucks and dumpers.

b. **Automated drilling machines and tunnel-boring systems**
    These are used in open-pit mining, tunnelling, and exploration activities. One operator can monitor up to five machines from a remote monitoring station. The remote operator needs only an interface with the machine to tell in what order the drill pattern should be drilled. The tunnel-boring machines significantly reduce the time, cost, and risks involved to build and expand an excavation underground, either a mine or the tunnel. These machines are likely to reduce the drilling and blasting time to half.

c. **Geographic information systems (GIS) and Global Positioning Systems (GPS)**
    GIS is now commonly used in almost all aspects of mining, from initial exploration to geological analysis, production, sustainability, and regulatory compliance. Over time, however, as the use of GIS becomes more evenly dispersed on a global scale, old procedures for mine surveying are becoming redundant. These days, the Digitalglobal positioning system (DGPS) to control *illegal mining* is common in India. The automated GPS can manage and improve the safe operation of heavy equipment such as dozers, drills, excavators, loaders, scrapers, graders, soil compactors, off-road trucks, and light vehicles.

d. **Long-distance belt conveyors and haul trains**
    Technologies are available that allow conveying of raw material in bulk by conveyors at a much steeper gradient and over long distances (long-haul trains). These are cent per cent automated from the mine site to the port and fully or partially controlled from a remote control room.

e. **Longwall plough and shearers**
    This technology is being implemented in the coal mining sector. Before automation, workers manned the longwall roof supports on hydraulic jacks called shields. Their automation and remote operation make the workplace safe and less hazardous (Case Study 6.3). Europe and its allied countries have achieved success in longwall mining compared to Asian countries. The American's innovation-led approach to sustainable mining, Future Smart Mining, and a willingness to collaborate with industry partners have enabled it to achieve a major milestone in 100% machine automation in longwall operation.

f. **Tele-remote loaders**
    Fitted with video cameras, thermal imagers, lasers, and sensors, tele-remote loaders are operated from a control room with a line-of-sight view. This type of automotive technology is useful for underground mines. The height and dimension of the loaders that include the cabin and control room are designed by the manufacturer as per the industry requirement (Figure 1.2).

Figure 1.2 Underground loaders (LHDs) fitted with the tele-remote operation.
Source: Caterpillar – R2900G, USA. https://www.cat.com/en_US/products/new/equipment/underground-hard-rock/

g. **Shovels, excavators, and semi-autonomous rock breakers**

These machines reduce the size of large-sized rocks and scoop up the ore at the actual place of mining. The powerful motor of the excavator/breaker and the larger swing angle of shovels make the digging and extraction operation more efficient and productive.

h. **Energy-efficient larger capacity crushers, grinding mills, and flotation cells for mineral processing**

The *mobile crusher* and *in-pit crusher* perform two tasks simultaneously as it transfers the crushed rock directly for processing via conveyors, eliminating the need for haul trucks within a mine.

Advances in metallurgical engineering can provide larger capacity and efficient grinding mills and flotation cells for mineral processing.

i. **Programmable logic controllers (PLCs) and control systems**

Modern-day heavy Earth-moving machines and equipment that have been in use in mines for various operations are fitted with various mechanical, electrical, and electronic systems to make them automatic systems. Such equipment has programmable logic controllers (PLCs) and control systems (CS) to make them energy efficient. PLCs are the requirement of nearly every machine.

Box 1.1: PLCs and Control Systems

Programmable logic controllers (PLCs): Flexible PLCs are digital computers that typically automate industrial electro-mechanical processes and replace relays, timers, counters, and sequencers. They are an enabling tool for improved process control. Once installed, they can be reprogrammed to improve the control of processes across the full spectrum of industry activity. This technology is the most crucial in the automation revolution and arguably the most important. It replaces the skilled and semi-skilled onsite manpower with the machine.

Control systems: These are of two types: onsite and offsite. As mines become automated, either mechanical or electrical control systems are common in use. Both control systems are used as per the site conditions encountered. Offsite rooms are comparatively bigger and more complex. Today, many mining companies with the most advanced technology have electronic control systems for their operational convenience.

j.   **Automatic monitoring**

For various types of machinery and equipment, using many different technologies, from cameras and thermal imaging to self-aware machinery, it is now possible to watch its progress and performance. Thus, monitoring automatically is extremely important these days. It is a part of the 'operation' as well as 'maintenance' required for the workforce of any mine site.

k.   **Improved and better chemistry to improve processing recoveries**

Undoubtedly, improved mineral processing and good chemistry enhance the metal recoveries from the ore. On a turn-key basis, such advanced metallurgical and chemical solutions are offered and made available to MDOs. On case-to-case basis for different mineral grades and types, these processes can be tried and implemented in the field.

The equipment and technologies described above, most of which are in use now, are indicative of technological advances in mining. Better mining productivity, automation, and remote operation for safety are the keys to these technologies at once implemented. In the long run, probably new technologies or practices such as 'deep-sea mining', 'asteroid mining', 'mining on the moon/space', and 'microbe mining' will radically change the global scenario of mining science. With uncertainty and risks involved in the excavation and mineral extraction process, it is rather difficult to predict how mining will be done in future. It is far sure that new technological advances in mining will play a major role.

1.6   Mining and mineral industry at a glance with ancillary industries and stakeholders

Ancillary industries that have close linkages with mines are power, cement, iron and steel, ferro-alloys, and fertilizer industry. Besides these, other industries, namely, building and infrastructure/construction (roads, rails, bridges, ports, airport terminals, and building materials such as sand, gravel, aggregate, brick clay, chalk, granite, marble, and gypsum), chemical, sugar, refractory, glass and ceramics, aluminium, and agriculture (lime), have an intimate connection because minerals are raw material feed for them. The manufacturing industry, trade transport, real estate, utilities, and other services have both direct and indirect involvement with the mines and mineral industry. Several ancillary industries generate employment through en-route mines, as they are partially dependent on them. In India alone, the mining sector with an employment elasticity of 0.52% creates 13 times more jobs than agriculture and 6 times more jobs than the manufacturing sector (Figure 1.3). According to the report of the high-level committee of the Indian National Mineral Policy, 2006, one direct mining

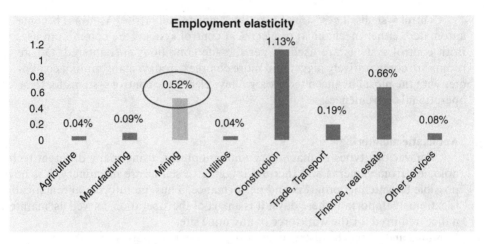

Figure 1.3 Sectoral employment in India including mining.
Source: FIMI, 2020; Page No. 26.

job creates ten additional jobs in the economy thus employment multiplication factor is 1:10 for the mining sector (FIMI, 2020). Similar is the condition with other countries which are dependent on minerals for developing their respective economy.

1.7   National and international organizations related to the mining industry

For the mega-sized mining and mineral industry, several functional and performing organizations right from policy to production then environment protection and social welfare are needed at different levels. World over, these organizations with no uniformity in their structures perform their roles at different locations in different countries. Sharing various responsibilities for the dedicated industry of mining and mineral sector, all across largely, these organizations are listed below:

- *Production Organizations/Mining companies* – With mining/mineral Production as a core business and with mining/mineral production as a diversified business.
- *Organizations for geological survey and mineral exploration* – Exploration company and geological survey organization of respective country, e.g. USGS, BGS, and GSI.
- *Organizations for environmental and ecology protection, e.g. UNEP and EPA.*
- *Organizations for mineral conservation (government agencies in different countries).*
- *Policy organizations* – National policy-making organization of respective country including planning arm organization.
- *Organizations for international cooperation, trade, e.g. (ICMM), export and import promotion* & safe, fair and sustainable mining and metals industry.
- *Organizations for statuary compliance* – Federal (Central) and state-level ministries in the concerned areas.
- *Academic institutes and R&D Organizations* – IISD *(Box 1.1)*, National mining research laboratory, National mineral processing laboratories, Engineering colleges, and polytechnic institutes imparting various degrees from graduate to post-doctoral, including professional diploma, and R&D cells of company/industry.

Box 1.2: International Institutes and Organizations

International Council for Mining and Metal (ICMM)/CRIRSCO: ICMM is an industry group that addresses key priorities and emerging issues for the mining and mineral sector. It seeks to play a leading role by promoting good practice and improved performance internationally and across different mineral commodities. In October 2009, the Committee for Mineral Reserves International Reporting Standards (CRIRSCO) became a strategic partner of ICMM.

ICMM provides a platform for industry and other key stakeholders to share challenges and develop solutions based on sound scientific practices and the basic principles of sustainable development. Its vision is for the mining and metals industry which is widely recognized as being essential for society. The ICMM currently has 28 member companies, representing most of the mining sector, 50+ countries of operation, and over 35 member associations, which include regional mining associations and regional and global commodity associations all across the world. The full range of its activities can be found at www.icmm.com.

International Institute for Sustainable Development (IISD): The International Institute for Sustainable Development (IISD) is one of the world's leading centres of research and innovation. The Institute provides practical solutions to the growing challenges and opportunities of integrating environmental and social priorities with economic development. IISD prepares reports on international negotiations and shares knowledge gained through collaborative projects, resulting in more rigorous research, stronger global networks, and better engagement among researchers, citizens, businesses, and policymakers. IISD is registered as a charitable organization in Canada and has legal status in the United States. IISD receives core operating support from the Government of Canada, provided through the International Development Research Centre (IDRC), and from the state Province of Manitoba. The Institute receives project funding from numerous governments inside and outside Canada, United Nations agencies, foundations, the private sector, and individuals.

- **Organizations for data and statistics:** *Collection and dissemination of mining and mineral industry-related data.*
- **Labour organizations:** International Labour Organization (ILO) and Trade unions.
- **Organizations related to safety:** Mine safety organizations in different countries.
- **Organizations for valuable metals:** International Council on Mining and Metals (ICMM), World Gold Council, Association for Iron & Steel, Diamond, Zinc, Silver, Copper, Lead, Tungsten.
- **Organizations for education promotion:** To promote geoscience education internationally at all levels, to encourage developments raising public awareness of geosciences, particularly amongst younger people, to work for the enhancement of the quality of geosciences education internationally of the International Geoscience Education Organisation (IGEO) is working.
- *Organizations for manufacturing and other ancillary industry for support to mines.*
- **Organizations for social welfare:** NGOs.

Concerning India and for reference examples, the short names of mining and mineral sector organizations are listed in Table 1.3.

PART II

1.8 Rudiments of mining

The word 'rudiments' explains a basic principle or element or a fundamental skill of a subject. It is first taught to or acquired while commencing the study or practice of a branch of knowledge. Rudiments of anything means learning the simplest or the most essential aspects of it. Concerning mining engineering, which involves 'coal mining' as well as 'metal mining', these first principles are described broadly right at the beginning, i.e. in this first chapter. All those points which are dealt with in this sub-section could be elaborated subsequently, either through the same book or through other reference books, and the reader will acquire deeper knowledge and understanding of mines, mining, and mineral-getting operations. Thus, for a mining engineer, these essential aspects will help to lay the firm foundation.

1.8.1 Mining as a business

Mining may be considered as second of humankind's earliest endeavours given that agriculture was the first. The two industries ranked together as the primary or basic industries of early civilization. Little has changed in the importance of these industries since the beginning of civilization. If fishing and lumbering are considered a part of agriculture and oil and gas production as part of mining, then agriculture and mining continue to supply all the basic resources used by modern civilization. Since civilization began, people have used mining techniques to access minerals on the Earth. In the earliest days, mining was slow-going and dangerous. However, as time has progressed, geologists and mining engineers have developed safer and more accurate methods for locating and uncovering minerals found in the Earth.

There are many terms and expressions unique to mining that characterize the mineral extraction process from Mother Earth. The following terms are closely related:

- **Mine:** An excavation made on the Earth to extract minerals.
- **Mining:** The activity, profession, and industry concerned with the extraction of minerals.
- **Mining engineering:** The practice of applying engineering principles to the planning, development, operation, closure, and reclamation of mines.

Mining and mining engineering can be distinguished as embracing the processes, the occupation, and the industry involved in the extraction of minerals from Mother Earth. Whereas mining engineering is the art and science of mining processes and the operation of mines inclusive of minerals uses. Mining engineers are the professionals who connect two fields so that they can help to locate and prove minerals, to design and develop mines, and to exploit and manage mines. The prime task for

Table 1.3 National Organizations & Companies Related to the Mining Industry in India

Production/mining companies	Geological survey and mineral exploration	Environmental and ecology protection	Mineral conservation	Data and statistics	Labour and safety
CIL & all its subsidiaries, SCCL, NMDC, MOIL, BGML, HCL, HZL, SAIL, TISCO, ESSEL Mining, UltraTech Cement	GSI MECL CMPDIL	MOEFCC CPCB and SPCB	IBM	MOM IBM MOSPI (Ministry of Statistics and Programme Implementation)	Min of labour DGMS

Trade promotion, export and import of minerals	Policy and Statuary compliance	Manufacturing and other ancillary industry	Research and development	Basic mining Education and its promotion	Social Welfare
MMTC, STC and Ministry of Industry & Commerce - Department for promotion of industry and internal trade	MOM MOEFCC Niti Aayog State govt. organizations & IBM and FIMI	BEML, Explosive manufacturing firms, e.g. IEL, IDL, Solar Explosive, Navbharat Explosives	1 - CIMFR (National Mining Lab) and other research institutions in India including CSIR.	3-IITs; 4 NITs; Engineering Colleges, Management Institutes and Polytechniques, SCMS (Skill Council for Mining Sector)	Trade Unions like INTUC, AITUC, HMS and Several NGO's and CMAI (Cement Manufacturers Association of India)

Note: For acronyms of organizations, see the common abbreviations list.

mining is to build entries (opening/drives) for an excavation that are used to enter the mineral deposit from the surface to below ground while passing through the soil and the rock to extract minerals from the treasure troves of Mother Earth. These entries may lie wholly on the surface or be made underground, which defines the locale of the mine. Normally, these openings are meant to provide access, make a development entry, and then excavate the deposit which has been formed for millions of years by the nature. Every mine and each mining method adopted by them for digging minerals is specified by the layout, procedure, equipment, and system involved. Moreover, its application is strongly related to the geologic, physical, economic, environmental, and legal conditions (Hartman, 2002). From concept to decommissioning, nearly all mines are business entities meant to produce and earn profit out of minerals. The essence of mining in extracting mineral wealth is to develop it as an industrial business for our present-day needs. All mines with large CAPEX, either on the surface (in plains or hills) or belowground, are principally directed towards a single goal – dig the mineral and utilize it as a resource sustainably.

A mineral can be found or traced in small or large quantities as the natural condition warrants. Accordingly, words like 'mineral deposit' (large) and 'mineral occurrences' (small) are framed for proper use by the industry. Note that when the economic profitability of a mineral deposit has been established with some confidence, the ore deposit (or ore) becomes feasible for exploitation and is called a 'commercial deposit'. However, coal and industrial mineral deposits are often not so designated, even if their profitability has not been firmly established. The field of boreholes (drilling) including digging operations is associated with mines extensively. At times, boreholes are used to extract the mineral values from the Earth.

The field of mining uses its unique vocabulary. Some common words, used technically, for indicating function, location and operation of mines, and mining operation of the real fields have been covered and described below as the rudiments concerning mining. These terms help to distinguish various types and categories of mined minerals in the geological domain, identify the nature of mineral deposits concerning mining, and help to find the exact (correct) economic differences among many mineral types used by mankind. These most common terms have intricate meanings, to differentiate many applications or uses, in the mining/mineral industry. The following are the industry-specific and most common terms used:

- **A mineral** is a naturally occurring inorganic element or compounds having an orderly internal structure and characteristic chemical composition, crystal form, and physical properties.
- **Rock** is a naturally formed aggregate of one or more types of mineral particles.
- **Ore** is a mineral deposit that has sufficient utility and value to be mined at a profit.
- **Gangue** is the valueless mineral particles within an ore deposit that must be discarded.
- **Waste** is the material associated with an ore deposit that must be mined to get ore and then discarded. Gangue is also a particular type of waste.
- **Metallic ores** are the ores of the ferrous metals category (iron, manganese, molybdenum, and tungsten), the base metals category (copper, lead, zinc, and tin), the precious metals category (gold, silver, the platinum group of metals), and the radioactive minerals category (uranium, thorium, and radium).

- **Non-metallic minerals**, also known as non-fuel minerals/ores of the industrial mineral category, are not associated with the production of metals. These include phosphate, potash, halite, dolostone, sand, gravel, limestone, and sulphur.
- **Fossil fuels** are also known as *mineral fuels*. *T*he organic mineral substances can be utilized as fuels, such as coal of all forms, i.e. peat, lignite (brown coal), bituminous, and anthracite.

> Note: The petroleum, natural gas, coal-bed methane, gilsonite (natural bitumen), and tar sands (natural asphalt) are not directly associated with fossil fuel but belong to the fuel category.

- **Mineral deposits** are of different types – Flat, dipping, sheet type, pipe type, vertical and horizontal deposits, massive deposit, disseminated deposits, regular and irregular deposits, oxidized deposits, lateritic deposits, stratified deposits, leached/weathered deposits, and low-grade/high-grade deposits, etc. Likewise, coal seams (thick and thin), beds, veins, and lodes are also classifications of mineral deposits of various types.

A mining engineer is a person directly associated with the mining and extraction of nearly all those resources belonging to the mineral category. However, the production of petroleum and natural gas has evolved into a separate industry with a specialized technology of its own. *Locating and exploring* a mineral deposit fall in the general purview of geology and the Earth sciences, whereas *mine* and *mining* enter into the realm of engineering and management. If the excavation used for mining is entirely open to the sky or operated from the surface, it is termed a 'surface mine' (see Section 1.9). If the excavation consists of openings for human entry below the Earth's surface, it is called an 'underground mine'.

Mining is never properly done in isolation, and it is impossible to develop a 'safe mine' if a proper sequence, layout, and proven method are adopted for mining. To get minerals, the extraction has to be planned scientifically. It is preceded by geologic investigations that locate the deposit and economic analyses that prove it financially feasible. Following extraction of the mineral, run-of-mine (ROM) material, comes the generally cleaned or concentrated ore/coal. The preparation or beneficiation of the minerals into a higher-quality end product is termed *mineral processing*, which falls next. Thus, the journey from raw mineral/ore to metal (as an end product) finishes with the smelting, refining, and concentration processes that provide a consumer with usable products. The end step in converting a mineral material into a useful product is marketing.

Quite frequently, excavations have been made to get minerals, but besides this, mining excavations are also useful for civil and military applications in which the objective is to produce a stable opening of the desired size, orientation, and permanence, e.g. vehicular, water, and sewer tunnels, plus caverns, underground storage facilities, waste disposal systems (underground), and military installations. These excavations, which were produced using standard mining technology, are non-mining applications.

Professionally, mining is a business endeavour that is large in size and capitalistic in nature. The mining of minerals and mineral industries are linked with mankind for centuries and are associated with human civilization. It has its development history which we had delineated in a separate section as 'history of mining'.

1.8.2   Mining stages

The overall sequence of activities in mining is often compared with the five stages during the life of any mine: prospecting, exploration, development, exploitation, and closure. Prospecting and exploration, precursors to actual mining, are linked and sometimes combined. Geologists and mining engineers often share responsibility for these two stages – geologists are more involved with the former, whereas mining engineers are more involved with the latter. Likewise, development and exploitation are closely related stages. They are usually considered to constitute mining proper and are the main province of the mining engineer. Mine closure has been added to these stages since the first edition, but a little later to reflect the times. These days, closure and reclamation of the mine sites have become a necessary part of the mine life cycle because of the demands of society for a cleaner environment and stricter laws regulating the abandonment of a mine. The overall process of developing a mine with the future scope for land uses in mind is termed *sustainable development*.

Prospecting: This is the first stage in the utilization of a mineral deposit, which is the search for ores or other valuable minerals (coal or non-metallic). Both direct and indirect prospecting techniques are employed. The direct method of searching minerals is normally limited to surface deposits, and it consists of a visual examination of either the exposure (outcrop) of the deposit or the loose fragments (float) that have weathered away from the outcrop. Geological studies of the entire area augment this simple, direct technique. Utilizing aerial photography, geologic maps, and structural assessment of an area, the geologist gathers evidence by direct methods to locate mineral deposits. Precise mapping and structural analysis plus microscopic studies of samples enable the geologist to locate the hidden as well as surface mineralization.

The most valuable scientific tool employed in the indirect search for hidden mineral deposits is Geophysics, the science of detecting anomalies using physical measurements of gravitational, seismic, magnetic, electrical, electromagnetic, and radiometric variables of the Earth. The geophysical method(s) are applied from the air, using aircraft and satellites; on the surface of the Earth; and beneath the Earth, using methods that probe below the topography. *Geochemistry*, the quantitative analysis of soil, rock, and water samples, and *Geo-botany*, the analysis of plant growth patterns, can also be employed as prospecting tools.

Exploration: The second stage in the life of a mine, exploration, determines as accurately as possible the size and value of a mineral deposit, utilizing techniques similar to but more refined than those used in prospecting. The line of demarcation between prospecting and exploration is not sharp; in fact, a distinction may not be possible in some cases. Exploration generally shifts to surface and subsurface locations, using a variety of measurements to obtain a more positive picture of the extent and grade of the ore body. Representative samples may be subjected to chemical, metallurgical, X-ray, spectrographic, or radiometric evaluation techniques that are meant to enhance the investigator's knowledge about the mineral deposit. Samples are obtained by chipping outcrops, trenching, drilling, and exploratory tunnelling (driving an exploratory adit). In addition, core logging and geologic and structural mapping may be provided to study the formation of the mineral deposit. Rotary, percussion, or diamond drills can be used for exploration purposes. However, diamond drills are favoured because the cores they yield provide knowledge of the geologic structure. The core is normally split along its axis with one half being analysed and the other half being retained intact for further geologic records. An evaluation of the samples enables the geologist or mining engineer to calculate the tonnage and grade, or richness, of the mineral deposit. This is useful in estimating the mining costs, evaluating the recovery of the valuable minerals, determining the environmental costs, and assessing the other foreseeable factors to conclude the profitability of the mineral deposit. The crux of the analysis is the question of whether the explored property contains a mineral of importance as a geological deposit or has an orebody. For an ore deposit, the overall process is called 'reserve estimation', that is, the examination and valuation of the mineral/ore body. After this stage, the project is developed, traded to another party, or abandoned.

Development: In the third stage, development, the work of opening a mineral deposit for exploitation is performed with the goal of starting the actual mining of the deposit, which contains the mineral, coal, or ore. Access to the deposit must be gained either (a) by stripping the overburden, which is the soil and/or rocks covering the deposit, to expose the near-surface ore for mining or (b) by excavating openings from the surface to access more deeply buried deposits to prepare for underground mining. In either case, certain preliminary planning work, such as acquiring water and mineral rights, buying surface lands, arranging for financing, and preparing permit applications and an environmental impact statement (EIS), is generally required before any development takes place. When these steps have been achieved, the provision of several requirements – access roads, power sources, mineral transportation systems, mineral processing facilities, waste disposal areas, offices, and other support facilities – must precede actual mining in most cases.

Stripping of the overburden will then proceed if the minerals are to be mined at the surface. Economic considerations determine the stripping ratio, the ratio of waste removed to ore recovered (in m^3/in tonne) and may range from as high as 20 for coal mines to as low as 0.8 in metal mines. Some non-metallic mines (limestone) have no overburden to remove; the mineral is simply excavated at the surface or from the hills.

Development for underground mining is generally more complex and expensive. It requires careful planning and layout of access openings for efficient mining, safety, and performance. The principal openings may be shafts, inclines/declines, or adits; each must be planned to allow the passage of workers, machines, ore, waste, air, water, and utilities. Many metal mines are located along steeply dipping deposits and thus are opened from shafts, while drifts, winzes, and raises serve the production areas. Many coal and non-metallic mines are found in nearly horizontal deposits. Their primary openings may be drifts or incline entries, which may be distinctly different from those of metal mines.

Exploitation: Exploitation, the fourth stage of mining, is associated with the actual production of minerals in magnitude (quantity) from the Earth. Although development may continue, the emphasis at this stage is on production and productivity, exclusively. Usually, sufficient development is done before exploitation starts to ensure that production, once started, can continue uninterrupted throughout the mine life. The mining method selected for exploitation is determined mainly by the characteristics of the mineral deposit and the limits imposed by safety, technology, environmental concerns, and economics. Geologic conditions, such as the dip, shape, and strength of the ore and the surrounding rock, play a key role in selecting the method. Conventional exploitation methods are divided into two broad categories based on the depth of coal or mineral deposit: surface and/or underground. Surface mine/mining operation could be either manual or mechanized and has variants such as an open-pit and an opencast (strip mining) and aqueous methods such as placer and solution mining. Underground mines are broadly classified into different categories – unsupported/supported mine, caved/non-caved mine, and backfilled underground mine that adopt methods likewise. The haulages, hoisting, underground transportation, mine support, mine ventilation, and underground mine drainage are other supportive ancillary services related to such underground mines, executed by the mine management.

Closure/Decommissioning (Reclamation): Any mineral deposits, located either at or below the surface of the Earth, after or during exploitation generate the degraded land which should be brought to its near original shape/stage and reclamation is the solution for this. Hence, this stage is the most essential of all stages, and it has plenty of scope for planning, surveying, developing, mapping, etc. Its importance is felt significantly at the closure phase (*Decommissioning stage)* of the mine. The mining engineer is deeply involved with this end phase (closure phase), and reclamation of the mine lease or mineral property equally shares those duties concerning the environmental fields.

The field of extracting metal from the ore – including processing, refining, and fabricating – is assigned to metallurgy, although there is often some overlap in mineral processing with mining engineering. These stages delineate the whole life of a mine; however, further details of these rudimentary principles will be explained in the subsequent section such as a basic method, unit operations, and economics of mining operation.

1.9   The basic methods

a.   **Surface mining:** Surface mining is the predominant exploitation method worldwide including the US, Australia, and India. About 80%–90% of all minerals, excluding petroleum and natural gas, are mined by surface mining methods. In open-pit mining, a mechanical extraction system (a machine mining method) is generally deployed to mine a thick deposit (or coal seam), to be mined by forming benches (or steps), although thin deposits may require only a single bench or mining face. Open-pit or opencast mining is usually employed to exploit a near-surface deposit or one that has a low stripping ratio (Altiti et al., 2021). Surface mining necessitates a large capital investment often resulting in high productivity, low operating cost, and good safety conditions in general.

Surface mining results in the development of an open-pit mine, a hill mine, and a small-scale mine (an artisanal mine) and is different from normal quarrying as it selectively attracts ore rather than aggregate. The open-pit mining method is usually non-selective and used for the extraction of coal, metallic and non-metallic ores for commercial purposes (Figure 1.4). It is applied to mine marble, dimensional stones, as well as high- and low-grade ores for a higher mining production to the tune of thousand tonnes per day. Basic elements of opencast mining are

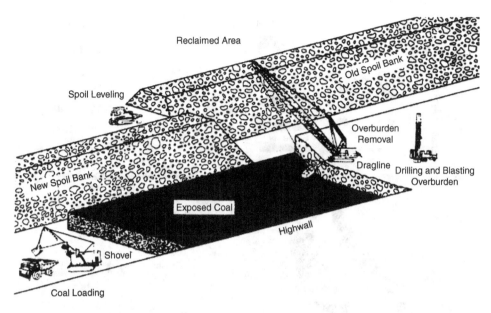

Figure 1.4 A typical open-pit mining of coal.

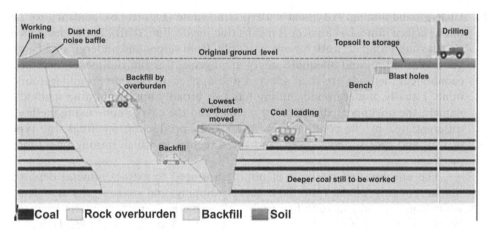

Figure 1.5 Opencast mining with its elements.

Source: https://www.open.edu/openlearn/nature-environment/environmental-studies/energy-resources-coal/content-section-2.5

open benches, heavy digging machines, drilling rigs, blasting equipment, specified waste dumping sites for overburden, and tailing disposal (Figure 1.5). The open-pit mine operation has backfilling, high-wall failures, mine pit slopes stability, and reclamation as parameters for the daily production cycle. The rudiments of surface mining are high production, low cost of the mining operation, and safe working condition on day-to-day basis.

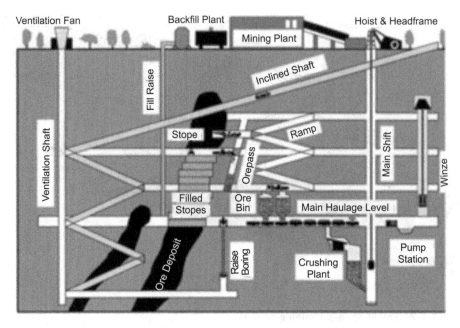

Figure 1.6 A typical underground mine.

b. **Underground mining:** A typical underground mine (Figure 1.6), contrary to the
    open surface mine, has an access means (incline/decline, shaft), a well-developed
    underground road network for ore extraction from stopes and underground faces,
    good ventilation, and adequate safety arrangement as the rudiments. Safe and
    economical methods are then selected and applied to operate an underground
    mine. Usually, underground mining has two broadly applied mining methods,
    namely, the unsupported/supported method and the caving/non-caving method,
    differentiated by the configuration and size of production openings, the type
    of wall and roof supports used, and the direction in which mining operations
    progress.

    The unsupported methods of mining are used to extract mineral deposits
    that are roughly tabular (plus flat or steeply dipping) and are generally associated
    with strong ore and surrounding rocks. These methods are termed unsupported
    because they do not use any artificial pillars to support the openings. However,
    generous amounts of roof bolting and localized support measures are often used.
    Room-and-pillar mining/Bord and pillar and *longwall mining* are the most common
    unsupported methods that are used primarily for flat-lying seams or bedded de-
    posits like coal, rock phosphate, limestone, and salt. Support of the roof is pro-
    vided by natural pillars of either the mineral or waste rocks that are left standing
    in a systematic pattern.

    A stope is a production opening in a metal mine and *Stope-and-pillar mining* is
    a method used in non-coal mines where thicker, more irregular ore bodies occur;
    the pillars are spaced randomly and located in low-grade ore so that the high-
    grade ore can be extracted. Two other methods applied to steeply dipping deposits

are also included in the unsupported category. In *shrinkage stoping*, mining progresses upward, with horizontal slices of ore being blasted along the length of the stope. A portion of the broken ore is allowed to accumulate in the stope to provide a working platform for the miners and is thereafter withdrawn from the stope through chutes. *Overhand sublevel open stoping* is also a most common method where a mineral is extracted through intermediate-level openings in the deposit. *Sublevel stoping* differs from shrinkage stoping by providing sublevels from which vertical slices are blasted. In this manner, the stope is mined horizontally from one end to the other. Shrinkage stoping is more suitable than sublevel stoping for stronger ore and weaker wall rock.

Supported mining methods are often used in mines with weak rocks. *Cut-and-fill stoping* is the most common of these methods and is used primarily in steeply dipping metal deposits. The cut-and-fill method is practised both in the overhand (upward) and underhand (downward) directions. As each horizontal slice is taken, the voids are filled with a variety of fill types (backfilling) to support the walls. The fill can be rock waste, tailings, cemented tailings, or other suitable materials. Cut-and-fill mining is a popular method used for vein deposits and has recently grown in use. *Square-set stoping* method involves backfilling mine voids; however, it relies mainly on timber sets to support the walls during mining. This mining method is rapidly disappearing in the world because of the non-availability of timber and the high cost of labour. However, it still finds occasional use in mining high-grade ores. *Stull stoping* is a supported mining method using timber or rock bolts in tabular, pitching ore bodies. It is one of the methods that can be applied to ore bodies that have dips between 10° and 45°. It often utilizes artificial pillars of waste to support the roof.

Caving methods are varied and versatile and involve caving the ore and/or the overlying rock. Subsidence of the surface normally occurs afterwards. Longwall mining is also used with caving, which is particularly well adapted to horizontal seams, usually coal, at some depth. In this method, a face of considerable length (a long face or wall) is maintained, and as the mining progresses, the overlying strata are caved, thus promoting the breakage of the coal itself. In some cases, Room and pillar/Bord and pillars are also used with caving after pillar extraction. A different method, *sublevel caving*, is employed for a dipping tabular or massive deposit. As mining progresses downward, each new level is caved into the mine openings, with the ore materials being recovered while the rock remains behind. *Block caving* is a large-scale or bulk mining method that is highly productive, low in cost, and used primarily on massive deposits that must be mined underground. It is most applicable to weak or moderately strong ore bodies that readily break up when caved. Both block caving and longwall mining are widely used because of their high productivity.

c. **Special methods:** In addition to these conventional methods, innovative methods of mining are also evolving. They apply to unusual deposits or may employ special techniques, processes, equipment, or a typical tailor-made unconventional methodology, e.g. deep-sea mining, placer mining (dredging), underground gasification, solution mining, hydraulic mining (hydraulicking), and liquefaction. These days, such methods are automated with rapid excavation progress but can only be applied to limited categories of mineral deposits.

1.10   Unit operations of mining

At different stages of mining, remarkably similar unit operations are normally employed to develop a mine – a storehouse of mineral/ore/coal. The unit operations of mining are the basic steps used to produce minerals from the geological deposit, and the auxiliary operations that are used to support them. The direct steps, contributing to mining operations, constitute the cycle of mineral extraction operations. The ancillary steps that support the production cycle are termed *auxiliary operations*. The production cycle employs unit operations that are normally grouped into *rock breakage* and *materials handling*. Breakage generally consists of drilling and blasting, and materials handling encompasses loading or excavation and haulage (horizontal transport) and sometimes hoisting (vertical or inclined transport). Thus, the basic production cycle consists of these unit operations.

Although production operations tend to be separate and cyclic, the trend in modern mining (including tunnelling) is to eliminate or combine functions and to increase the continuity of extraction. For example, in coal and other soft rock mines, *continuous miners* perform rock-breaking and loading operations together in the mineral winning process, thus eliminating essential drilling and blasting unit operation; boring machines (TBM) perform the same tasks in medium-hard rocks. The cycle of operations in surface and underground mining differs primarily by the scale of the equipment. Specialized machines have evolved to meet the unique needs of the two regimes. In modern surface mining, blast holes with a diameter of 75–380 mm are drilled by rotary or percussion drills for the placement of explosives when consolidated rock must be removed. The explosive charge is then inserted and detonated to remove the overburden or ore/coal to a size range suitable for excavation. The broken material is loaded by shovel, dragline, or wheel loader into haulage units generally dumpers/trucks for transport. Rail wagons are also used for haulage, and belt conveyors are often deployed for handling the continuous supply of crushed material after excavation operation. Overburden/waste rocks of mines, including soil and coal/ore, are often moved in the same manner, although blasting is sometimes not mandatory and unnecessary. In the marble mining and quarrying of dimension stone, the blocks are often freed without blasting using *wire saws* or other mechanical devices.

In underground mining, the conventional production cycle is similar, although the equipment used may be scaled down in size. Smaller drill holes, low-profile loaders, low-profile dumpers/trucks, continuous miners, and shuttle cars are used in underground mines. Load-Haul-Dump (LHD) machines and belt conveyors are common in coal production. Salt, potash, and talc are often mined without the use of explosives or mined after undercutting the face to reduce the consumption of explosives.

In addition to the main unit operations of the production cycle, certain auxiliary operations must be performed in many cases. In surface mining, the primary auxiliary operations include those providing slope stability, pumping, power supply, maintenance, waste disposal, and supply of material to the production centres. Underground mining usually includes roof support, ventilation and air-conditioning, power supply, pumping, maintenance, lighting, communications, and delivery of compressed air, water, and supplies to the mine working sections.

1.10.1 Economics and mineral-getting operation

The uniqueness of minerals as economic products accounts for the complexity of mineral economics being described here in short, considering mining as a business. Minerals are unevenly distributed, and unlike agricultural or forest products, cannot reproduce or be replaced. A mineral deposit may therefore be considered a *depleting asset* whose production is restricted to the area in which it occurs. These factors impose limitations on a mining company in the areas of business practices, financing, and production practices. Because mineral assets are constantly being depleted, a mining company must discover additional reserves or acquire them by purchase to stay in the mining business. Other peculiar features of the mineral industry are associated with the operations performed in the mineral-getting operation. Production costs tend to increase with depth and declining grades. Thus, low-cost operations are mined first, followed by the harder-to-mine deposits. In addition, commodity prices are subject to market price swings in response to supply and demand, which can make the financial risk of long-term minerals projects quite high. A change in mining or processing technology can also drastically alter the economic landscape.

The economics of mineral-getting operation is very wide and elaborate involving various parameters. Some minerals, such as precious metals (gemstones) and high-value minerals (gold and strategic minerals of rare category), have different cost-economics compared to those of low-value minerals (limestone, dolomite, clay). Likewise, those minerals which have an acute demand by the industry, non-recyclable and recyclable minerals (iron and most of the base metals can be recycled economically), have altogether different economics. This means demand, supply, mineral value, and mineral reuse all have a significant contribution to mineral economics. Considering these facts, the markets for freshly mined minerals and metals are affected and a trend and practice that is favourable for the future has been set. Even though fluctuations in the economic value of minerals and market price(s) are inevitable, they have to be dealt with and learnt periodically.

As a raw material feed, all industrial minerals, including coal, are great contributors to the exchequer that have an impact on the economy. Hence, substitutes for a particular mineral have been developed particularly in context with the price of the mineral and metals, e.g., aluminium, iron, and copper are substituted with plastics and glass for several applications. The pattern of mineral usage in terms of intensity of use (kg/capita) and total consumption of metals on the world market for non-ferrous metals shows that the intensity of usage of many of these minerals and metals continues to go up and down, while the overall consumption dwindles according to the demand and supply. Any swing in the intensity of use due to substitution or recycling can greatly affect the mineral/metal market and its prices. Therefore, mining companies are asked to keep their economics under control and prices low through further productivity improvements. In brief, mining economics is a major and extremely intricate area that should be dealt with at various levels to avoid great economic hardships and thus create a stable market price.

1.10.2   Reclamation/restoration/revegetation of the mines and degraded mining areas

The final stage in the operation of most mines is *reclamation,* after the process of extraction or before closing a mine. Recontouring, revegetation, and restoring the water and land values form an essential part of the reclamation process. The best time to begin the reclamation process of a mine is initiated with the start of the first excavations. In other words, mine planning engineers should plan the mine so that the reclamation process is considered at the operational stage itself and the overall cost of extraction plus reclamation is minimized, not just the cost of mining itself. The concept that evolved in the mining industry is *sustainability*, that is, the meeting of economic and environmental needs of the present while also enhancing the ability of future generation's needs. In planning for the reclamation of any given mine, many concerns must be addressed. The first of these are the environmental protection needs, secondly the environmental health, and thirdly the safety of the mine site or mining area, which are addressed by the abbreviated word **SHE** (safety, health, and environment). The removal of plants and buildings, processing facilities, transportation equipment, utilities, and other surface structures must generally be accomplished before the reclamation work is started. The mining company is then required to seal all mine shafts, adits, and other openings that may present physical hazards.

Yet another major issue to be addressed during the reclamation of a mine site is the restoration of the land surface, the water quality, and the waste disposal areas so that long-term water pollution, soil erosion, dust generation, or vegetation problems do not occur. The restoration of native plants is often a very important part of this process, as the plants help build a stable soil structure and naturalize the area. Heavy metals that pollute streams form a part of the reclamation. The successful completion of the reclamation of a mine enhances the public opinion of the mining industry and keeps the mining company in the good graces of the regulatory agencies. The final concern of the reclamation ends with subsequent use of the land after mining is completed. Reclaimed mine sites (old) should be converted to wildlife refuges, eco-friendly parks, pit lakes, underground storage facilities, solid waste disposal areas, and other miscellaneous uses such as real estate developments, museums, shopping areas, golf courses, and airports that can benefit society.

1.11   Life cycle of the mine

Mines are the storehouse of minerals that are naturally occurring. The mine and mining operations are complex and meant to produce minerals (coal and iron ore) and deliver refined metal commodities like gold, silver, copper, zinc, and aluminium. In its life cycle, the mine has the following different stages:

1.   **Exploration stage:** Search of mineral deposits in various deposit forms either over the Earth or underneath the Earth (orebody assessment and delineation).
2.   **Planning and design stage**: For a mine or a mine site.
3.   **Construction stage**: Development of a mine/underground workings.

4. **Exploitation stage:** Production of coal/mineral/ore either by surface mining or by underground mining (depillaring). 4 (i) Milling/Minerals processing stage – a stage in addition (for coal, it is washing, and for a metallic ore, it is to recover metal from the ore).
5. **Decommissioning stage:** Mine closure and reclamation.

The life of a mine is in several years depending on the quantity of ore/coal (mineral) available for exploitation. All mines and mining projects are capital-intensive projects involving millions of dollars. Mining projects and mine complexes provide subsistence to many needy during execution and development. When made fully operational, they bring about significant local and regional changes in the economics of the area. The environmental and social changes mark the region.

1.12 Mining consequences (impacts of mining)

Mining along with the ore beneficiation processes, i.e. milling, affects the environment considerably. It is a well-known fact that the impacts/effects are adverse as well as positive. Plenty of widely scattered literature is available on this subject area being discussed. However, we are dividing the impacts of mining into two broader categories: *the impact of mining on society* and the *impact of mining on the environment*. Health and safety impacts, caused by mining, will be covered under these two broad categories. These impacts may be briefly summarized as follows:

* *Impact of mining on society*
    The positive impacts of mining on society can be listed as follows:
    * Development of infrastructure facilities and asset creation such as roads, schools, buildings, health centres, electric supply lines, hand pumps, and solar panels in and around the mines either in urban areas or in the village localities.
    * Development of literacy and skill development such as dairy development, poultry farming, pisciculture, vegetable farming, beekeeping, car driving, TV repairing, carpet weaving, tailoring, goat and pig rearing, and cycle maintenance.
    * Community activities such as the formation and support of youth clubs, organization of sporting events, cultural programmes, and health camps.
    * Reduced gender gaps and improved lifestyle.
    * Improved quality of life (QOL) in the mines and mining areas.
    When a materialistic culture is developed in the mining society, it is quite obvious that ill effects, such as theft cases, crime against women, and vandalism, also flourish and get momentum. These are the visible negative societal impact of mining on society. Cultural changes may be observed in society in the long term and can be both positive as well as negative.
* *Impact of mining on the environment*
    In general, the impacts of mining on the environment called *pollution* are negative. Air, water, and land including flora and fauna are disturbed. Noise is created as a result of mining activity which is under the industrial category and leads to

noise pollution creating health and safety impacts. In short, these environmental impacts can be described below:

i.   **Air pollution**: Mining causes air pollution and it can be divided into two different groups: *ambient air pollution* and *workplace pollution*. The airborne particulate matter is the main air pollutant contributed by mining and mineral dressing processes. The fine-sized solid particles (fugitive dust) generated during drilling, blasting, loading and transportation of minerals/waste in mines get spread in the atmosphere due to wind action. This causes particulate matter generation. The mineral dressing processes, especially dry crushing and grinding, may also result in air pollution. The particulates that are most hazardous for human health are having a diameter of less than 2–10 μm. Also, the smaller the diameter of the particulate, the longer will be the airborne time and the more potential it has for spreading to a wider area. Fugitive dust and ambient airborne air pollution both have health effects or negative impacts, either long term or short term. The most hazardous air-polluting particulates are those which are toxic to both plants and animals, mainly those that contain heavy metal particles such as lead, zinc, copper, nickel, cadmium, arsenic, and aluminium. In plants, the toxic effects are caused mainly by the destruction of chlorophyll, thus disrupting the process of photosynthesis. In coal mining areas, spontaneous combustion of the carbonaceous matter contained in the waste rock dumps results in atmospheric pollution with smoke and noxious and hazardous gases (sulphur dioxide, nitrous oxide, carbon mono oxide, phenols, and creosols) emanation. Thus, air pollution both in the ambient environment and at workplaces is the impact of mining on the air we breathe.

ii. **Water pollution**: One of the most common water pollution problems caused by mining and ore beneficiation processes is the effluent discharge from mines, deposition of suspended solids in adjacent watercourses/areas/water bodies, wash-off from waste dumps and tailing ponds, effluent discharge resulting from the metallurgical processes, such as flotation, and leaching from the waste dumps and tailings areas. The most common and direct form of water pollution is turbidity and hardness changes of water (a physical parameter for rivers, lakes, and irrigation canal pollution). Water turbidity results in decreased light penetration which affects the life cycle and food chain mechanism of the aquatic ecosystem. It has a direct serious effect on fish and other water-living organisms.

One of the most serious water pollution problems associated with the mining industry is caused by toxic metals and by acid mine drainage (AMD) from both underground and opencast mines. This form of pollution results from the oxidation of the primary minerals being mined and/or of the other minerals in the ore, particularly those containing sulphide minerals. Oxidation causes acid formation in coal mines and pyrite mines, and the release of metals and other complex chemical compounds into the underground and surface water causes pollution to occur.

iii. **Land degradation**: Land degradation is the most visible and serious environmental impact of mining. Surface mining creates scars on the Earth's surface and green hill areas rendering the land unsuitable for other uses unless it is restored or rehabilitated. The impact of mining on land is more severe in the case of surface mines developed either as a pit mine or a hill mine. Land degradation reduces the land use potential and the degree of damage to the land varies with the topographic setting in which the surface mine or the opencast mining is carried out. In the case of underground mining, surface subsidence is the most common environmental impact which is a hazard too from a safety angle.

The fallouts of land degradation, being negative, have a linkage with the terrain type, climatic conditions (rainfall), and the quantity of land/soil handled (dimensions of the deposit mined and the overburden tackled). All these variables influence the management of land while doing the reclamation/restoration/revegetation of land.

iv. **Noise pollution**: It is an industrial impact of mining and associated activities, particularly drilling, blasting, loading, transportation, and crushing/grinding, that are essential for mineral excavation. Noise and vibrations result in increased fatigue for the miners and other persons involved. Continuous noise exposure and high noise levels have serious health impacts on human habitation located adjacent to the mine, crushers, ore processing plants, major haulage roads, etc. Noise pollution is detrimental to human health.

v. **Loss of vegetation**: Mines are located in remote, densely vegetated areas like jungle/forest areas. For opencast mining, the land has to be cleared of any vegetation growth or forest. Likewise, for the infrastructure development, i.e. approach road preparation, necessary dump, or tailing storage, certain amount of damage has to be done to the flora or the vegetation involved. Thus, the impact of mining on flora is quite apparent. The loss of vegetation depends upon the location of the countryside and the mining area, scale of operations, mining method, degree of mechanization to be adopted, etc.

vi. **Impact on fauna/wildlife**: As a result of mining, the disturbance to the wildlife and animals is caused apparently. Obstruction, destruction, and migration of wildlife habitats (wildlife corridors) are serious impacts of mining on fauna.

Health and safety impacts are extremely important in mines and mining operations. While the former is largely associated with the person, the latter involves both personal and the workplace. The general health risk arises from dust/dusty working conditions in mines which may lead to respiratory problems. Radiation exposure, wherever applicable, also poses both health and safety issues for the miners. Underground mines are generally more hazardous compared to their surface counterparts because of poor ventilation, visibility, and risk of roof fall. In the case of surface mines, the safety risks are associated with the use of heavy machinery and equipment.

The adverse mining impacts on the air quality, water quality, and other natural resources can be either reduced to a minimum or kept contained, utilizing the mining policies and key scientific management practices, but not fully eliminated. Sustainable development goals (SDGs), which aimed to meet the growing needs of future generations, can lessen the negative impacts of mining thereby maintaining an equilibrium in the life cycle of non-renewable mineral resources.

1.12.1 'GO' and 'NO-GO' areas for mining

Government always prefer economic development over environmental protection. It is a well-known fact that mining causes negative impacts. To reduce the mining and environmental impacts, the concept of 'go' and 'no-go' areas was evolved by a typical clash between the 'green lobby' and the 'development lobby'. The purpose of demarcating the go/no-go areas, on a map, is an exercise similar to the preparation of the mineral map. It is meant for the coal-blocks allocation in the coalfields and thereafter smooth coal production in a sustainable manner. This has been applied in India in 2009–2010, particularly for coal mining areas, and it is helpful in decisions for the union ministry and coal company.

PART III

The financial aspects form an important part of the mining engineering devoted completely to the backdrop of planning, production, and management of a mine. The importance of the financial factors in the entire value chain system is extremely essential and exclusive for an improved mining operation (excavation). With a change in the nature and characteristics of each industry, the outlook of financial handling and management takes an altered shape barring the basic rules. Concerning the discussed industry as well, financial changes are inevitable because of the higher degree of risk and uncertainties (Section 1.14.1) involved. *Mining* and its co-brother *metallurgy* when paid due attention, as it should be, the purpose of the mineral excavation is fulfilled completely.

1.13 Introduction

Because mining is a capital-intensive industrial activity, a mining engineer should know important aspects of finances and investment. Although useful for senior executives, its basic knowledge is necessary for a budding engineer as well from a practical perspective. This cost-oriented knowledge has been discussed and described in this section of the chapter forming an integral part of undergraduate-level education. Compared to other industrial businesses, mining-related finances and investment are undoubtedly risky as it has a lot of variability with uncertainties involved at many stages. Minerals are a wasting asset and *mining economics*, which includes cost accounting, mining taxation, mine cost engineering, and mine valuation, all are directly linked to the cost of mineral production per ton. Indeed, such linkages of production and economics provide improved solutions and decide the investment as well as the financial health of the mining enterprise and mining operations.

1.14 Examination of mineral properties

Before getting started on the mine and mining operation, the examination of mineral properties is essential. It is done in stages starting through reconnaissance survey, prospecting, exploration, and reserve estimation (Box 1.3). This requires different *feasibility studies*, which mining companies often undertake while deciding whether to develop a mining project. Hence, *scoping study, pre-feasibility study, feasibility study, and*

Box 1.3: Reconnaissance and Exploration

Reconnaissance: Judging the location of a mineral deposit, its accessibility, rough interpretation from specific geological evidence, and knowledge about anticipated quantity, grade, geological characteristics, continuity, etc. of the identified mineral resource are known.

   Exploration: It is the search for natural resources, the testing of some places for identification of the resources through drilling or boring, and the examination of samples for possible mineral/metal content in the deposits. Exploration aims at locating and confirming the presence of economic deposits and establishing their nature, shape, and grade.

bankable-feasibility study are all comprehensive forward analyses of a project's economics to be used to assess the creditworthiness for project financing and investment. These studies vary in the depth of information on a case-to-case basis and are based on geological data, reserve data, and cost data. Having examined and evaluated the mineral property for commercial exploitation, the involved risks and uncertainties are dealt with.

   A mineral resource can be both economical and uneconomical, which are extracted during the normal course of mining. In this regard, two words are extremely important: *reserves (mineable)* and *resources*. The United Nations Framework Classification (UNFC) for reserves and resources, as applicable in the industry, is mostly referred to for this purpose **(Annexure 2.1)**.

   Reserves are those parts of the ore body and mineral resources that have reasonably good prospects for economical extraction. In this context, mineable reserve (quantity and quality) is more valuable as it is a material of intrinsic economic interest. The reserve of a deposit could be either a 'proved reserve' or 'probable reserve' and 'measured reserve' or 'indicated reserve' **(Annexure 2.1)**. As their name explains, these reserve types must include adequate information on exploitability (mining), processing (metallurgical aspects), and other relevant economic factors that demonstrate and give a technical justification or explanation for its economic extraction at the time of reporting. A complete description of the ore body is provided when the mineral properties are examined. The ore, metal content, and by-product information are all included in the examination report because the *ore* is a mixture of valuable minerals and gangue minerals from which at least one of the minerals can be extracted economically. Similarly, the by-products are secondary or additional products recovered in the extraction process, e.g. molybdenum is a common by-product of copper. In the examination of the mineral properties, their grouping in a ferrous and a non-ferrous group of metals could also be done accordingly (Figure 1.7).

1.14.1   *Risk and uncertainties*

Mining is an economic activity that involves risk and uncertainty in its various modules to become an everyday reality. Therefore, the explanation of uncertainty and risk must become a constant concern for the mining fraternity. Uncertainty and risk both

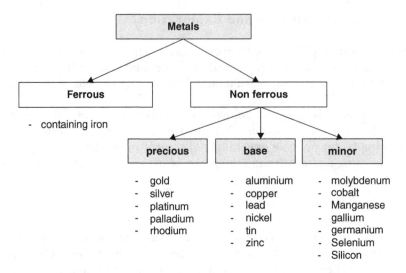

Figure 1.7 Ferrous and a non-ferrous group of metals.

look like relatively identical terms, but differ widely in economic theory, How? Let us define and explain risk and uncertainty:

- Risk is defined as the 'possibility of loss or injury' or 'someone or something that creates a hazard'. Using this definition of risk, you could classify nearly everything as risky because it's such a broad definition. For example, there's a risk that the battery inside the computer someone is using right now could spontaneously combust and blow up in his/her face. Thus, the risk is the probability of that happening. Right now, it is low and the impact that it would have on life is also low, but it could be more as well if combust happened on some occasion. Thus, it is apparent that there is risk in everything we do, whether the same is large or small.
- Uncertainty is defined as: 'not known beyond doubt' or 'not having certain knowledge'. In another way, uncertainty can also be defined as *incalculable or unquantifiable risk* because it cannot be measured. Uncertainty is the complete absence of quantifying the outcome of an endeavour.

Thus, uncertainty is different from risk. The major difference between risk and uncertainty is that the risk is quantifiable and uncertainty lacks quantification completely. A good modern-day example of risk and uncertainty, particularly in financial terms, is the 'crypto currency' and 'block chain'. In summary, we all unconsciously think about risk and uncertainty on a daily basis and at greater depths as a means of human instinct and survival. Risk and uncertainty are necessary and interesting as well for the hazardous mining industry.

An example: Suppose you have a jar with 50 red and 50 black marbles inside and you are asked to reach in and grab a red marble while blindfolded. If you grab a red marble, you'll win a cash prize of $5000. However, to play the game, you have to pay a certain amount of money. At this point in the experiment, the majority of those tested agreed to pay a reasonable amount of money for the chance to win $5000. Now, suppose you have a jar with the same number of marbles, but you don't know how many are red or black. If you grab a red marble, you'll win a cash prize of $5,000. At this point in the experiment, nearly all of those tested did not agree to pay to play the game for obvious reasons. When it was easy to calculate the probability of winning or losing, more people were inclined to pay for the chance to win a much larger prize. However, the opposite occurred when the risk of winning or losing could not be quantified.

1.15 Mineral economics/mining economics

Mineral economics, also referred sometimes as 'mining economics', is that branch of 'applied economics' which deals with the economic aspects of minerals exclusively. Since the problems and applications of the mining and mineral industry require different approaches, an approach to analyse facts and problems about minerals is needed. Defining it further:

> Mineral Economics is the academic discipline that investigates and promotes the understanding of economic and policy issues associated with the production and use of mineral commodities.

'Mineral Economics' is a complex subject quite apart from pure economics and its specialities are:

– The mineral industry and all activities related directly or indirectly to the economical extraction of minerals have a **long gestation period**.
– A high amount of **risk and uncertainties** in terms of quality, quantity, and usability of ore/mineral because of hidden occurrences below the ground.
– Prospecting/exploration is unique to mineral resources because an **in-depth and prior understanding** is needed for the mining of minerals.
– Minerals are **finite resources and non-renewable** independent of political boundaries. Their siting on Earth's crust is beyond human control and indispensable for the survival of mankind. The minerals are always supply-dependent.

'Mineral Economics' emerged as a separate academic field only after World War II because of the threat of supply interruptions for strategic and critical supply needs. All countries from the developed to developing and from the Middle East to the Soviet Union and Africa have fears of their smooth availability, as a raw material for industries, which has forced them to develop these non-renewable commodities (i.e. minerals) into an economic commodity, including applications, for the benefit of various aspects of the mineral industry.

1.15.1 Welfare economics

Welfare economics is a branch of economics like mineral economics that uses microeconomic techniques to evaluate well-being at the grass-root level. Welfare economics is the study of *how the allocation of resources and goods affects social welfare.* Together, they strengthen the institutional mechanism on problems and prospects related to the mineral sector, hence making it practically useful. Both for the company's financial and developmental works as well as regulatory needs, these economic principles are necessary for the long-term planning of mineral resources at the central level and the state level and are needed for proper resource use. Welfare economics, when applied, improves social welfare, i.e. overall well-being of people, income, and efficiency affect the local economy. Using social welfare functions and tools, through the government plan and schemes in the public interest, brings up an upliftment of societal status, thereby benefiting the masses directly.

Amartya Sen, the Indian economist and 1998 Nobel Prize winner in economic sciences, who was awarded the prize for his contributions to welfare economics and social choice theory is said to be the father of welfare economics. The problems of the poorest members of society can be tackled with an abiding interest in welfare economics. In welfare economics, many theories of welfare are developed by different economists. All of them are not agreeing with a single view that can be used for measuring social welfare; instead, there exists heterogeneity in the view on welfare.

Mining is a labour-intensive activity and labour force participation is an important component of the mining activity at most levels. Thus, the question arises: Is welfare good for the labour-oriented mining society and economy? Certainly, the answer is, Yes. Welfare measures help to reduce poverty and raise income, primarily through an increase in earnings of the poor families that has subsistence from the mines. It includes welfare programmes and welfare activities, such as medical aid, security, nutrition assistance, child's health, insurance programmes for future subsistence, housing assistance, and the tax credit (loan), required for temporary assistance to needy families. Some of them are supplemental and some attempt to apply for support, either for individuals or for the mining community.

Thus, knowledge of welfare economics is helpful in smooth mine operation, evaluating the costs and benefits of changes to the economy and guiding public policy towards increasing/enhancing the quality of life for the overall good of society.

1.16 Mine cost engineering

Mining projects have many different risks and uncertainties, as described in another section of this chapter depending on the specific situation of the project. There exists geological risk, ore-quality risk, economic and financial risk, geopolitical risk, and social risk. To get their work started, the mining and metal extraction firms need a large infrastructure as well as investment. Considering this fact, it is quite apparent that the commerce terms like cost, expenses, asset, loan, depreciation, credit, debt financing, salvage cost, valuation, and financial management become extremely important for any MDO of a mine be it an underground mine or a surface opencast mine. To know them, one has to take the help of 'mine cost engineering'. Mining companies may have different connotations for it but all led to the fact that they have to manage their operations as well as finances properly and effectively in order to become productive and viable. Both the immediate operational period (short) and long gestation period can be handled in a better way through scientific approaches in an organized and planned manner till the decommissioning phase is reached. In this interdisciplinary 'mine cost engineering' section, all these aspects are described in limited pages. Only important and key aspects have been covered here, broadly.

Mine cost engineering is a combination of different financial aspects right from cost assessment to financial valuation for a mine or mining enterprise that is undertaking the excavation of minerals. Different approaches and methods that are applied in financial management are also applicable to yield improved performance, efficiency, and better financial health of the mine or mining enterprise. It is necessary to have a basic knowledge of mining engineering, accounting procedure and finances/investment, and safety risks concerning that mine.

For corporate finance and accounting, many types of costs are involved. These are categorized as *direct costs & indirect costs, fixed costs & variable costs*, and *capital costs & operating costs.* For simplified description, two of the most common costs involved in mining have been described: (a) *the capital cost* and (b) *the operation cost.* A short description is given here under:

- Capital expenditures (CAPEX)

    Capital expenditures (CAPEX) are the major purchases a company makes that are designed to be used over the long term. Examples of CAPEX include physical assets, such as buildings, equipment, machinery, vehicles, computers, plant, and installations. CAPEX includes the cost of all infrastructure build-up.

    Capital expenditures are listed on the balance sheet **(Box 1.4)** under the fixed asset cost under the property, plant, and equipment section. CAPEX is also listed in the investing activities section of the cash flow statement. Fixed assets are depreciated over time to spread out the cost of the asset over its useful life. In this way, the word 'depreciation' comes into the picture which will be defined and explained later.

Box 1.4: Balance Sheet

A *balance sheet* is a financial statement that reports a company's assets, liabilities, and shareholders' equity at a specific point in time, which also provides a basis for computing rates of return and evaluating its capital structure. In other terms, a balance sheet is a summary of all the business assets and liabilities that the business owes. At any particular moment, it shows you how much money you would have leftover if you sold all your assets and paid off all your debts.

On one side, if there is an asset of a company, and on another, the liabilities of the company both must be kept balanced for the venture to run smoothly. The name '*balance sheet*' evolved from the fact that 'assets' will equal 'liabilities' and 'shareholders equity', every time. Accordingly, it is said that the balance sheet is (+ positive) or (– negative).

An annual balance sheet shows at a glance the business record for the asset, liabilities, and income statement (tax etc.) of the company and it is the most commonly referred document of the senior company officials. It is a snapshot of the company's business plan depicting the financial health of the company. Knowing the balance sheet is essential for the mining engineer when he/she occupies the senior responsible position in the company hierarchy.

CAPEX can be financed externally, which is usually done through collateral or loan (debt financing). Companies issue shares and bonds, take out loans, or use other debt instruments to arrange as well as increase their capital investment. Shareholders who receive dividend payments pay close attention to CAPEX numbers, looking for a company that pays out income while continuing to improve prospects for future profit.

- Operating expenses (OPEX) are the day-to-day expenses a company incurs to keep its mining business operational. The following are common examples of operating expenses:
  - Wages and salaries
  - Rent and utilities
  - Overhead costs and administrative expenses
  - Property taxes
  - Business travel
  - Interest paid on debt
  - Accounting and legal fees
  - Expenses for research and development (R&D)
  - Expenses for the cost of goods sold.
  - If a company chooses to lease or rent a piece of equipment instead of purchasing it as a capital expenditure, the lease cost would be classified as an operating expense.

  OPEX are necessary for a normal business whether in a mine or in any other industry. Any mining company reports OPEX in their 'income statement' and OPEX is subjected to the taxes for the assessment year in which the expenses were incurred.

Since all types of mining business costs can be organized under these two broad categories, we must also know the differences between them. The key difference between CAPEX and OPEX is that capital expenditures are major purchases that will be used beyond the current accounting period in which they're purchased. Whereas the OPEX are short-term expenses and are typically used for evaluation of the accounts in the concerned financial year. This means that they are paid weekly, monthly, or annually. CAPEX costs are paid upfront all at once. The returns on CAPEX take a longer time to realize, for example, machinery for a new project, whereas the returns of OPEX are much shorter, such as the work that an employee does daily to earn their wages. The goal of any company is to keep OPEX minimum to maximize the output. Because of their different attributes, each is handled separately for the company's efficiency over time.

1.16.1 Some cost engineering terms

Assets: The term 'asset' signifies all kinds of resources that help generate revenue as well as receivables. Assets are resources that often help to reduce expenses, enhance profitability, and generate robust cash flow as they help convert raw materials or can be converted into cash or cash equivalents. Further, due to their economic value, they can be quickly sold or exchanged. Notably, assets are reported by the mining company as a resource that is maintained/involved in the commercial practice. It is mandatory to make mention of the asset in the company's balance sheet.

Liabilities: The term liability signifies all types of account payables. It can further be defined as a financial obligation that individuals must meet. Usually, liabilities tend to play a significant role when it comes to financing expansion or ensuring the smooth processing of everyday operations of commercial practices. Further, depending on the

Table 1.4 Asset and Liabilities

Assets	Liabilities
• Fixed assets and current assets	• Current liabilities
• Tangible and intangible assets	• Non-current liabilities
• Operating and non-operating assets	• Contingent liabilities

type of company, such liabilities can either be limited or unlimited. In the case of the former, owners are not entirely obligated to compensate or pay off for the venture's liability, whereas in the latter, the resulting liability is solely the responsibility of the owners. The liability of the **commercial** company is equally important as that of the asset since it is an obligation and must be reported in the balance sheet of the company.

Different **types of assets and liabilities** are given in Table 1.4.

When a company is sold or purchased, both assets and liabilities tend to play a vital role. Similarly, when it comes to ensuring the profitability of a business or its long-term viability, both **assets and liabilities** are vital components for cost considerations. They are to be managed effectively and judiciously for the sustainability of a commercial venture.

Book value: It is the amount of a property shown in the books after allowing necessary depreciation year-wise. The book value is independent of market value.

Market value: It is defined as the value that a property can fetch when sold out in the open market. This value is variable, depending upon the will to buy or sell.

Working capital: In accounting terms, the working capital is the difference between current assets (inventory, cash and accounts receivable) and current liabilities (accounts payables, short-term debt and debt due within the next year). From a cash flow perspective, it is the difference between non-cash current assets (inventory and accounts receivable) and non-debt current liabilities (accounts payable).

Liquidity: Liquidity refers to the efficiency or ease with which an asset or security can be converted into ready cash without affecting its market price. Cash is universally considered the most liquid asset because it can most quickly and easily be converted into other assets.

Annuity: The return on capital investment in the shape of annual instalments (monthly, quarterly, half-yearly, and yearly) for a fixed number of years is known as an annuity.

Present value: The present value of the building can be found using any of these two methods, viz., Value depending upon the Plinth Area and Value from detailed measurement. The *plan area* is multiplied by either the plinth area or the area measured manually.

Gross or net income: Total amount of the income received from a property during the year, without deducting outgoing. Outgoings are the incurred expenses such as taxes, repairs, management expenses, insurance premiums, and other miscellaneous fixed by town planners or municipalities before doing any construction.

Marginal cost and incremental costs: The 'marginal cost' is the increase in cost as a result of a unit change in output. In other words, marginal cost is the rate at which total cost changes with the change in output. This cost is the additional cost incurred resulting from the change in the existing output or production of coal or ore.

'Incremental costs' may be defined as the addition to all costs resulting from a contemplated change in the nature and the level of business activity including policy. Examples of the change are change in the level of investment, production/output level changes, product line or product mix changes, adding or replacing a machine or equipment, and change in the advertising policy of the company. While taking a decision, it should be ascertained that incremental revenue generated (revenue resulting from a change) should exceed the incremental cost. Besides produced minerals, goods and services have a marginal and incremental costs too.

In the case of mines, the per ton production cost of mineral/ore is significantly influenced by these two costs practically because the volume of material handled is in terms of million tonnes. Since there exists a very small (hairline) difference between these two additional costs, the difference must be clearly understood. The main difference that is significant to note between these two costs is that, while the marginal cost refers to the change in total cost (resulting from producing an additional unit of output), the incremental cost comes into the picture when there is an overall change, e.g. new grade or type of ore is added for excavation (as a result of the policy decision or as a result to expand output, a new product line or new cost centre is added).

1.16.2 Expenditure and cost accounting in mining

Mining businesses have a variety of expenses, from the lease they occupy, the rent they pay for their offices, colony and factories installations, wages they pay to their workers, the cost of the raw materials for their products, etc. In professional business terms, all costs involved, including expenditure, are accounted for in cost-accounting terms and essential to know the overall financial health as well as the growth of the operative mining company.

Cost accounting is an accounting process that measures both fixed and variable costs associated with coal/ore/mineral production, including goods and services. Both together can make the mine/industry, running, growing, and established. The purpose of cost accounting is to assist mine management in decision-making processes to optimize mining operations based on efficient cost management. Thus, assessing a company's cost structure allows management to improve the way it runs its business and therefore improve the value of the mining company. Cost management and cost accounting help the mining firm throughout its life be it the establishment stage or during the process of establishing.

1.16.3 Depreciation and amortization

The assets a company acquires come at a cost. Most assets don't last forever, so their cost needs to be proportionately expensed for the period they are being used within. The method of depicting/portraying (prorating) the cost of assets throughout their useful life is called 'depreciation' and 'amortization'. Thus, in brief, *depreciation and amortization* are the two methods of calculating the value of business assets over time.

Depreciation and amortization are income statement items. There are many different terms and financial concepts incorporated into income statements. Two of these concepts – depreciation and amortization – can be somewhat confusing, but they are essentially used to account for *decreasing value of assets over time*.

Depreciation: Expensing a 'fixed asset' over its useful life cycle is called 'depreciation'. Buildings, equipment, land, machinery, office furniture, and vehicles all have a depreciation value. It is the cost that should be included in the account records. As an example, an office building can be used for several years before it becomes run down and is sold. The cost of the building is spread out over its predicted life, with a portion of the cost being expensed in each accounting year. Similarly, a vehicle is worn out with time and it has a *'depreciation cost'* to be included over years during its life cycle. Any running vehicle of any type is an example of 'accelerated depreciation' meaning that it has a larger portion of the asset's value expensed in the early years of the asset's life cycle. Thus, the main cost plus depreciation covers the entire cost of an asset on a balance sheet.

Amortization: The practice of spreading an intangible asset's cost over the asset's useful life cycle is called amortization. Intangible assets are not physical assets. Intangible assets that are expensed through amortization include organizational costs, franchise agreements, proprietary processes like copyrights, trademarks, and patents, as well as the cost of issuing bonds to raise capital for the company. Amortization is typically expensed on a straight-line basis, meaning the same amount is expensed in each period over the asset's useful life cycle. Assets expensed using the amortization method usually don't have any resale or salvage value, unlike depreciation. The word amortization carries a double meaning, so it is important to note the context in which someone is using it. An amortization schedule is used to calculate a series of loan payments of both the principal and interest in each payment as in the case of a mortgage. So, the word amortization is used in both accounting and lending with completely different definitions.

The main difference between depreciation and amortization is that depreciation is used for tangible assets, while amortization is used for intangible assets. All assets, liabilities, depreciation, taxes, and income are portrayed in the company's balance sheet every year **(Box 1.4)**.

Method for calculation of depreciation: There are four main methods for depreciation calculation:

1. Straight-line method
2. Declining balance method
3. Sum-of-the-years' digits method and
4. Units of production method

Generally, the method of depreciation to be used depends upon the patterns of expected benefits obtainable from a given asset. This means different methods would apply to different types of assets in a company. However, in reality, companies do not think about the service benefit patterns when selecting a depreciation method. In general, only a single method is applied to all of the company's depreciable assets.

Depreciation is calculated each year for tax purposes and is expressed in monetary terms such as dollars/pounds/rupees. The first-year depreciation calculation is:

Depreciation value = Cost of the asset – Salvage value/Years of useful life

Each year, use the prior year's adjusted cost (depreciation value/cost) for that year's calculation. The next year's calculation is based on the previous year's total. Here, we must understand what is the 'salvage value'. Salvage value is the amount that an asset is estimated to be worth at the end of its useful life. It is also known as **scrap value** or residual value and is used when determining the annual depreciation expense of an asset. Thus, *Salvage Value* is determined for the property/asset without being broken into pieces, i.e. as on whereas on basis. After being discarded at the end of the utility period and sold, the amount realized by the sale or the value of the asset is recorded on a company's balance sheet, while the depreciation expense is recorded on its income statement.

Among the four methods mentioned above, the straight-line depreciation method is a very common method as it is being the simplest. In this depreciation method, the expense amount is the same every year over the useful life of the asset.

Example: Consider a piece of equipment that costs say $25,000 with an estimated useful life of 8 years and a $ 0 salvage value. The depreciation expense per year for this equipment (Table 1.5) would be as follows.

Here, only one basic method of depreciation is described for explanation and the rest is left at the readers' discretion. Interested readers can refer to the literature for it.

1.16.4   Mining cost estimation and production/productivity linkages

In a mining project, the mining cost estimates matter a lot. Reliable cost estimation reduces financial risk. If the cost, budgets, and financing have been fixed properly, the other planning can be set accordingly. The cost estimation provides accurate CAPEX and OPEX estimates recognized by financial institutions globally. These are required for new mine projects, mine project expansion, plant upgrades, retrofits, greenfield projects, and brown-field expansions.

Table 1.5 Depreciation by Straight-Line Method

Year		1	2	3	4	5	6	7	8
Straight-line method									
Opening book value	–	25,000	21,875	18,750	15,625	12,500	9,375	6,250	3,125
Depreciation	8	3,125	3,125	3,125	3,125	3,125	3,125	3,125	3,125
Ending book value	25,000	21,875	18,750	15,625	12,500	9,375	6,250	3,125	0

Note: All values in the dollar ($) or currency used.

Capital cost (Cc) and Operating cost (Oc): The cost estimate for *underground mining* and *surface mining* differs because the components involved are different. The estimation of the capital cost and operating costs revolves around the various production costs in the case of a mine. For a surface mine, *pit development cost; excavation cost, i.e. drilling blasting and loading; mineral/coal/ore transportation cost;* and *crushing/sizing cost* are considered, whereas for an underground mine, *cost towards roadway development/drifting; haulage cost; stope development and depillaring, i.e. drilling, blasting, and mucking; HW/FW or roof support cost;* and *ventilation cost,* based on the mining method selected for mining, are considered. For instance, the access to the ore/deposit (shaft/incline-decline for an underground mine), mineral processing cost for metal recovery from the ore, the capacity and efficiency of the processing plant, and the grade and percentage of recovery are involved. Therefore, in cost estimation, it is required that all costs involved, i.e. infrastructure costs, mining (excavation) costs, and mineral processing costs, must be included.

Largely, capital cost (also termed as 'ownership cost', hailed with entrepreneur or company for the assets owned) is fixed, but the operating costs keep on varying. The operation cycle of mining may continue round the clock and the machines deployed may or may not be used. This decides the operating cost; hence, it is referred to as the variable. If the equipment is new, the maintenance cost is low, and if it is old, the reverse holds good.

Capital and operating costs can be estimated as follows. For both surface mine (OC) and underground mine (UG), they are given separately. **Annexure 1.3** may be referred to for details of the calculation.

- Capital Cost (Cc) estimation:

    Capital Cost$_{OC}$ = Pit development cost (Roads and permanent installations) + Asset installation cost + HEMM cost + Infrastructure cost + Interest on capital + depreciation.

    Capital Cost$_{UG}$ = Incline/shaft drivages cost with permanent support + Mine roadway development cost + Cost of Conveyor belt installation + Bunker cost + Cost for winder and haulage with track and trolleys + Interest on capital + depreciation

- Operating cost (Oc) estimation

    Operating Cost$_{OC}$ = Excavation cost (i.e. drilling, blasting, loading) + transportation cost for mineral/coal/ore + crushing/sizing cost + maintenance cost for equipment and machinery + manpower cost and tax plus depreciation.

    Operating Cost$_{UG}$ = Digging cost (drilling and blasting with cost of explosive) + loading cost + transportation cost + support cost + ventilation cost + maintenance cost for equipment and machinery + manpower cost and tax plus depreciation.

Cost analysis: Inherently, underground mining cost exceeds the surface mining cost (Figure 1.8). An expert/financial consultant assistance may be taken for cost analysis because production and productivity of the mine are intricately linked with the per ton cost of ore/mineral produced from the mine. All these tools and procedures are helpful to mitigate financial risk and obtain the expected return on investment.

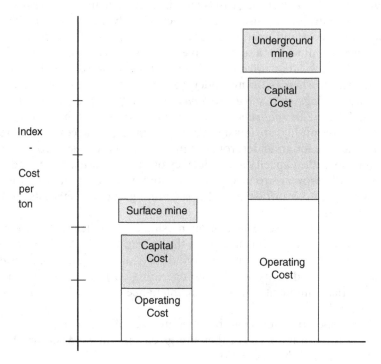

Figure 1.8 Cost comparison: underground vs surface.

1.17　Mine valuation

If a question is asked by a mining engineer, what is the valuation of your mine and how the mine valuation is done technically? This section answers the same.

What is mine valuation?

- A 'valuation' is an objective search for the 'true value' of a property.
  OR
- Valuation means the fixation of cost or returns expected of a property, building, or engineering structure in a project either owned by govt. or private at present-day rates.

Since, the cost or returns of property, evaluated in monetary terms, dwindles with time and varies with its upkeep and maintenance. The assessed value may be either more or less from the original depending upon the present utility of the property. For example, a house/building, made at a cost of say @ 10,000 dollars in the year 2000 may have the assessed/evaluated cost of 20,000 dollars in the year 2021 if the building is well maintained or else also fetch 5,000 dollars if it is in depilated condition.

Generally, the necessity of valuation arises for a man-made property, e.g. plant, building, and machine. Buying and selling, for acquisition, for rent, for the mortgage, and for insurance policies of property necessitate the valuation of the property. The role of an engineer in valuation is felt when an engineering structure is to be valued.

1.17.1   Mining companies, valuation, and corporate considerations

A good valuation indeed provides a precise estimate of the mine. It is comparatively difficult to assess a mine in terms of its value, but generally, the geological reserve of the mine is a major parameter for mine valuation. With common sense, one can say easily that amine, which is a storehouse of minerals with ample reserve at the start of the mining, may have a higher value than a mine that has its ore completely or partially exhausted.

There is no precise *valuation method* and most valuations are biased. Practically, all valuations of mining companies today differ from each other, not only because the companies are different but also because different people with different knowledge and backgrounds do the valuations. Uncertainties, especially in valuing a mine, are more prevalent because estimating the production for the coming years in advance is quite difficult and unpredictable. The price forecasts of the metals (extraction from ore/minerals), as underlying commodity, e.g. copper, zinc, coal, and manganese, are also very difficult to assess even between professional analysts. Also, other forecasting parameters like discounted factors and taxes (tax on the mineral resources keep on changing the world over frequently) that are considered as input in the valuation methodology vary from analyst to analyst. Because of these reasons, i.e. possible forecast and uncertain conditions, no valuation method for a mine (or a mineral mining company) can be said to be right, but no method is wrong either.

In brief, mine valuation is a task for corporate considerations by the mining companies that remains seriously distorted in both short and long terms. If a mining project is running as a corporate entity, various company-related issues arise that can have a significant impact on the company's market value. Mining companies with a strong balance sheet, adequate working capital, and reasonable leverage generally occupy a premium space in the capital market. Mining companies with higher capitalization and greater liquidity generally attract a premium in the market. Conversely, mining companies with high debt levels and financial liquidity generally face trading difficulties and attract less attention. Depending on the market's outlook about the mineral being produced (high-value/low-value mineral), the mining company may attract a significant premium or discount in the capital/share market. It is significant here to note that the mining valuation is inter-linked with the mining investment.

1.18   Mining taxation

Minerals are a national asset. To carry out the mining activities, all mineral extraction companies, either a public limited organization or a private company, have to pay a cost usually referred to as 'mining taxes'. Prospecting and exploration of minerals also attract taxation and fall under this category. These taxes are of different types and generate a significant amount of revenue for the government. A government-controlled mechanism exists for such tax/revenue collection that includes a geological search of minerals as well. This section of the book describes *mining taxation* citing the example of India which makes us understand its importance necessary for the industry in general, and the concerned government of the nation in particular. For a mineral-rich country, the revenue from the mines and the minerals extraction is a helpful resource for national development.

The taxes or levies payable by the mining company (mine) differ in quantum and nature depending on the country and states where it is located. Principally and broadly, two types of taxes apply to the mining industry: (a) Direct taxes, such as corporate tax (or minimum alternative tax) and (b) Indirect taxes, such as customs duty, service tax, value-added tax; stamp duty; water tax; forest-related taxes (for the forest produce removed from forest areas). In this sub-section, taxes as applicable to the mines located in India have been described. The taxes are as per the prevalent laws involving public policy and national policy that promote the sustainable development and conservation of minerals.

a.   **Royalty:** Royalty is the most common form of tax or a charge levied on the mineral produced from mines. The federal government specifies the royalty rates (payments) for each mineral and the state government collects the royalty on mining. Royalty in most cases is charged differently for the various mineral produced from the mines and can be a fixed payment on a per ton basis or an ad Valorem basis as a percentage of the price notified by the government. Enhancement to the royalty can be made through government intervention periodically, maybe once every 3 years.
b.   **Dead rent:** A mining rights holder is liable to pay either royalty or dead rent in respect of a mining area, whichever is higher. Dead rent is, therefore, meant to be paid when the mine is closed or is being underexploited. Dead rent is fixed by the federal government and is collected by the state. Any enhancement to the dead rent can only be done once in 3 years.
c.   **Cess:** It is a fixed tax levied on mineral ores under various legislation.
d.   **Surface rent:** The mining rights holder (lesse) has to pay, wherever applicable, *surface rent* to the surface rights owners.
e.   **Mining concession fees:** This fee, fixed by the government, remains payable by the mines or the mining organizations holding the mineral rights to the government for the mining.
f.   **Other taxes:** Goods and Service Taxes (GST), Income taxes, Exports taxes, Sales taxes, Excise taxes, Property/Estate taxes, Gift taxes, and Road taxes are some other taxes types, applicable for the mining areas and for miners who live in the mining areas. GST for various commodities and services are applicable at different rates ranging from 5% (minimum) to 28% (maximum).

Besides this, application fees for the licence or lease that are fixed by the federal government and collected by the state or agency involved are also a form of tax payment applicable when a mining lease comes into existence. While running the mine, NMET fund and DMF contributions are the indirect taxes or a form of tax levied on the mining companies meant for societal development and revenue generation. A mining rights holder has to pay a sum equal to 2% of the royalty as a contribution to the NMET fund. DMF contributions are to be fixed by the federal government but cannot exceed one-third of the royalty specified.

1.18.1   NMET fund

'National Mineral Exploration Trust' is abbreviated as NMET. For carrying out regional and detailed exploration for minerals, a fund is created called 'National Mineral Exploration Trust (NMET) fund'. This fund has been augmented under the National

Mineral Exploration Trust Rules, 2015, in exercise of the powers vested with the Central Government of India, under the Mines and Minerals (Regulation and Development) Act, 1957.

The NMET fund money comes from the mineral royalty and comprises of payment of 2% equivalent of royalty payable by the holders of the mining lease or prospective licence-cum-mining lease under the MMRD Act. Thus, contribution to the fund comes from the holder of a mining lease or prospecting licence-cum-mining lease through the mineral royalty route.

The NMET fund is kept in the 'Public Account' of the concerned state under the specific head (Consolidated Fund of India) booked for the purpose. The NMET fund is a non-lapsable and non-interest-bearing fund. The concerned state government provides information regarding the amount collected and the amount transferred to the Consolidated Fund of India to the Indian Bureau of Mines monthly. This information shall be provided not later than the tenth day of any particular month or the succeeding month in which the royalty is deposited. Since the Indian Bureau of Mines (IBM) is a custodian of royalty payment, it maintains an updated record of the amount transferred from the respective state along with a database of royalty payments. IBM provides such information to the government periodically for utilization.

1.18.2   District Mineral Foundation (DMF)

District Mineral Foundation (DMF) is a mechanism to drive developmental work in mining-affected areas. It includes both money/funds as well as powers to execute. State government official – 'District Collector' – is the principal authority for controlling DMF. The DMFs were introduced in January 2015 by the Government of India in all districts in the country affected by mining-related operations. The provision was brought through the Mines and Minerals (Development and Regulation) Amendment Ordinance, 2015 and hailed as a golden pill for the upliftment of such areas.

The exact composition and functions of the DMFs were to be decided by the state governments, but the periodical amendment of the law had specified that the use of this fund has to be in line with the provisions of *the Panchayats Act, 1996* and *Forest Rights Act, 2006*. In 2015, the government notified the rates of contribution payable by miners into the DMFs, and according to that, all mining leases executed before 12 January 2015 require mine leaseholders to contribute an amount equal to 30% of the royalty payable by them, while in the case of mining leases granted after that, the rate of contribution is 10% of the royalty payable. Following this, by now, the DMFs have been constituted in 574 districts in 21 states with about Rs. 41,650 crores (Rs 416 billion) fund collected till September 2020.

With a focus on livelihoods based on natural resources to improve the local economy and community, the DMF is a contribution to infrastructure strengthening, societal development, and environmental protection. DMF and its mechanism seem sound at first look, but it is marked with many notable deficiencies. For example, (a) The use, misuse, and abuse of the DMFs are very high, e.g. Civil society leaders and experts complain about the poor implementation of the DMF funds, the delivery of services, and the involvement of local communities. (b) It has a lot of political interference and hence is not transparent in fund handling. (c) Poor participation of mining-affected communities affects the ground truth regarding execution and implementation. The DMF fund flow is not to the extent expected and may not be reaching the mining-affected communities. In many mining areas across India, the lives of the affected mining communities, local people, and biodiversity including forest and water bodies have undergone a transition with the help of DMF.

1.18.3 Green fiscal regime

Environmental challenges arising from mining are immense. While framing the basic objectives of taxation, the idea of the 'polluter pays principle' is evolved. In consonance with this, a new term took birth while designing, planning, and framing the financial system for the mining sector called the 'Green Fiscal Regime'.

A green fiscal regime for the mining sector is defined as a regime that uses the tax system which influences environmental outcomes from mining activities. Such a regime would tax pollutants, water, and energy used in the mining project while reducing taxes on non-polluting inputs. Such an approach can be an alternative or a complement to traditional regulatory approaches. By integrating environmental concerns into the tax system, better environment protection is possible because the mining activity can lead to significant environmental and social impacts ranging from landscape degradation to air and water pollution, leading to long-lasting consequences for the local environment and health of the population. The mining sector is also water and energy-intensive, with the use of fuel oil and diesel making mining a major emitter of greenhouse gas (GHG) emissions. Thus, placing environmental protection alongside revenue generation (taxation) and investor returns marks a major change in the current fiscal practices of the mining sector.

1.18.4 Multiplicity of mining taxes

An industry review reveals that the mining industry in various parts of the globe is inflicted with the problem of multiplicity of taxes. This persistent problem of the mining industry requires a relook to improve further.

The principal objective of explaining the mining taxation and related fund here is to explain to readers the financial resources available for actual use and the inter-linkages between mining and exploration which are in practice practically.

1.19 Investment decision

When a mine is opened and becomes operational, even thereafter it continues to attract many different excavation activities, e.g. shaft or adit drivages, winder installations, overburden bench sizing and development, sump construction, cross-cut development and approach for stopes/new mining district underground, during the operational phase. All these require a large size capitalistic investment practically. These big sizes of work and services we are referring to here as a 'major investment', barring routine production activities of the mine.

Major investments in an ongoing mining project amounting to millions of dollars require a firm decision that starts with the preparation of a Detailed project report (DPR) supported by a base feasibility document referred to as a 'Techno-Economic

Feasibility Report (TEFR)'. All major investment proposals require feasibility assessment, checking, and in-depth evaluation of technical as well as financial parameters before being implemented into the field. The mining company, either on their own or through an expert agency, prepares it.

The DPR/TEFR is a report-type document based on multidisciplinary engineering and economics input, comprising all details that include introduction, the proposal, geology of the mine, exploration details, ore reserves, i.e. total reserves, level-wise reserves, and shaft-wise reserves. In addition, mining details describe the details of production (year-wise and grade-wise clean ore/coal production), production schedule, development details in a designated period, i.e. current and projected, operating cost of mine, and how access to the different underground levels is in real existence at the mine and their status. Justification and need for the mining installation, i.e. new or existing excavation, be it a vertical shaft, an incline/decline including stopes, depillaring district, or the development panel, all the important salient information form part of the report description. TEFR alone contains complete technical cum commercial information for the work schedule, quantity, and cost estimate for the proposed investment, execution plan, coal/ore sale cost, and year-wise investment in monetary terms. The financial estimate of the project, as well as major raw items such as steel, iron rails, concrete, pipes, and cables, has been detailed in tabular form and explained with drawings and explanatory annexures. Having prepared the TEFR, it is placed before the decision takers who are the project experts and the specialist for examination, evaluation, and implementation. In brief, TEFR is a document based on which an investment decision can be implemented into practice. Once an investment decision has been taken and the funds allocated, it is often cumbersome as well as difficult to justify the project in both technical and economic terms if it has not been proved scientifically. Therefore, it is essential as well as important that major-size investment proposals are properly evaluated and funds are correctly disbursed. In this context, third-party examination (an independent agency other than the DPR/TEFR preparation agency and the project funding organization), scrutiny/in-depth analysis, and vetting of the investment proposal are the further steps involved in the real-time execution of the project.

Mining is a risk-prone industrial investment, be it in exploration or mine development. Geological uncertainties and site potential are the characteristic features of all mining activities in different world regions. Therefore, the key factors determining investment decisions are fiscal and socio-political besides the uncertainties. One of the most important criteria that companies consider in making investment decisions is related to taxation they have to consider. The financial evaluation of investment is done considering the prevalent tax structure at the prevailing rate. Value-Added Tax (VAT), Goods and Service Tax (GST), Service Tax (ST), and Cess are some common tax types applied in financial evaluation. The type of taxes applicable differs from country to country, e.g. Indian tax system adopts GST into practice, whereas VAT is applicable in the UK. An effective tax system provides an attractive investment and makes the investment rather easy given the diversity in the mining sector operations (Mitchell, 2005).

Many a time, international investors are involved in major mining investment decisions, e.g. shaft sinking. In such selected cases, a combination of technical expertise, strategic partnership, and political will pave the way for international cooperation thereby attracting investors for making mining investments. Looking at the commercial prospects and potential profitability of the project, these investments with higher risks have been undertaken (Pritchard, 2005).

To help for the quick economic evaluation of the risky mining investment, mathematical planning tools and computer program have been developed, e.g. OPT-MINEINVEST – an optimum investment decision model with a computer program that provides the values of key economic parameters in terms of revenue, thus enhancing its application for all mineral types (Mohnot and Singh, 2014). The correlations between optimum mine investment vs revenue and optimum mine operating cost vs revenue are useful to guide towards incurring more profits in mining ventures.

1.19.1   Methodology for evaluating investment proposals

Mining – A risky business with uncertainties about ore bodies underground, fluctuating metal prices, volatile exchange rates, environmental issues, legislation hurdles, and political risks makes capital investment a puzzle for every decision-maker throughout the industry (Mirakovski et al., 2009). Thus, the major challenges of project evaluation and investment decisions are: How to evaluate it and deal thereafter with risks and uncertainties? Several methods, namely, Discounted Cash Flow (DCF), Decision Trees (DT), and Monte Carlo Simulations (MCS), are commonly used for the evaluation of mining investment projects. It applies to mines and other ancillary investments in major investment categories equally.

DCF method is the most commonly used method of investment evaluation and has been around for many years. NPV methods and IRR methods, a part of the DCF method, are those methods for project evaluation, being applied in a mining company. NPV method and IRR method of project appraisal/evaluation consider the payback period and cash flow. It is superior to other methods because the value of money invested at that time is duly considered. Other methods are not popular because of their inherent disadvantages. There are four major important heads of evaluating investment proposals and they are:

  i.   Accounting rate of return (ARR)
 ii.   Payback period (PP)
iii.   Net present value (NPV) method
 iv.   Internal rate of return (IRR)

The NPV of a project is defined as 'the present value of all future cash flows produced by an investment, less the initial cost of the investment'. This method discounts the future cash flows associated with the investment project using the cost of practical as the appropriate discount rate. If the NPV of the discounted cash flows is positive, one should accept the project as viable and the decision can be taken in favour. As mentioned earlier, NPV and IRR methods are considered superior to the ARR and PP methods because of the advantages offered by them. The most important advantages of the NPV method are:

•   It takes account of the time value of money by discounting the cash flows arising in the future.
•   It takes account of all relevant cash flows.

- It is consistent to maximize profit (i.e. shareholder wealth), which is assumed to be the primary objective of a business.
- It provides a clear decision concerning the acceptance/rejection of a project.

The *IRR method* is yet another method similar to the NPV method. The IRR of a project is the discount rate at which the present value of future revenues is less than the present value of future costs. In other words, IRR is the method of discounting future cash flows, and it will normally give the same accept/reject decisions and rank investment projects in the same way as the NPV method. However, the IRR method has difficulty in handling unconventional cash flows and does not address the issue of 'wealth maximization' clearly as explained by the NPV method. Thus, from a theoretical viewpoint, the IRR method is inferior to the NPV method. IRR is expressed in percentage (%) and NPV is expressed in rupees (Rs). However, many prefer the IRR method to the NPV method because it provides an answer in % which is easily understandable by the reader. The IRR and NPV are inversely proportional to 'capital costs' and directly proportional to profit (revenue less operating costs).

In the abovementioned methods, two related terms, namely, 'Profitability Index (PI)' and 'Payback Period', are usually encountered.

The *profitability index*, also known as the *profit investment ratio* and *value investment ratio*, is the ratio of payoff to the investment of a proposed project. It is a financial tool that tells us whether an investment should be accepted or rejected. PI greater than one indicates that the present value of future cash inflows from the investment is more than the initial investment, thereby indicating that it will earn profits. Its calculation is shown in the next sub-section below. For the proposed investment to be supportive and feasible so that it can be implemented into practice, the PI should be close to 1 (i.e. ≈0.7, 0.8, 0.9) and estimation is done *before tax* and *after tax*. If mine production is increased, the *profitability* is also enhanced and PI will reach nearly or equal to 1 which is indicative of a feasible/viable investment.

The *payback period* refers to the amount of time it takes to recover the cost of an investment. In more simple words, the payback period is the length of time an investment reaches a break-even point. The payback period, expressed in terms of years/months, should be less as far as possible. Shorter paybacks mean more attractive investments. The desirability of an investment is directly related to its payback period.

The choice of the NPV, IRR, ARR, or any other method to arrive at a proper decision depends on the kind of proposal being assessed (feasibility assessment) and the financial evaluation sought. The availability of data also matters. Therefore, the theoretical aspects of 'mine economics' must be diligently understood which covers both economical and technical parameters. Here, only calculation details are described below. However, for the theoretical details, readers are advised to refer mine economics book.

1.19.2 Calculation details for the techno-financial investment

- **NPV Calculation:** NPV is the sum of all cash flows at a discount rate. It is calculated using the following formula which represents an investment potential:

$$\text{NPV} = \text{Cummulative cash flow} \times \text{Discount rate} \tag{1.1}$$

To calculate the NPV, one will discount each cash flow with the discount rate keeping in mind the time-lapse. NPV is calculated in currency (dollars/pounds/rupees) and estimation is done *before tax* and *after tax.*

- **PI Calculation:** The PI is calculated as follows:

$$\text{Profitability Index (PI)} = \frac{\text{Present value of cash flow}}{\text{Present value of cash outflows}}$$

$$\text{i.e.,} = \frac{\text{Discounted cumulative cash flow}}{\text{Investment proposed}}$$

- **IRR Calculation:** IRR is one of the major and good parameters for judging an investment. The bigger the IRR, the better the proposal. In brief, the IRR is the interest rate that makes the NPV zero. IRR is represented in % and it is impossible to understand IRR without knowing the NPV. The IRR is calculated using an MS excel worksheet quite that is easily available online. Like NPV, IRR can be calculated/estimated for various values, namely, before tax and after tax, with cumulative cash flow or with net cash flow.

1.19.3 Mining valuation vis-à-vis mining investment

For a mine, investment and valuation are two interrelated terms. Every asset of mine has an intrinsic value that can be evaluated, based on its financial characteristics, e.g. in terms of value, cash flows, growth, and risk. In a *discounted cash flow valuation (DCF)*, the value of an asset is the present value of the expected cash flows on the asset. To obtain the present value using DCF, one needs to know the life of the asset, the cash flows during the life of the asset, and the discount rate that applies to these cash flows. The mining investment, on the contrary, fulfils the financial need of the mining company which is later evaluated. Markets are assumed to take care of both investment and valuation in pricing assets across time and are assumed to correct themselves over time as new information comes out about the assets. When forecasting the future growth of a mining company, it is important to assess the working capital needs, investment planning, and cash flows. At the same time, the valuation of the asset and liabilities needs to be done judiciously because it enables us to assess fiscal growth and the effects of such growth on the company's overall financial health.

PART – IV

From a policy perspective, there are two broader domains: firstly the *National Policy* and secondly the *Industrial Policy.* Inter-linked to these two is the 'Organization Policy' which is a prime mover for the mine or the company. The organization policy is a must for every company because it gives focus to the work and priorities. Better the organization policy, the best will be the execution and implementation turning the company as profitable. Each of the countries has its own policies in place that are driven by a large number of national & international constraints and local & regional factors required for national development. Consumer satisfaction, enterprise resource planning, and progressive system improvements in the company and projects, through monitoring mechanisms, are the policy targets, to be achieved.

When one (or a single) sector is focused on the industrial policy, it is termed as the industry-specific policy for that specific sector. For mineral-rich countries, mining or mineral-oriented industrial policy is desirable. Indeed, generalization of the policy, either national, industrial, or mineral, is not a correct approach. Considering these aspects, the following are framed.

1.20   Mineral policy

In general and for those countries where minerals are excavated in large magnitude and the minerals industry contributes to the economy significantly, a country-level mineral policy is framed and kept in place. It is also true that all countries may not have such policies. One such mineral-dominant country is India. In India, National Mineral Policy (NMP), 2019 was formulated by the Government of India, Ministry of Mines for non-fuel and non-coal minerals. 'Dept. of Coal' is functional at the mines ministry to look after the policy matters of coal separately. The reproduced vision of the Indian NMP, 2019 document says that:

> Minerals are a valuable natural resource being the vital raw material for the core sectors of the Indian economy. Exploration, extraction and management of minerals have to be guided by national goals and perspectives, to be integrated into the overall strategy of the country's economic development. The endeavour shall be to promote domestic industry, reduce import dependency, and feed into the Make in India initiative. Natural resources, including minerals, are a shared inheritance where the State is a trustee on behalf of the people and therefore the allocation of mineral resources must be done in a fair and transparent manner to ensure equitable distribution of mineral wealth to sub-serve the common good. Mining needs to be carried out in an environmentally sustainable manner keeping stakeholders' participation, and devolution of benefits to the mining-affected persons with the overall objective of maintaining a high level of trust between all stakeholders. It shall also be ensured that the regulatory environment is conducive to ease of doing business with simpler, transparent and time-bound procedures for obtaining clearances. Since mining contributes significantly to state revenues, there is a need for an efficient regulatory mechanism with high penetration of e-governance systems to prevent illegal mining and value leakages. Mining contributes significantly to employment generation, thus, there shall be a keen focus on gender sensitivity in the mining sector at all levels. An endeavour shall be made to set up a unified authority at the national level for mineral development and coordination to fulfil the objectives of this policy.
>
> (NMP, 2019)

The national policy on minerals provides power to encompass important topics such as regulation of minerals, the role of the state in mineral development, prospecting and exploration, a database of mineral resources and tenements, mining and mineral development, foreign trade and foreign investment, fiscal aspects, research and development, inter-generational equity, and an inter-ministerial mechanism for sustainable development, covering most of the practical aspects of the mining and mineral sector. The outcome expected from the national policy document on minerals exclusively is that the policy will bring holistic development of the mineral sector on a sustainable basis to fulfil the demand of downstream industries dependent on mineral/ore supply. Effective implementation of the national-level policies causes cost reduction in production and supports the manufacturing sector (industry) in its growth so that the future needs of the economy can be met. It will help to felicitate the *import* and *export* of minerals in bulk, required for foreign exchange (external cash) as well as for in-house consumption. The policy is expected to take the country on a loftier development trajectory.

1.21 Organizational policy

A company-owned policy is called 'Organizational Policy'. In a company whose core business is mining, the focus of the organization policy remains on the two most important aspects, namely, the production of minerals from mines and the environment. Every mining company, be it the world's largest coal producer, namely, the Coal India Limited (CIL), the state-owned SCCL (Singerani Collieries), the manganese producer (MOIL Ltd.), and the iron ore producer (NMDC Ltd. or Steel Authority India Limited as mining PSU's or big private players like HZL and UlraTech Cement), has its policy to run and govern their mines and industrial establishments as per the law of the land. With distributed responsibilities in the various hierarchy, the policy is subjected to periodical update.

The organizational policy of the company ultimately gives direction to the productivity and consistent growth of the company. Clean technology projects, acquisition of assets for the company, project relevance, strategic importance of the project(s), conservation of minerals and energy, and care for the environment, including environmental management, are some of the important areas linked to the organizational policy. The organizational policy has to be in tune with the national policy and should touch the peoples' welfare at the grass-roots level besides the industrial safety /workplace safety and future outlook.

1.22 Industry–institution interaction

To create a win-win condition, better interaction between a mining-affiliated institute and the mineral industry is the need of the hour (Figure 1.9). This will have a great bearing on the overall industrial growth of the mining engineering sector.

With the advent of globalization and liberalization in the economic policies in several parts of the world, competition among industries has become stiff. To solve their engineering problems, the industry looks up now to engineering consultants, research organizations, or academic institutions of importance in that field. There is a cross-country flow in expert consultant and know-how transfer. Similarly, to prepare new budding engineering students for jobs in multinational companies, a need is always felt for industry–institution interaction. Such interaction gives exposure to them in new areas of engineering, technologies, and management. By bridging the gap between industry and institutes, particularly academic institutes required objectives can be achieved successfully. To promote industry–institute interaction, roles of both

Figure 1.9 Win-win negotiations.

industry and academia are important which we will be describing in the following sub-sections. In this regard, several schemes have been planned and undertaken.

1.22.1   Role of industry

The following are the role of the mining industry:

1. Establishment of industry–institute partnerships through MOUs.
2. The establishment of an 'interaction cell' at the corporate level in the industry is similar to an *environment cell* for pollution control of mine.
3. Organizing workshops, conferences, and symposia with the joint participation of the industry and other institutions.
4. Encouraging engineers from the industry to visit engineering institutions to deliver technical lectures.
5. Professor chair/scholarships/fellowships to be instituted by the industry.
6. Periodical visits of industry executives and practising engineers to the academic institute on regular basis for observing ongoing research work in laboratories, discussions, and delivering lectures on industrial practices, trends, and experiences.
7. Financial support from the industry for establishing R&D laboratories (sponsorship by the industry for the institute).

1.22.2   Role of academia

In mining engineering education, the following are the role of academic institutes that impart mining engineering education and prepare/build mining engineers for the industry:

1. Encouraging scholarships and fellowships instituted by industries at the institute for students.
2. Part-time degree courses for practising engineers from industry, particularly at the postgraduate level or higher like Ph.D.
3. Instituting collaborative degree programmes (like dual degrees at the postgraduate level, i.e. B.Tech./B.E. and M.E./M.Tech. on specific names of the industry).

4. Participation of industry experts in curriculum development of the academic institute.
5. Professional consultancy and industrial testing by the faculty and technicians are to be done either at the field site or in the laboratory for the industries.
6. Joint research programmes and field studies by faculty and people from industries.
7. Arranging visits of academic faculty and staff members to various industries and vice versa.
8. To send students for practical training and project work in industries like vocational training, thesis work, summer camps, and survey camps.
9. Human resource development (HRD) programmes in consultation with industry for the faculty and practising engineers.
10. Short-term assignment to faculty members in industries.
11. Visiting faculty positions from professor level or above from institute to industries.

Liaison between the industries and academic institutions provides support for identifying new opportunities, providing initiatives to collaborate, and facilitating institutional association, which is good for industry promotion and improving the institute's image through an exchange. Enhanced and involved roles between the academic institute and industry have the strength to bring the two sides emotionally and strategically closer thereby encouraging the overall growth of the industry.

1.23 Mineral inventory (MI)

Mineral inventory is the inventory of mineral resources covering mineral deposits in leasehold and freehold areas. This is the data bank that provides an overview of the geological and techno-economical status of mineral resources, necessary for planning and programming mineral development. At the national level and for a country, such mineral inventory becomes a single and sole listed entity of minerals, of course as a resource, and termed as 'National Mineral Inventory (NMI)'. The NMI covers the location of Mineral deposits/lease, nature of the land, infrastructure, geology, exploration, physical and chemical properties, beneficiation, resources, reserves, the status of freehold/leasehold, etc., along with the primary source of information. The information generated from the NMI database is utilized specifically for policy planning by the government and for the benefit of mining and mineral industries. The NMI provides reserves/resources of minerals at a glance as per UNFC classifications, which is a three-digit code. This code is a handy ready reckoner to unfold the broad scenario of mineral resources in the country and is updated periodically. The mineral resource data of NMI is also disseminated through various publications and is being used by government and industry.

In the Indian context, it has been observed that a fairly large variety and adequate quantity of mineral deposits are endowed, with exception of a few deficient minerals. Indian Bureau of Mines, Nagpur (India) maintains a fully computerized database of NMI for more than 75 ores and minerals of India. To present mineral resources on the global platform, the NMI is quite a helpful inventory. Considering the ever-increasing industry and mineral demand for the future, the NMI preparation is a continuous process being updated periodically, e.g. NMI in India was initiated in the year 1968 and successively updated till April 2015. Rare Earth Elements (REE) were included in NMI in 2015. Thus, the journey of mineral inventorization and exploitation goes hand in hand. For NMI knowledge, the national mineral policy of the government and mines ministry is essential suggested reading.

Note

1 Flint is a hard microcrystalline quartz, a stone that is typically called 'chert' by geologists. This sedimentary rock breaks with a conchoidal fracture.

References

Altiti, A. H., Alrawashdeh, R. O., and Alnawafleh, H. M. (2021), Open Pit Mining, In: *Mining Technique – Past, Present and Future*, edited by A. Soni, London: InTech Open, pp. 1–23. DOI: 10.5772/intechopen.92208.

Allchin, B., and Allchin, R. (1982), *The Rise of Civilization in India and Pakistan*, Cambridge: Cambridge University Press, p. 396.

Gregory, C. E. (1980), *A Concise History of Mining*, Oxford: Pergamon, p. 259.

Hartman Howard, L., and Mutmansky Jan, M. (2002), *Introductory Mining Engineering*, Student Edition, New York: Wiley.

Lacy, W. C. and Lacy, J. C. (1992), History of Mining, In: *SME Mining Engineering Handbook*, Section 1.1, 2nd Edition, edited by H. L. Hartman, Littleton, CO: Society for Mining, Metallurgy, and Exploration, pp. 5–23.

Lewis, R. S. and Clark, G. B. (1964), *Elements of Mining*, Third Edition, New York: John Wiley & Sons, Inc. p. 768.

Mirakovski, D., Boris, K., Aleksandar, K., and Filip, P. (2009), Mine Project Evaluation Techniques, University Goce Delcev UGD Academic Repository, p. 7, https://eprints.ugd.edu.mk/3735/

Mohnot, J. K. and Singh, U. K. (2014). 'OPTMINEINVEST' – A Model for Optimum Mine Investment Decision. *International Journal of Engineering Science and Innovative Technology*, Vol. 3, Issue 2, pp. 350–357.

Paul, M. (2005), Taxation and Investment Issues in Mining, Advancing the EITI in the Mining Sector, pp. 27–31. http://citeseerx.ist.psu.edu/viewdoc/citations;jsessionid=A6F9DC29A22C-DA134C310AF202E23776?doi=10.1.1.182.9749; Accessed on 01/02/2021.

Paul, T., Huber, S., and John, H. (2021), Points on Plug and Play. *Global Mining Review (GMR)*, Vol. 4, Issue 4, pp. 33–35.

Pritchard, R. (2005), Safeguards for Foreign Investment in Mining, In: *International and Comparative Mineral Law and Policy*, edited by E. Bastida et al., The Hague: Kluwer Law International, p. 21.

PwC (2020), Mine 2020-Resilient and Resourceful, June, p. 31. PwC.com.

Raymond, R. (1984), *Out of the Fiery Furnace*, Melbourne, Australia: Macmillan, p. 274.

Rickard, T. A. (1932), *Man and Metals*, Vols. 1 & 2, New York: McGraw-Hill, p. 1068.

Sheo, S. R. (2020), Digital Transformation for Improving the Productivity of Mining – An Approach, *MGMI News Journal, A Quarterly Publication of the Mining, Geological and Metallurgical Institute of India (MGMI)*, Vol. 46, Issue 3, pp. 34–41.

Soni, A. K. (2020), History of Mining in India, *Indian Journal of History of Science (IJHS)*, Vol. 55, Issue 3, pp. 218–234. DOI: 10.16943/ijhs/2020/v55i3/156955.

Stoces, B. (1954), *Introduction to Mining*, London: Lange, Maxwell and Springer.

Stuart, John, M. (1879), *Mining: Its Theory and Practice*, New York, p. 67.

Annexure 1.1

(Refer to Chapter 1, Section 3)
India: Mining and mineral industry in India at a glance

Introduction

Primary raw materials to industries, such as iron & steel, ferro-alloys, aluminium, cement, thermal power generation, refractories, ceramics, glass, chemicals like caustic soda, soda ash, calcium carbide, paint, and pigment, are minerals. Nearly, all minerals required for industrial use are mined in India, barring a few. In all, India is producing 95 principal minerals (4 fuel minerals, 5 atomic minerals, 10 metallic minerals, 21 non-metallic, and 55 other minor minerals) from across the 29 states and 7 union territories. Coking coal (low ash content), non-coking (high ash content), and lignite among mineral fuels; bauxite, copper, chromite, iron, lead, zinc, manganese, ilmenite, and rutile among metallic minerals; limestone, dolomite, chalk, kaolin, clay, flintstone, salt and gypsum among non-metallic category, and almost all industrial minerals (Wollastonites, soapstone, kyanite, sillimanite fluorite, Apatite, etc.) are mined in India. Precious and semi-precious minerals, namely, gold, diamond, and gemstones are also mined in India. India is a producer of 'mica' sheets too which is a compound of both metals and non-metals in crystal form. Thus, India continued to be wholly or largely self-sufficient in minerals.

The mineral sector is critical to India's economic and social well-being and today contributes about 2% of the gross domestic product (GDP) and provides daily direct employment to about 1 million people. According to an industrial analysis, India continues to be an importer of higher value and scarce minerals and metals, whereas it exports iron ore and mica. Though exports from the mineral sector as a whole account for over 20% of the total value of all merchandise exported.

Mineral reserves and their status

Geological reserves, either coal or ore, when economically mineable become mineral reserves. All geological reserves and mineral resources that are geologically suitable and feasible for exploitation (mining) have been classified as per the UNFC classification. **(Annexure 1.2)**.

- **Coal reserves:** As per the Geological Survey of India (GSI), the geological reserves of coal in India as of 01.04.2020 was 3,44,020.84 Million Tonnes. The Reserve of coking coal (prime, medium, and semi-coking) was 35,004.12 Million Tonnes and non-coking coal was 309,016.72 Million Tonnes (MOC, 2021).

- **Mineral reserves:** As per the National Mineral Inventory (NMI) prepared by the Indian Bureau of Mines, Ministry of Mines, Govt. of India, the mineral reserves position for various minerals (70 minerals other than coal/fuel and atomic minerals) has been prepared and is given in Table A1 (National Mineral Inventory (NMI) 2021). The NMI contains the latest position of mineral reserves and resources of minerals feasible for mining. It contains details about the deposit-wise inventory of freehold deposits and inventory in respect of different lease areas of India spread across many states. The said information is not available freely but is available on a payment basis from IBM. The schedule of charges has been fixed in both Indian Currency and Foreign Currency for NMI.

Table A1 Mineral Wise Reserves/Resources (as of 01.04.2010)

Sl. No.	Mineral	Unit	Reserves	Remaining resources	Total resources
1	Alexandrite	-	N. E.	N. E.	N. E.
2	Andalusite	000' tonnes	-	18,450	18,450
3	Antimony		-		
	Ore	tonnes	-	10,588	10,588
	Metal		-	174	174
4	Apatite	tonnes	2,090,216	22,138,530	24,228,746
5	Asbestos	tonnes	2,510,841	19,655,762	22,166,603
6	Ball clay	tonnes	16,777,842	66,615,662	83,393,504
7	Barytes	tonnes	31,584,128	41,149,746	72,733,874
8	Bauxite	000' tonnes	592,938	2,886,682	3,479,620
9	Bentonite	tonnes	25,060,508	543,306,838	568,367,346
10	Borax	tonnes	-	74,204	74,204
11	Calcite	tonnes	2,664,338	18,281,110	20,945,448
12	Chalk	000' tonnes	4,332	585	4,917
13	China clay	000' tonnes	177,158	2,528,049	2,705,207
14	Chromite	000' tonnes	53,970	149,376	203,346
15	Cobalt (Ore)	Million tonnes	-	44.91	44.91
16	Copper				
	Ore	000' tonnes	394,372	1,164,086	1,558,458
	Metal		4,768.33	7,518.34	12,286.67
17	Corundum	tonnes	597	740,194	740,792
18	Diamond	carats	1,045,318	30,876,432	31,921,750
19	Diaspore	tonnes	2,859,674	3,125,144	5,984,818
20	Diatomite	000' tonnes	-	2,885	2,885
21	Dolomite	000' tonnes	738,185	6,992,372	7,730,557
22	Dunite	000' tonnes	17,137	168,232	185,369
23	Emerald	-	N. E.	N. E.	N. E.
24	Feldspar	tonnes	44,503,240	87,832,212	132,335,452
25	Fire clay	000' tonnes	30,104	683,415	713,519
26	Fluorite	tonnes	4,712,316	13,501,588	18,213,904
27	Fullers Earth	tonnes	58,200	256,593,879	256,652,079
28	Garnet	tonnes	19,324,793	37,638,032	56,962,824
29	Gold				
	Ore (Primary)	tonnes	24,124,537	469,570,375	493,694,912
	Metal (Primary)		110.54	549.30	659.84
	Ore (Placer)		-	26,121,000	26,121,000
	Metal (Placer)		-	5.86	5.86
30	Granite (Dimension Stone)	000' cum	263,692	45,966,608	46,230,300
31	Graphite	tonnes	8,031,864	166,817,781	174,849,645
32	Gypsum	000' tonnes	39,096	1,247,402	1,286,498

(Continued)

Table A1 (Continued) Mineral Wise Reserves/Resources (as of 01.04.2010)

Sl. No.	Mineral	Unit	Reserves	Remaining resources	Total resources
33	Iron Ore Magnetite	000' tonnes	21,755	10,622,305	10,644,060
34	Iron Ore Haematite	000' tonnes	8,093,546	9,788,551	17,882,097
35	Kyanite	tonnes	1,574,853	101,670,767	103,245,620
36	Laterite	000' tonnes	24,714	446,119	470,833
37	Lead and zinc				
	Ore	000' tonnes	108,980	576,615	685,595
	Metal Lead		2,245.01	9,304.38	11,549.39
	Zinc		12,453.26	24,211.64	36,664.90
	Lead +Zinc		0	118.45	118.45
38	Limestone	000' tonnes	14,926,392	170,008,720	184,935,112
39	Magnesite	000' tonnes	41,950	293,222	335,172
40	Manganese ore	000' tonnes	141,977	288,003	429,980
41	Marble	000' tonnes	276,495	1,654,968	1,931,463
42	Marl	tonnes	139,976,150	11,704,870	151,681,020
43	Mica	kg.	190,741,448	341,495,531	532,236,979
44	Molybdenum				
	Ore	tonnes	-	19,286,732	19,286,732
	Contained MOS_2			12,640	12,640
45	Nickel (Ore)	Million tonnes	-	189	189
46	Ochre	tonnes	54,942,176	89,319,089	144,261,265
47	Perlite	000'tonnes	428	1,978	2,406
48	PGM (Metal)	tonnes of metal content	-	15.7	15.7
49	Potash	Million tonnes	-	21,816	21,816
50	Pyrite	000' tonnes	-	1,674,401	1,674,401
51	Pyrophyllite	tonnes	23,275,451	32,807,451	56,082,902
52	Quartz/Silica Sand	000' tonnes	429,223	3,069,808	3,499,031
53	Quartzite	000' tonnes	86,599	1,164,649	1,251,248
54	Rock Phosphate	tonnes	34,778,650	261,505,701	296,284,351
55	Rock Salt	000' tonnes	16,026	-	16,026
56	Ruby	kg	236	5,112	5,348
57	Sapphire	kg	-	450	450
58	Shale	000' tonnes	15,331	580	15,911
59	Sillimanite	tonnes	4,085,052	62,902,385	66,987,437
60	Silver				
	Ore	tonnes	187,558,668	279,426,291	466,984,959
	Metal		8,039.57	19,588.68	27,628.25
61	Slate	000' tonnes	0	2,369	2,369
62	Sulphur (Native)	000' tonnes	-	210	210
63	Talc/Steatite/ Soapstone	000' tonnes	90,026	178,996	269,022
64	Tin				
	Ore	tonnes	7,131	83,719,066	83,726,197
	Metal		1,132.43	101,142.41	102,274.84
65	Titanium minerals	tonnes	22,030,223	371,965, 694	393,995,917
66	Tungsten				
	Ore	tonnes	-	87,387,464	87,387,464
	Contained WO_3		-	142,094.35	142,094.35
67	Vanadium				
	Ore	tonnes	410,955	24,307,933	24,718,888
	Contained V_2O_5		1,602.72	63,284.45	64,887.17
68	Vermiculite	tonnes	1,704,007,803	803,003	2,507,010

(Continued)

Table A1 (Continued) Mineral Wise Reserves/Resources (as of 01.04.2010)

Sl. No.	Mineral	Unit	Reserves	Remaining resources	Total resources
69	Wollastonite	tonnes	2,487,122	14,082,751	16,569,873
70	Zircon	tonnes	1,347,470	1,786,482	3,133,952

Source: National Mineral Inventory (NMI) (2021).
Note: (1) Reserve figures rounded off.
(2) N.E., not estimated.

Production status

Two major sectors, i.e. coal and non-coal sectors, contribute to the mineral production in India. Its summary is as follows:

Coal

In the year 2019–2020, the total production of raw coal in India was 730.874 MT, whereas it was 728.718 MT in 2018–2019. Thus, in 2019–2020, production of coal increased by 0.30% in comparison to 2018–2019. In the year 2019–2020, production of lignite was 42.096 MT against 44.283 MT in 2018–2019, thus in 2019–2020 lignite production decreased by 4.94% against 2018–2019 (Table A2). The contribution of the Public Sector and Private Sector in the production of raw coal in 2019–2020 was as follows (Table A3):

The production of coking coal in 2019–2020 was 52.936 MT, whereas it was 41.132 MT in 2018–2019, thus registering a growth of 28.70%. In 2019–2020, the production of non-coking coal was 677.938 MT, whereas it was 687.586 MT in 2018–2019, showcasing a negative growth of 1.40% (MOC, 2021). The production trend of coal by type of mines, i.e. opencast mines and underground mines, and for two major coal companies, i.e. Coal India Limited (CIL) and Singareni Coal Company Limited (SCCL), has been depicted in **Figure A.1**.

Table A2 Company-Wise Raw Coal Production in India

Coal Production Year: 2019–2020; All values in million tonnes (MT)

Company	Coking	Non-coking	Total
ECL	0.026	50.375	50.401
BCCL	25.945	1.784	27.729
CCL	20.027	46.862	66.889
NCL	0	108.053	108.053
WCL	0.178	57.458	57.636
SECL[a]	0.250	150.296	150.546
MCL	0	140.358	140.358
NEC	0	0.517	0.517
CIL	46.426	555.703	602.129
SCCL	-	64.044	64.044
Other Public	0.300	31.751	32.051
Total Public	46.726	651.498	698.224
Total Private 6.210	26.440	32.650	
ALL INDIA	52.936	677.938. 730.874	

Source: MOC (2021).
[a] Including Production from SECL (GP-IV/1) and (GP-IV/2&3) coal blocks.

Table A3 Production of Raw Coal in 2019–2020

All values in million tonnes (MT)

Sector	Coking	Non-coking	Total	% increase from 2018–2019
Public	46.726	651.498	698.224	
Private	6.210	26.440	32.650	
All India	52.396	677.938	730.874	28.70% (Coking Coal) (−) 1.40 % (Non-coking)

Source: MOC (2021).

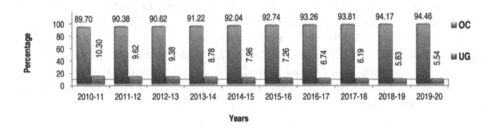

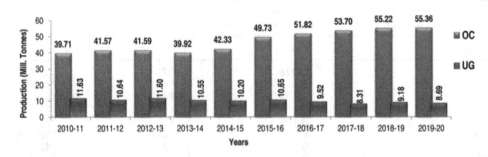

Figure A.1 Production by type of mines for CIL and SCCL companies.

State-wise, in 2019–2020, Chhattisgarh state registered the highest coal production of 157.745 MT (21.58%), followed by Odisha 143.016 MT (19.57%), Jharkhand 131.763 MT (18.03%), and Madhya Pradesh 125.726 MT (17.20%). In 2019–2020, Tamil Nadu was the largest producer of lignite and produced 23.516 MT (55.86%) followed by

Gujarat 10.357 MT (24.60%) and Rajasthan 8.223 MT (19.53%). Coal India Limited produced 602.129 MT (82.38%) and SCCL 64.044 MT (8.76%) of coal in 2019–2020. In that year, the main producer of lignite was Neyveli Lignite Corporation which produced 24.864 MT (59.06%). In this paragraph, figures shown in the bracket in % are the share percentage of total all India production, i.e. 730.874 MT (MOC, 2021).

Non-coal

In India, the non-coal sector includes both major and minor minerals of the metallic and non-metallic categories. Overall, the production of minerals in the non-coal sector registered an increase compared to the earlier years. Both private sector and public sector mines witnessed it indicating the growth trajectory of the industry. However, it should also be noted that the *production data* of minerals keeps on dwindling and the authenticity of data is one big problem faced by the reader. As far as the gross revenue from mineral production since 2000 is concerned, according to the Indian Bureau of Mines, publications an upward increasing trend has been observed mainly by the public sector mines (IBM, 2020). The total value of mineral production (excluding atomic & fuel minerals) during 2020–2021 has been estimated at Rs.129,950 crores, which shows a decrease of about 11.35% over that of the previous year which is due to the COVID-19 pandemic and countrywide lockdown. During 2020–2021, the estimated value for metallic minerals is Rs.49,285 crores or 37.93% of the total value and non-metallic minerals including minor minerals is Rs.80,664 crores or 62.07% of the total value (Table A4) (Annual Report).

Key developments, recent amendments, and future takeaways

Broadly, some key developments took place in India in the mining sector very recently, e.g. non-coal minerals auction was introduced after 2015 as per the basic MMDR Act, 1957 and its amendment act of 2021. Though, not very new, private companies involved in coal extraction (for the mining operation exclusively) were opened first. Laws related to minor minerals excavation, auction, and transportation (through state government) have been made more transparent by several state governments with technological interventions like DGPS and online auctions. The 'captive mine' and 'non-captive mine' concepts have been modified for mining productivity (see box below).

Any lessee may, where a mineral is used for the captive purpose, sell mineral up to 50% of the total mineral produced in a year after meeting the requirement of the end-use plant linked with the mine in such a manner as may be prescribed by the central government and on payment of the such additional amount as specified in the sixth schedule which makes mention of an amount for bauxite, Chromite, iron ore, limestone, and other minerals. In the sixth schedule, an amount for captive coal and lignite mines that were auctioned and allotted with a condition allowing the sale of coal up to 25% of annual production has also been done.

In India, the basic mining laws of 1957 [Mines and Minerals (Development and Regulation) Act or MMDR Act, 1957] have been amended in 2015–2016 and 2020. This has provided a new legal framework to categorize mineral and metal resources overseen by the state and central governments, including mining leases and methods, to ensure the welfare of the local population residing in and around the mining area. Favourable incentives from the state and central, coupled with an investment environment from private players, play a key role in the growth of the Indian mining industry for the foreseeable future. With new rules and regulations, India aims to improve the potential of the mining/mineral sector to match that of resource-rich countries, including Australia and South Africa. The massive Indian infrastructure and manufacturing sectors, specifically the foreign direct investment (FDI), are the direct beneficiaries of these changes. The GDP growth of the country is bound to get acceleration and the government's ambitious target of reaching a US$5 trillion economy by 2025. The mining industry will get stimulation by these policies cum economy-related steps. Selected points that need review, description, and pondering for the Indian mining and mineral sector are given below (Anuj Mudaliar, 2021):

1. Mineral exploration activities in India need to be incentivized. With reforms, such as the commercial coal block auction, emphasis has been given to the importance of opening the sector to private players.
2. The modern-day industry requirements of *additional jobs* & *superior technologies* in the mining sector, through new reforms, proposed and implemented by the government for the mines located in India are partially fulfilled as this has become the need of the hour.
3. Changes in regulations and the introduction of the MMRD Amendment Bill, 2021, will notably contribute to other ancillary and downstream industries either directly or indirectly.
4. For the country's energy security, which has more than 70% of its power generation from coal, the amendments will be critical to the country's energy security, and consequently, the Indian mining industry.
5. The auction process for mines and mineral blocks/mineral deposits; open acreage licensing policy for allocation of mining rights; powers of the government in managing the NMET funds and the District Mineral Foundation (DMF); revenues and royalties-related issues; issues of relocation, rehabilitation, and compensation; risk of abuse on local tribal communities and the environmental impacts are the leading concerns in India, addressed by the amendments. The industry experts felt/opined such steps as desirable for the Indian mining industry.
6. Imports for minerals and raw materials such as copper, steel, iron, and coal in India are likely to be reduced with new laws in force. This will be giving a boost to the self-reliance initiatives preferred and boosted by the Indian government both at the centre and state.

These amendments to the MMDR Act which are new have set the ground for the development of the Indian mining industry. The legislative changes will bring about several improvements that have the potential to revolutionize the Indian mining industry in the coming years. It is felt that the key provisions of the legislation will allow the government organizations and statuary government bodies in their routine work improvement. Longer and extended mining lease duration, by 10 years from the present, will improve the efficiency of mine and better utilization of the exploited mineral resources.

Public sector enterprises, state governments, central government, and other private mines can gain from the brought about changes for mines as well as the industry as a whole. The changes to the act will allow the central government to remove the classification between captive and non-captive mines. Now, captive mines can sell up to half of all their ores in the open market, provided that they pay additional charges to the government for the same. The government can allow sales to go above the 50% cap if necessary. The District Mineral Foundation (DMF) primarily operates to optimize the benefits for locals in the mining areas. The amendments have given the power to the central government to direct the usage of funds from the foundation for the development of the locality. The amendments have removed the need for the conventional non-exclusive licence permit required by the mining companies to reconnoitre an area to assess mineral potential. The new act allows for only a 2-year permit for mining operations to start after the issuance of a licence, which makes the issuance of permits difficult owing to time constraints while pushing for mines to become active faster. The new amendments provide licence clearances for long-term mining operations before providing an opportunity for the lease to be transferred to a new bidder. The government has also pushed amendments to keep mines from becoming inactive. Consequently, in a case where a state government is unable to auction off a mine, the central government can now step in to conduct the auction procedures. New amendments have removed regulatory differences between non-captive and captive mines for statutory payments as well. Also, the NMET (National Mineral Exploration Trust) is set to become an autonomous organization.

The socio-political scene in India plays a key role in the development of the mining industry in India while meeting the mineral security requirements of the country for the foreseeable future. However, future takeaways from the 2020–2021 amendments include measures such as: (i) Mining/mineral producing companies need not take out exploration and mining licences separately, which until now, the companies had to obtain stepwise. (ii) Setting the standards of mineral exploration for the future. (iii) A greater potential for metal and mineral exports through domestic and foreign investments and associated infrastructure. (iv) To allow the transfer of mining leases through routes other than auctions. (v) To minimize obstacles associated with mining concessions. (vi) Distinction between captive and non-captive mines with stressed assets to be used for domestic industrial consumption.

Until now, only government agencies including CMPDIL, GSI, and MECL have been involved in mining activities. With the amendments, for the first time in India, private organizations will be allowed to take part in mineral exploration activities.

Discussion and analysis

Considering the changes for the developments in the mining/mineral industry on policy fronts, ease of doing business (economic liberalization), environment, and the discoveries of new deposits that have occurred in the last decades, the growth trajectory of mineral production is likely to continue. The degree of self-sufficiency in respect of various principal minerals and metals/ferro-alloys in 2020–2021 was highly appreciative and conducive (Table A3).

The review and analysis of the Indian mining and mineral industry are possible in many ways, but we have kept our focus on the following key points described in brief relating to the textbook content:

- **Procedural delays** – An analysis of the Indian mining and mineral industry on the procedural delays, which reviews the following major steps that have been taken up recently for improvement:

    i. Digitization in mining has been adopted and encouraged.
    ii. Private participation in the coal sector has been permitted.
    iii. Single window clearance (for prospecting and reconnaissance permit) to reduce the red-tapism.
    iv. Many legislative measures in terms of simplification of rules and regulation have been introduced.

    As regards the procedural delays, the most significant are the 'environmental clearance' and 'forest clearance' for the mining projects. The former can be partially tackled with the abovementioned improvement, but for the latter one, the explanation and solution are as given hereunder.

    It is known to everyone that mines are located in forest areas and not in urban localities. In India, the forest policy and mineral policy are not in tandem, there exists a long-standing discord between the conservation of forest resources and the exploitation of mineral resources. Forest clearance for mining seems to be a formidable obstacle to the speedy development of mineral resources. In this context, overlaying of forest map and mineral map and then implementing the Go and No-go areas concept will categorize forests into different categories that are worth mining. Accordingly, forests for other uses can be separated from the mineral-rich forests having geological reserves for exploitation may be now or in near future.

- **Labour laws:** Mining Industry in India is labour-intensive. Labour makes an impact on both employers and employees. It causes labour problems in mines which are a form of industrial unrest being faced by mine management. The labour laws on wages, industrial relations, social security and occupational safety, health, and working conditions have been used/misused with the promise of making the industry modern, progressive, vibrant, and better suited to the needs of changing business environment. Covering different categories of labourers as employees of mines and mining enterprises, it is analysed that to expand the social security net, regulate conditions of employment, and streamline conditions relating to health and security at the workplace, some new changes have been introduced in the industry (Amanda et al., 2020; Anuj, 2021; New Labour Codes). They are:

Table A4 Production of Selected Minerals, 2016–2017 to 2020–2021 (Excluding Atomic & Fuel Minerals) (Value in Crore)

	Unit	2016–2017 Qty	2016–2017 Value	2017–2018 Qty	2017–2018 Value	2018–2019 (P) Qty	2018–2019 (P) Value	2019–2020 (P) Qty	2019–2020 (P) Value	2020–2021 (E) Qty	2020–2021 (E) Value
All Minerals			110,845.58		131,585.41		144,883.77		146,592.21		129,949.60
Metallic			39,759.61		50,975.52		64,042.45		66,084.02		49,285.22
Bauxite	th. tonnes	24,745.49	1,486.55	22,786.11	1,578.42	23,687.72	1,716.84	21,823.79	1,578.56	21,239.97	1,629.02
Chromite	th. tonnes	3,727.78	3,193.75	3,480.94	3,203.70	3,970.69	3,583.61	3,929.26	3,332.66	1,250.25	712.72
Copper Conc.	th. tonnes	134.79	650.61	141.99	770.66	155.44	939.52	124.69	844.58	100.94	796.47
Gold	kg	1,595.00	436.24	1,650.00	476.98	1,664.00	524.17	1,724.00	643.10	1,193.00	589.08
Iron Ore	M. tonnes	194.58	25,229.18	201.42	34,713.10	206.45	45,184.14	246.08	48,107.41	188.63	34,508.61
Lead Conc.	th. tonnes	268.05	966.93	306.40	1,142.94	358.37	1,631.68	351.27	1,807.28	405.47	1,889.39
Manganese Ore	th. tonnes	2,395.14	1,624.84	2,599.81	1,990.75	2,820.23	2,270.25	2,904.37	1,941.64	2,120.20	1,494.40
Zinc Conc.	th. tonnes	1,484.24	4,338.56	1,539.66	4,979.93	1,457.17	5,608.38	1,446.82	6,023.12	1,655.46	6,637.94
Other Met. Minerals			1,832.95		2,119.04		2,583.86		1,805.67		1,027.59
Non-Metallic Minerals			8,029.19		8,855.46		9,214.90		8,881.77		9,037.96
Diamond	Crt	36,491	63.96	36,491.00	37.41	36,491.00	58.11	36,491.00	39.81	36,491.00	30.04
Garnet (abrasive)	th. tonnes	85.413	78.73	158.28	161.89	123.40	156.82	0.55	0.47	1.14	0.51
Limeshell	th. tonnes	12.344	3.48	14.77	5.14	7.54	2.78	4.60	1.87	0.00	0.00
Limestone	M. tonnes	314.669	7,387.84	340.42	8,099.57	379.05	8,484.11	359.33	8,312.02	353.47	8,340.91
Magnesite	th. tonnes	299.149	74.93	195.06	59.37	146.58	39.66	97.68	35.03	73.55	28.62
Phosphorite	th. tonnes	1,124.44	299.67	1,515.65	366.83	1,284.58	354.76	1,400.19	431.91	1,648.00	572.24
Sillimanite	th. tonnes	68.131	53.59	81.64	67.17	69.03	55.98	13.24	3.63	17.52	4.31
Wollastonite	th. tonnes	166.186	15.88	153.05	12.60	184.06	17.40	124.66	11.91	116.14	10.81
Other Non-Met. Min.			51.10		45.48		45.30		45.12		50.52
Minor Minerals			40,976.35		52,810.07		52,810.07		52,810.07		52,810.07

Source: Annual Report. Ministry of Mines, Govt. of India, Page No. 221. (a) MCDR Minerals: MCDR returns; (b) Minor Minerals: State Governments (data repeated in case of non-availability).

Note: Includes 31 minerals declared as minor minerals vide notification dated 10.02.2015. The data for these minerals for 2014–2015 onwards is included in minor minerals. M.Tonnes, million tonnes; th.tonnes, thousand tonnes; kg, kilogram; (P), Provisional; (E), Estimated figures.

- **Layoffs:** Industries that employ more than 100 workers have to obtain the government's permission to temporarily refuse to work or terminate workers.
- **Reducing trade union conflicts:** One of the many reasons for industrial unrest during the 'licence raj' era was certain provisions relating to trade unions and strikes which brought industrial activities to a grinding halt for months together. Till now, large industrial establishments could have more than one trade union to protect the interests of workmen. As trade unions rely on collective bargaining with employers, at times differences arose between the mine management and trade unions. To address such conflicts, the new reforms allow recognition of any union as a 'negotiating union' in an industrial establishment if it receives the approval of 51% of workers or more. This will reduce labour unrest.
- **Setting minimum wage:** Presently, central and state governments prescribe minimum wages for workers based on their skills and experience. But there's no uniform minimum amount. To address this, a national floor wage has been introduced on wages. The amounts are yet to be notified, but once implemented, state governments won't be able to prescribe wages below that.
- **Extending social security net:** A gig worker is defined under the new law as a person who earns out of activities that fall outside of the employer–employee relationship. Also, a person who undertakes work through online platforms for providing services outside the employer–employee relationship is considered a platform worker. The law on social security enables the government to extend social security benefits to gig workers as well as health, security, and safety.
- **Recognizing fixed-term labour**: Loosely, they are also referred to as 'contractual labour'. Contractual practices and no regular recruitment in the government sector have increased the number of these workers in several organizations. So far, fixed-term contracts have been permissible only for a limited period with no rights in many organizations. Neither they are allowed regular employment through the normal absorption process nor given their legitimate rights. These workers face the agony of omitting and avoiding. This leads to an increased set of challenges like worker rights, due diligence, and liability issues. In recent years, lot of changes in the labour laws have made the hiring of contractual labour by companies much easier if the company follows the rule. This addresses industrial relations on its own (New Labour Codes). The applicability of these labour laws will apply to mines, mineral processing plants, and factories that employ more than 100 workers at a time.
- **Fiscal regime:** To make financial health sound and healthy, a 100% foreign direct investment in the mineral sector is allowed by the Indian government. For the exploration and mining of diamonds, 74% of FDI is allowed through an automatic route (FIMI, 2020).

Enhancing private investments in the mining and mineral sector is the biggest need for India today to generate employment. The coal sector of India, which was restricted to the government-owned public sector, has been opened for private sector participation through the selling of new and virgin coal blocks on open competitive bidding. Since the mining sector is one of the highest employment-generating sectors, other ancillary sectors, such as mining machinery, steel, aluminium, commercial vehicles, rail transportation, ports, shipping, and power generation, that are closely linked to the mining sector have recorded an automatic improvement in the employment front and fiscal regime. Thus, the Indian economy got a boost inevitably.

Banking and finances are the major needs for running the mines and plants of the capital-intensive mining and mineral industry in general. Several new reforms in recent years have been witnessed by the financial sector institutions, i.e. banks and money lending companies, thus providing ease in doing business and raising the capital as required for smooth industrial operations. On review of the Indian mining industry, it is analysed that this biggest support is fully extended for the industry enabling the smooth running of the Indian mines.

The green fiscal regime for the mining sector which was lacking has been introduced in the industry slowly and in a phased manner (New Labour Codes).

- **Legislative measures:** An overhauling of the National Mineral Policy and MMRD Act of 1957 is the biggest legislative measure taken up recently in the mineral and mining sector in India (The Mines and Minerals (Development and Regulation) Amendment Act, 2021).
- **Sustainable development:** Mineral resources are finite and technology provides the basis for innovative solutions in a range of areas. Mining in a sustainable manner is the mantra of the Indian mining industry and is being done by adopting different tools and techniques of 'best-practice mining'. Environmental legislation, as applicable to the mining industry in India, is strong and elaborate that paves the way for sustainable development together with mineral exploitation.

An alarming increase in greenhouse gas (GHG) emissions worldwide, presently over 415 ppm, is an area of concern for sustainable development. Reducing Carbon footprints is a positive way by which the coal mining industry can contribute to the reduction of GHG and CO_2 sequestration, through the unworkable coal seams and the mining voids. R&D work is continuing in the Indian coal mining industry.

References

Amanda, Guimbeau, James Ji, Nidhiya Menon, and Yana van der Meulen Rodgers. (2020), Mining and Gender Gaps in India, Bonn, Germany: IZA Institute of Labour Economics, November 2020. http://ftp.iza.org/dp13881.pdf; Accessed on 07/12/2021.

Anuj, Mudaliar. (2021), *Creating a Favourable Future, Global Mining Review*, September 2021, Palladian Publications, pp. 13–16.

Annual Report. Ministry of Mines, Govt of India, p. 264. https://mines.gov.in

FIMI. (2020), Indian Mining – A Synopsis, Federation of India Mineral Industry (FIMI), New Delhi, p. 55. https://www.fedmin.com/

MOC. (2021), Coal Directory of India 2019–2020, Ministry of Coal (MOC), April 2021, Government of India, Coal Statistics Compiled by Coal Controller's Organisation, Kolkatta, p. 213. https://www.coal.nic.in/

IBM. (2020), *Indian Minerals Yearbook (IMBY) -2019*, Vol. I, II and III, 58th Edition, Advance Release (September), Indian Bureau of Mines (IBM), Nagpur: Ministry of Mines, Govt. of India. https://ibm.gov.in/?c=pages&m=index&id=107&mid=24021

National Mineral Inventory (NMI), Indian Bureau of Mines, Ministry of Mines, Govt of India. https://ibm.gov.in/?c=pages&m=index&id=354; Accessed on 21/06/2021.

National Mineral Policy (NMP). (2019), Government of India, Ministry of Mines, p. 14. https://mines.gov.in/admin/storage/app/uploads/64352887bcfa41681205383.pdf

New Labour Codes: Top 5 Changes That Will Impact Employers and Employees. https://green-fiscalpolicy.org/blog/green-fiscal-regime-for-mining/

The Mines and Minerals (Development and Regulation) Amendment Act, 2021, The Gazette of India Extraordinary. https://mines.gov.in/writereaddata/UploadFile/mmdr28032021.pdf; Accessed on 03/28/2021.

Annexure 1.2: United Nations Framework Classification (UNFC) for reserves and resources

The UNFC classification is a three-digit code-based system. It has three axes X, Y, and Z and four digits (1 to 4) as Economic Viability, Feasibility Assessment, and Geological Assessment. The economic viability axis represents the first digit, the feasibility axis, the second digit, and the geologic axis, the third digit. The three categories of economic viability have codes 1, 2, and 3 in decreasing order. Similarly, the three categories of the feasibility study have also codes 1, 2, and 3, while the four stages of geological assessment are represented by 4 codes: 1 (detailed exploration), 2 (general exploration), 3 (prospecting), and 4 (reconnaissance). Thus, the highest category of resources under the UNFC system will have the code (111) and the lowest category, the code (334).

Explanation: The process of geological assessment is generally conducted in stages of increasing details. The typical successive stages of geological investigations, i.e. reconnaissance, prospecting, general exploration, and detailed exploration, generate resource data with a clearly defined degree of geological assurance. These four stages are, therefore, used as geological assessment categories in the classification. Feasibility assessment studies form an essential part of the process of assessing a mining project. The typical successive stages of feasibility assessment are geological study as the initial stage followed by pre-feasibility study and feasibility study/mining report. Similarly, the degree of economic viability (economic or sub-economic) is also well-defined. It is assessed in the course of pre-feasibility and feasibility studies. A pre-feasibility study provides a preliminary assessment with a lower level of accuracy as compared to that of a feasibility study which assesses the economic viability in detail.

The UNFC classification most applicable in the industry has defined various terms. Their definitions in brief are as follows:

a. **Mineral occurrence:** A mineral occurrence is an indication of mineralization that is worthy of further investigation. The term mineral occurrence does not imply any measure of volume/tonnage or grade/quality and is thus not part of a mineral resource.

b. **Uneconomic occurrence:** Materials of estimated quantity that are too low in grade or for other reasons are not considered potentially economic. Thus, uneconomic occurrence is not part of a mineral resource. If quantity and quality are considered worthy of reporting, it should be recognized that an Uneconomic occurrence cannot be exploited without major technological and/or economic changes, which are not currently available.

c. **Total mineral resources:** Total mineral resources comprise the reserve plus remaining resources. Thus, the total resource minus reserve gives the remaining resource which is in addition.

d. **Mineral reserve:** Economically mineable parts of measured and/or indicated mineral resources are called mineral reserves. The following are two sub-classes:

    i. **Proved Mineral Reserves (UNFC CODE: 111)**
       Economically mineable part of measured mineral resource.

    ii. **Probable Mineral Reserves (UNFC CODE: 121 and 122)**
       Economically mineable part of indicated, or in some cases, a measured mineral resource.

e. **Mineral resource:** A mineral resource (remaining or additional) is the balance of the total mineral resources that have not been identified as a mineral reserve. The following six sub-classes as per UNFC classification exist:

    i. **Measured Mineral Resource (UNFC CODE: 331)**
       That part of a mineral resource for which tonnage, density, shape, physical characteristics, grade, and mineral content can be estimated with a high level of confidence, i.e. based on detailed exploration.

    ii. **Indicated Mineral Resource (UNFC CODE: 332)**
       Tonnage, density, shape, physical characteristics grade, and mineral content can be estimated with a reasonable level of confidence based on exploration, sampling and testing information, location of borehole, pits, etc.

    iii. **Inferred Mineral Resource (UNFC CODE: 333)**
       Tonnage, grade, and mineral content can be estimated with a low level of confidence inferred from geological evidence.

    iv. **Reconnaissance Mineral Resource (UNFC CODE: 334)**
       Estimates based on regional geological studies and mapping, airborne and indirect methods, preliminary field inspections as well as geological inference and extrapolation.

    v. **Pre-feasibility Mineral Resource (UNFC CODE: 221 and 222)**
       That part of an indicated, and in some circumstances, measured mineral resource that has been shown by the pre-feasibility study as not economically mineable or can become economically viable subject to changes in technological, economic, environmental, and/or other relevant conditions.

    vi. **Feasibility Mineral Resource (UNFC CODE: 211)**
       That part of a measured mineral resource, which after a feasibility study is economically not mineable.

Annexure 1.3: Detailed cost estimation by an example

Name of the mine: XY mine. Name of the mineral: Coal

Problem 1.1

Calculate the total per ton mining cost of XY mine assuming that the mine has both surface operation and underground operation.

XY Open-pit mine	Values (assumed)	XY Underground mine	Values (assumed)	General costs	Values (assumed)
• Production	6.5 MT per year	• Production	2.0 MT per year	• Annual fixed cost	13%
• Investment	25 million dollar	• Investment	65 million dollar	• Wages costs	$20,000 annual
• Workforce	130 person	• Workforce	275 person	• Energy cost	2 cents per KWH (0.02$)
• Energy consumption	10 KW per ton	• Energy consumption	25 KW per ton	• Indirect and direct costs	$ 1 per ton

Solution/Answer

Estimation of the per ton mining cost can be done as follows. The general cost will be applied equally for open and underground mines as both are at the same site and owned by one mining company.

Total per ton mining costs

For surface mine

Costs	Calculation	Million dollars per year
Capital	$ 25 × 0.13	$ 3.25
Labour	($ 20,000 × 130)/1,000,000	$ 2.60
Energy	6.5 × 10 × 0.02	$ 1.30
Other costs (direct & indirect)	6.5 × 1	$ 6.50
Total		$ 13.65
Per ton cost	13.65/6.5	$ 2.10 per ton

For underground mine

Costs	Calculation	Million dollars per year
Capital	$ 65 × 0.13	$ 8.45
Labour	($ 20,000 × 275)/1,000,000	$ 5.50
Energy	2.0 × 25 × 0.02	$ 1.00
Other costs	2.0 × 1	$ 2.00
(direct& indirect)		
Total		$ 18.95
Per ton cost	18.95/2.0	$ 9.95 per ton

Conclusion: **At XY colliery, the cost per ton by an underground method of mining is 4.7 times more than that of an open-pit mine.**

Problem 1.2

Estimate the operating cost and capital cost of five 85-ton dumpers deployed at XY mine assuming the following operating conditions and data:

  i. Make: Bharat Earth Movers Limited (BEML); diesel operated
 ii. Annual investment: $300,000
iii. Unit cost of dumper: $6,50,250
 iv. Freight, loading/unloading, and moving Charges: $27,712
  v. Economic life of dumper: 15,000 hours (or say approximately 7.5 years)
 vi. The operating period of dumpers: 2,000 hours/year
vii. Tyre cost one set (4 nos.): $ 32, 512 and Tyre life : 2,500 hours
viii. Diesel cost – $1 per litre; Consumption – 17 litres/hour
 ix. Labour/Operator wages – $12 per hour
  x. Tax rates: 18% (Value-added tax/Goods and service tax)
 xi. Operating condition – average; not very favourable for tyres

- Factor for tyre repair @ 17% (unfavourable condition)
- Lubrication factor @ 1/3 (as per standards)
- Factor for mining equipment repair & maintenance @ 45% (average)

A. Capital cost (Ownership cost)
   The price of one dumper is $6,50,250, while the expenditure for freight, moving, and loading/unloading is $27,712. Then,

   Cost/price of one dumper (at delivery) = 6,50,250 – 27,712 = **$5,25,000***

   * This price will be borne by the mining company.
   If the total economic life of the machine is 15,000 hours, then
   - **Depreciation** = 525,000/15,000 = $35 per hour
   - **Expenditure** (Charges, fixed and incurred) = 54,000/2000 = $27 per hour
   - Per annum investment (after tax) = Investment × tax = 300,000 × 0.18 = **$54,00**
   The salvage value will be 15% of the price. Thus, **Salvage Value** = 0.15 × 650,250 = **$ 97,538**

Therefore, **Total Capital cost (before tax)** = Depreciation + Expenditure
$$= \$ 35 + \$ 27 = \$ 62.$$
Total Capital cost (after tax) $= 62 + (62 \times 0.18) = 62 + 11.16 = \$ 73.16.$

B. Operating cost

To calculate the operating cost, we should know and determine the following:

- Cost towards repair and maintenance of dumper including the tyre replacement cost
- Cost towards lubrication of HEMM
- Cost towards fuel (hauling cost)
- Cost towards labour and wages

If the price of one set (4 nos.) of dumper tyres is $ 32,512 (say) and tyre life is 2,500 hours, then

Tyre cost = 32,512 / 2,500 = $13 per hour.

Considering the tyre repair factor as 17%,

Tyre repair cost = $0.17 \times 13 = \$2.21$ per hour

Since the mine has no favourable conditions, we will consider a 45% factor for an average condition and apply depreciation ($35) as calculated in the capital cost. Then,

Maintenance cost = $0.45 \times 35 \times 1.5 \, (15/10)^{\text{depreciaation}} = \$23.63$ per hour

For fuel consumption, the cost of running one dumper will be

Fuel cost = $17 \times 1 = \$17$ per hour

Other operating costs will include the cost of lubrication because the dumper is a movable machine and essential mining equipment for the transportation of minerals in the mine. Since 17 litres per hour is the fuel consumption and 1/3 is the factor for lubrication,

Lubrication cost = $17 \times 1/3 = \$5.67$ per dumper

Wages cost (for 8 hours shift and minimum 4 hours paid per day) = $48 per day

Therefore, the operating cost for one dumper will be

Operating cost = $2.21 + 23.63 + 17 + 5.67 + 48 = \$96.51$ per hour/per day.

Since the number of dumpers deployed varies on day-to-day basis, the running hours keep on changing at the XY mine. Thus, if 4 dumpers are deployed on all working days regularly and 1 is kept on standby for the breakdown/maintenance, then for an operating period of dumpers = 2000 hours/year or 5.5 hours/day, considering 360 days operation of the mine in a year. Then, the **total operating cost per day** of all dumpers can be calculated as:

Total operating cost for the fleet of 4 dumpers at XY mine:

= per day operating cost × working hrs × numbers deployed
= $96.51 \times 5.5 \times 4 = \$2,123.22$ per day

Cost estimation is extremely necessary to know the costs incurred in the mine development and project operation. For control, visibility, and financial planning of a project, these are the basic tools.

Chapter 2

Production and productivity in mines

Live a lifestyle that matches your vision and have enough surrender to see the truth as it is.

(Bhagwad Gita)

Today, all mining companies face the challenge of achieving growth and enhancing the production and profitability of their company. Every organization thus needs to arm its team with skills and tools that will help them understand and analyse its performance. Their engineers remain perturbed in searching for the exact customer requirements before they decide to improve. This means that, in a mining company, real-field conditions play a pivotal role as it varies from one site to another. Taking production and productivity to a higher level will yield better profitable results only when we have committed interaction, smart management as well as updated human resource development (training) that recognizes the strength and weaknesses. The mining company should take advantage of the expertise gained over the years in whatever shape and manner is required. This enables us to be effective in planning and management thereby harnessing a better future for the company to establish it as a leader in mineral production.

Practical experiences narrate that production and productivity have several linkages, many measures, and different ways. Every single mine is unique and the production company must rely on meaningful problem study, the correct approach, and tailor-made solutions to arrive at the best solution. We must develop both a long-term and short-term relationship with our industry partners and apply an integrated approach and know-how to unblock the hurdles whether tried and tested earlier or not. Considering a new or recently opened deposit, a virgin deposit, an operational (existing/ongoing exploitation) deposit, and an abandoned deposit (already exploited), production and productivity have altogether different connotations. Similarly, conventional and special methods of mining have their dimension towards productivity if their selection is done properly. This chapter is particularly devoted to the vital aspect of production and productivity concerning mining because our ultimate goal is to produce more minerals at a reduced cost.

DOI: 10.1201/9781003274346-2

2.1   Production and productivity

Production is the process of converting any kind of inputs to outputs, either tangible or intangible. In a mine, all mining unit operations and processes result in the production of coal, ore, or mineral in bulk quantity which we usually refer to as mine production (output). In general, *productivity* is a measure of performance or output. It is a measure of how effectively the business targets of mining companies are being met. Both production and productivity are intertwined. The main factors influencing productivity are training, the experience of the workforce, quality of management, investment in production, technology, equipment, facilities/infrastructure, and general level of awareness both social and technical. Especially, in underground mining, productivity implies the state of being safe and meeting the challenges of unit operations as well as auxiliary operations leading towards actual production, i.e. mining from an underground mine. If an underground mine can produce economically and progress with time, in terms of equipment and technology, its productivity is certified.

Production and productivity issues are of immense importance in the mining industry, and thus, every single effort must be put in place to enhance them. Better mine productivity, an increase in the mine output, i.e. coal or mineral production, is assured. At every mining stage, if a smooth and continuous flow of production-related activities is ensured, the profitability of the company is easy to be achieved if not guaranteed.

Many hitches arise during mineral production at several key operational points in the mines affecting production, e.g. drilling, blasting, rock fragmentation, mucking/loading, stoping, ventilation systems, ore dilution, haul road/rail, and other transport systems including its maintenance, ore crushing, sizing, handling, and upkeep of all those equipment deployed in mines. These key points and focus operation areas, if not up to the mark, shortfalls in the mine's production are the likely cause. Our practical experience says that to improve productivity as well as production, firstly, supervision of operations should be well-governed and maintenance needs to be optimized; secondly, technology innovations and upgradation are vital; and, thirdly, efficient implementation of strategic and tactical plans are necessary.

2.2   Measures to improve productivity in mines

Many managerial and operational factors improve the mine productivity broadly pointed out in the previous section. A conceptual framework of mine productivity parameters is depicted in Figure 2.1 and the details are explained below point-wise.

2.2.1   *Management and supervision*

If the management and supervision are target and delivery-focused, improved production and enhanced productivity both are bound to occur thereby ensuring higher profitability for the mine. It is mainly concerned with planning, organizing, coordination, and controlling a part of the operation, management, and supervision concerning the mine being dealt with. In the mining context, this managerial factor is dealt with by the mine management, either handled by a leader or a group of leaders or coordinating among one another. An able leadership can successfully turn inputs (materials, equipment technology, and human resources such as staff/workers) into better outputs

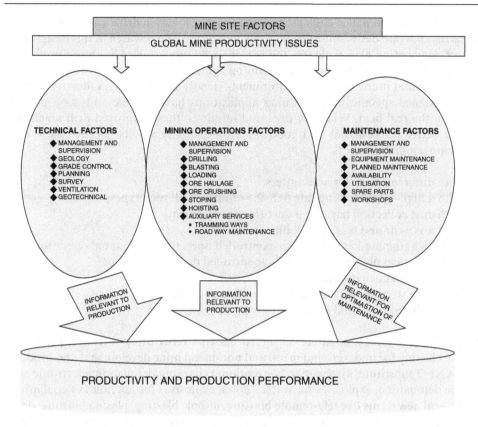

Figure 2.1 Conceptual framework of mine productivity parameters.
Source: After Jairo and Peter (2020).

(production of ore/coal/mineral) in an efficient manner. Everyday management, compliance, and supervision have to support the company's planned goals. There are five major groups of activities performed for the different aspects of management and supervision – (a) meticulous planning, (b) effective organizing, (c) timely execution and scheduled implementation of work, (d) interactive supervision, and (e) good coordination of all production-related activities/operations. Considering mine as a focal centre, these activities not only optimize cost but also protect assets and provide the best use of resources. Control over the production and productivity through the management and supervision route raises the mine performance significantly. For each mine, these aspects differ very widely depending on the organization's policy and the efficiency of the managers/supervisors.

2.2.2 *Operational factors*

2.2.2.1 *Drilling and blasting (D&B)*

D&B is a well-established unit operation of mining. Both of these areas have been extensively researched. Modern and advanced drilling equipment with good drilling rates and low maintenance costs are the latest technologies in drills and drilling operations that nurture better mine productivity outcomes. The research in this designated area of engineering also evaluates the reasons why modern rigs are more proficient and effective as regards the mines and mining operations. Based on drilling research, it is evolved that innovative and environment-friendly DTH drill rigs (down-the-hole drills) designed specifically for mining applications have reliable and good performance in the real field. Wherever deployed in mines, these improved drill machines achieve better performance and good mine productivity results because these machines are equipped

- with different diameter hole cutters;
- with a high-speed drilling rate (0.8–2.5 m/min) in all rock types;
- with dust collection bags for a dust-free environment;
- with less wear and tear of drill bit/cutter;
- with an automated and computer-controlled operation (no manual operation);
- with electrical power and hydraulic-controlled drilling operation;
- with easy manoeuvrability;
- with auto stop, auto wash, and anti-jamming functions;
- with reduced noise and less energy consumption.

In blasting technology, computer-designed blast patterns using software, perimeter control blasting for underground metal and non-metal mine development and production, ANFO substitute, alternate to conventional charging/detonation (electronic and electric detonators), explosives for watery holes, explosives for hot holes of coal mine, and several new terms like tele-remote blasting, in-hole blasting, plasma blasting, deck charging, AI-based programmable chargers, and image-based fragmentation software's for productivity enhancement, both for field and data analysis, have made their advent in the mining industry. The goal of all these devices, software, and equipment is to improve productivity in mines. Each of these terms requires an elaborate description for detailed understanding, but because of brevity, only the names of the related terms are mentioned here.

2.2.2.2   Stoping and stope development in hard rock mining underground

Underground hard rock mining for the extraction of metallic minerals (copper, zinc, manganese, etc.) is often practised the world over. Globally, *stoping* is the process of extracting the desired ore or other metallic minerals from an underground mine, leaving behind an open space known as a *stope*. In these metal mines, the surrounding rocks are hard rocks of different types that are strong enough to permit the drilling, blasting, and removal of ore without caving, although in most cases artificial support is also provided. As mining progresses, the stope is often backfilled with waste rock materials. In mining, stoping is considered a productive work to access the ore being extracted such as from levels, stopes, or underground chambers. It is seen that when the rock is sufficiently strong not to collapse into the stope, long-hole and in-hole blasting is practised. It enables an effectively controlled blast, modelling of blast designs, and projection of vibration and fragmentation in a scientific way (*Daunia Coal Mine, Queensland, Australia*).

Figure 2.2 Chambishi copper mine, Kitwe, Zambia.

In the mining industry across the globe, different mines use different stoping methods depending on the size and shape of the ore body, e.g. *Chambishi underground copper mine (Zambia)* practices open stoping with room-and-pillar to excavate steeply dipping ore bodies having reasonably firm ore (Figure 2.2). *Caylloma Mine in Peru* uses a cut-and-fill stoping method, which is applicable in the mining of steeply dipping ore bodies in stable rock masses. Canadian mining company (INCO) uses vertical crater retreat (VCR) which is used all over the world in areas with competent, steeply dipping ore and host rock (Goel, 1988).

The productivity of a stope in a metal mine is directly related to the stope size. Large stopes have high tonnages that can support high mucking rates for long periods, resulting in higher productivity rates, e.g. The *Homestake mine in South Dakota, North America* is a VCR mining operation. Their switch from the cut-and-fill method to VCR mining has been a major factor in their increased productivity since the 1970s. The next best method employed at Home stake is mechanized cut-and-fill (MCF), which has proven to be at a 48% cost disadvantage to the VCR method in terms of direct mining costs. The cost advantage mainly contributed to higher productivity and less required ground control. The productivity of a stoping method will vary based on the dimensions of the orebody, however, if the orebody meets the requirements, then VCR mining will provide high productivity rates when compared to other methods (Jairo and Peter, 2020).

2.2.2.3   Coal/ore handling and loading equipment and systems

Worldwide, mines have different ore handling equipment, loading systems, and method of extraction (mining). Both the open cast and the underground mine adopt different transport systems too for ore handling, materials, equipment, and personal handling. Selected and important equipment used in mines to handle/load rock, coal & ore in bulk quantity is

i.   Underground LHDs
ii. Dumpers and trucks (for use in open mines)
iii. Underground locomotives
iv. Electric mining locomotive
v.   Shovels and draglines (for use in open mines)
vi. Rippers and dozers

Concerning the cost and productivity, it is cardinal and has critical importance because of homogeneity, size, and equipment deployed for use, e.g. coal produced from one mine may be homogeneous in grade but underground ore tends to be very heterogeneous. Similarly, the ore and lump sizes matter may be large or small. These days, equipment manufacturer carries out a lot of research and development on the equipment they manufacture. Hence, tremendous scope exists for productivity enhancement be it a motor HP, drive control, or the type of power transmission for their running (diesel/electric). In general, size, capacity, capital cost, and manoeuvrability (mobility/flexibility) add to the production efficiency of mine haulage systems or material handling systems. Reliability, efficiency, and safety are the common requirements for underground ore handling.

2.2.2.4   Ore crushing/screening and sizing

Ore crushing is done in both opencast mines as well as underground mines. Crushing of ore is not always needed, but in underground mines, primary crushers are installed at different depth levels as per the requirement. In some mines, secondary crushing systems are put in place to further reduce the ore sizes. Extremely coarse ore is sometimes difficult to handle and causes choking/blockage in the ore stream due to its large size, e.g. at the Caledonia gold mine in Zimbabwe.

In captive opencast limestone mines of the cement plants, limestone of the required size is the requirement of the raw material feed for the kilns to produce cement. Sometimes, the ore is gravity fed from the ore bins onto the conveyor, which discharges the ore onto a vibrating grizzly feeder, which discharges the oversize ore/mineral requiring crushing. The ore crushing, screening, and sizing ensure that all the run-of-mine (ROM) ore is reduced to the required sizes as this provides for the optimization and greater cost efficiency in the bulk loading, handling, and hoisting operations in mines. This allows mining operations to continue without interruption.

2.2.2.5   Hoisting/transportation of mineral and waste

Open mines make use of the road, rail, aerial ropeways, and conveyors as their transport means for bulk handing of minerals and waste. In general, all underground mines worldwide make use of the winder as a hoisting system. All these transport means if maintained and used properly give cost-effective results. For the productivity of the mine, the transportation of equipment, personnel, mined ore, and waste must be quick, safe, and fast. The hoisting system (consisting of various components in which there are winders, ropes, and electric motors) and transportation equipment should have all the features essential for smooth production, e.g. the safety and reliability of the hoist depend on its design, so the proper and accurate design of a hoist is essential. A cost-effective hoisting system is a great call for all those mines that are using it to enable ore handling from the underground. With mining increasingly becoming cost-conscious, it has become necessary to yield greater hoisting benefits from improved electrical and mechanical appliances used for hoisting particularly in deep underground mines. The use of state-of-the-art hoisting and automation not only reduces the risk of accidents and injuries but also reduces manpower costs and spillage and enhances mine efficiency in terms of lowering loading and hoisting costs. An example of the *Blanket Gold Mine* operated by Caledonia mining in *Zimbabwe* can be given in this regard.

All these unit processes of mine operation improve the economic efficiency and the productivity of the whole mine. Proper machine utilization and equipment availability play an equal role in mine productivity. It is very important that practically applicable and technically feasible production plans,[1] in line with the requirements of the mining operations, must be kept ready all the time for profitability and productivity as it ensures the success of the production.

2.3 Ways and means of reducing per ton cost of mineral extraction and productivity enhancement

Economizing the unit operations of mining is always helpful in reducing the per ton cost of mineral extraction leading to productivity enhancement. Improving production by a few more tonnes through operational measures and ore recovery by even just 1% can mean a lot for a mine. It saves millions of dollars for the whole mine life.

For a mine, a tailored solution combining high-precision drilling, software-controlled blast design, and automatic blast monitoring (vibration, flyrock, and fragmentation analysis) is a practical measure. Backed by technical expertise and enterprise analytics, an integrated solution for higher productivity with increased profit is possible. For productivity solutions, an optimized *drill and blast cycle* is an easy way to manage and control. Some steps from a broader perspective are as follows:

Automation: Automation has the potential to liberate individuals and organizations from arduous, time-consuming repetitive tasks freeing them to be more strategic and productive. Repetitive works in any mining-related tasks can be automated, be it a unit operation of loading, explosive handling, repetitive blasting, survey, or analysis of samples for grade control. The mining operation, being hazardous, has an edge over other industrial activity as it can be controlled without compromising on man and machine safety. At every stage and for nearly all operations, automation in mining helps in reducing the per ton cost of mineral extraction. Thus, automation is overall advantageous for production and productivity enhancement.

Technology: Fortunately, a variety of advanced technologies have become affordable, cheap, and widely available over the past few years to help organizations. Technologies that execute effective work (mining) and carry out health and safety programme more efficiently are akin to mining tasks. Central to these efforts are environmental, health, safety, and quality (EHSQ) management. These powerful platforms – hardware as well as software solutions, available in traditional and digital forms, increasingly via the cloud computing model – allow important data to be captured electronically rather than through old and cumbersome school spreadsheets or even pen-and-paper methods. Many software solutions can support a variety of mining functions from map preparation to integrated management systems. Information sharing and dissemination through digital platforms are very common these days and are the new means of reducing the per ton cost of mineral extraction and productivity enhancement.

Good rock fragmentation: Globally, blasting is one of the key components of mining activities and its significant outcome is fragmentation. Fragmentation is the first result of blasting that is directly related to the mining costs. Unfortunately, mining operation needs secondary blasting that increases the cost per ton of mineral production. The effective utilization of the explosive energy assisted by employing optimized blasting practice can yield good fragmentation of the in situ rock meaning that the blasting operation is carried out in such a way that loader buckets should not have difficulties in grabbing the blasted material or ground.

In the fragmentation research, it was observed that some mine loaders more often fail or are not able to grab the problematic blasted material such that production has to standstill or is delayed. Poor rock fragmentation occurs as a result of either poor drilling practice or poor blasting practice or both. In this way, good rock fragmentation is a way to reduce the per ton cost of mineral extraction and productivity enhancement.

Machine simulators and virtual training: Safety and efficiency at a mine site are of supreme importance and the 'machine simulators' for training are vital tools in ensuring the productivity that should be achieved as desired. In virtual mode, the training is imparted without the mine having to take an actual machine out of production for too long as increased availability of the machine (drill) aids cost efficiencies. A machine from any original engineering manufacturer can be simulated. Some leading manufacturers, namely, Atlas Copco, Caterpillar, P&H, and Sandvik have been doing work on how to reduce the per ton cost of mineral extraction and improve the production efficiency and productivity of drill rigs, shovels, and dumpers through machine simulators. Example: (a) Drills used in mines as rigs are intricate machines and vital devices in mining operations. The *drill rig simulators* precisely imitate aspects, such as drill length and percussion, and train operators to use the machine safely and productively. (b) A virtual mine (Chapter 9) can impart training to beginners for improved efficiency and better production performances (productivity).

Cyber mine/Digital Mine: Digital mine, digitization, process-controlled mine, and cyber mine all are different terms but connected with a similar current trend of *digital opportunity*. Since mine is the nerve centre of mineral production, a digital transformation approach for the insight of the real-time data has enabled improved planning, control, and decision-making and support across the mining value chain in actuality (Rai, 2020). Integrated data platforms and well-governed big data support all mining processes and all time horizons. IOT/IIOT, cloud platforms, AI and ML tools, sensors, industrial communication networks (meshing) and reporting/analysis of real-time data, and insight gained from analysing trends, patterns, and opportunities for improvement lead towards better productivity and less per ton cost of mineral extraction. From experience and for the new developing/upcoming mining projects, the digital mine should be the future goal of the mining company.

2.4   Target capacity determination – an assessment

The target capacity determination, a part of the general mine planning, is a mining engineer's requirement for designing a new mine, an operational mine as well as upgrading an existing mine. Indeed, if integrated properly and explicitly with the technical issues, a safe and productive mine can be planned and designed.

A mine has to be designed considering physical, geotechnical, geological, and mineralogical factors and its production capacity per year together with the total target of coal/ore production should be designed accordingly. Many details go into the planning of a mine and the underlying mine economics principles are applied equally as they are of great concern and importance. This sub-section of the chapter addresses many of the factors to be considered for a mine during its various phases starting from the initial phase to the closing phase. Remember that planning an underground mine is different from those with a surface mine. Preliminary and detailed mine planning requires the following technical information largely to be gathered or generated from the primary and secondary resources:

Site Related

- Mining company's information
- Property and location
- Mining history of the property
- Land and Ownership
- Lease area and its size (length, width, and thickness)
- The overall operation areas, i.e. areas to be mined
- Regional and local surface features
- Power, access, water, and communication position
- Any important or protected zone or multiple areas

Exploration Related

- Review of exploration activities
- Mineral deposit type and geology
- Tabulation of geologic resource material
- Resource/reserve calculation based on geostatistics
- Coal seam-wise reserves/Ore deposit size and zone (area) wise reserves
- Geology and mineralogy related information obtained during exploration

Economics Related

- Economically appropriate or not for modem mechanized mines.
- Coal/ore grade and tonnage
- Related mining costs
- Related taxes as applicable
- Royalty and cess
- Capital to be invested and financial outlay

Technical

- HEMM and equipment: Selection, performance, versatility, acceptance, flexibility, and application conditions
- Material characteristics (the hardness, toughness, and abrasiveness of the material extracted, i.e. both ore and host rocks)
- Rock types and rock structures in established mining districts (for underground mines)

- Strength parameters of rocks, ore, coal, and overburden material
- Slope stability-related parameters (cohesion, angle of internal friction, and angle of repose)
- Factor of safety

2.4.1   *Targeted mine production and timing (an influencing factor)*

For any given ore body or deposit, the development required before the start of production is generally related to the size of the production as well as the mining method. The necessary stripping ratio (OB: mineral cover ratio) for a surface mine and the time for the development of an underground mine are time-dependent. This gestation period will be quite long if planning is not done properly. In combination, all of the related factors mentioned above could be substantially reduced. In the past, the time for the mine development has varied from 2 to 8 years. Two indirect economic effects could result from this:

a. Capital would be invested over a long time before a positive cash flow is achieved.
b. The inflation rate-to-time relationship is controlled. In some countries in the past, this is the main cause to push the costs upward by as much as 10%–20% per year (cost overrun), thereby eliminating the benefits of the economy of scale in the case of large-size mining projects.

The timing of mine development and the cost-timing are often more important than the amount of the cost. In a financial model, the timing of a development cost must be studied in a sensitivity analysis. In this respect, any development that can be put off until after a positive cash flow is achieved without increasing other mine costs should certainly be postponed.

It means this influencing factor should be checked to aid the mining engineer to develop a mine. In planning, the approximations of the time have tremendous importance in fixing production targets, cost, and capacity.

2.4.2   *Statutes, government policies, and the mine size considerations*

For mines that are at the stage of start/primarily development/lateral development, the speed of the development can vary considerably. Concerning this, the government policies and concerns related to legal & legislative statutes matter a lot for the mine development. Experience has shown that the mine's development only progresses when they are handled properly. Government attitudes, policies, and taxes generally affect all mineral extraction systems and should be considered as they impact the real execution at the ground level, i.e. mining methods, production, and the size of the mining operations. Some examples are (Bullock, 2011)

i. One can assume that a mine is being developed in a country and that the political scene is currently stable hence it is difficult if not impossible to predict beyond 5–8 years. In such a case, it would be desirable to keep the maximum amount of development within the mineral zones, avoiding development in waste rock as much as possible. This will maximize the return during a period of political stability. Also, it might be desirable to use a method that mines the better ore at an accelerated rate to obtain an early payback on the investment. If the investment remains secure at a later date, the lower-grade margins of the reserve might later be exploited. Care must be taken, however, that the potential to mine the remaining resource, which may still contain good-grade ore, is not jeopardized, whatever the stability situation. There is merit in carefully planning to mine some of the higher-grade portions of the reserve while not impacting the potential to mine the remaining reserve.

ii.   Some mining methods, such as room-and-pillar mining, allow the flexibility of delaying development that does not jeopardize the recovery of the mineral remaining in the mine. In contrast, other mining systems, such as block caving or longwall mining, might be impacted by such delays.

The number of HEMM/equipment or numbers of personnel required to meet the needs of all mines cannot always be given in absolute and precise terms. Some of the general problems that may be encountered are related to the mine site as well as size because actions to mitigate these problems are as big and large or small as the size of the machine/equipment.

As a result of a country's tax or royalty policies, sometimes established to favour mine development and provide good benefits during the early years of production; in later years, the policies change. This kind of situation might arise affecting the mine planning. Moreover, the flexibility of the mining system must be considered taking into account the mineral resource conservation as the mineral is a wasting asset. Once coal/ore is formed in nature, it takes several million years to form.

2.4.3   Planning the organization's workforce and production design

When planning the details of a mining workforce, it is necessary to consider several factors to sustain the production level dictated by other economic factors. Of all the items involved in mine design, the *organization of the workforce* is the most neglected and can be the most disastrous. Workforce and local community issues need to be investigated at the beginning when the property is being evaluated and designed. This will provide adequate time for specialized training, minimize unexpected costs, and also prevent economic projections based on implemented policies. The productivity and profitability of a mine or mining operation are linked to high morale and good labour relations. Where such parameters are poorly rated, industrial operation can be drastic. Such matters can make the difference between profit and loss.

For production design, it is extremely necessary to give a thought – Are the local people trained in similar production operations or must they be trained before production can achieve full capacity? Hence, for better production, design, workers' customization to a mining environment, and work schedule (shift workings) are essentially required. Though human resources (people) can be trained in such programmes, this item could cause a delay in designing, planning, and organizing a large-sized daily production of a mine to fall far short of its desired targets.

2.4.4   *Underground vs surface*

As far as the general basis of mine planning goes, the facilities, services, and utilities in the two broad types of mine categories are different hence the planning also differs. The underground facilities, such as underground pumping stations, power transfer points, computer-controlled installations for production and communications, transformer stations, underground offices (shift and maintenance foremen's offices, etc.), safety manholes and ore storage pockets, skip-loading stations, storage/warehouse space, lunchrooms, and refuge chambers, are unique and require different engineering study and design. Hence, these types of facilities are additional. More detailed information on these aspects is available in the mining literature.

2.5   Mine size and productivity concept, its measurement, problems, and suitability

Before we outline the mine size and productivity concept relation, let me make the reader aware and explain – What is Quarrying? and how we can define Artisinal mining or small-scale mining? Since these words have intricate relations with the size of a mine, they need to be defined first. Practically, the mine size and productivity are linked with each other. The productivity of a bigger size mine makes use of machinery and small mines/quarries are dependent on the available resources, which are mostly manual labour. Selective equipment and limited finances make them different from their bigger counterpart. Hence, productivity measurement and suitability come across some typical practical problems as well. This is the reason why this section is included here.

Definition of Quarrying: A quarry is an excavation or pit, usually open to the air from which the resources contained in it are obtained, usually rocks, for human use, be it for construction or other industry. Some form of quarrying is carried out virtually in every country of the world, if not for the extraction of construction materials, but for the extraction of different minerals that include coal, stone, and ores.

In general, the quarry is an excavation (usually below the ground surface but not in the case of hill quarrying) that plays a big role in the local employment generation. It has important economic, environmental, and social effects too for the country as a whole or regions where it is carried out. The quarrying operation requires an interdisciplinary team of prospectors, geologists, geophysicists, excavation specialists, and geochemists. How to run a quarry, founding a quarry, the various quarrying techniques (i.e. operation part), quarry wastes, quarry fines, and quarry products are some aspects that are dealt with in quarrying to put focus in entirety.

Definition of AM/SSM: Usually, 'artisanal mining' and 'small-scale mining' (abbreviated as AM or SSM) in international terminology are two synonymous terms. From country to country, these terms and their definition differ. In India, no separate policy decisions or legislative provisions have so far been made specifically for SSM even though it is an important segment of the overall mining industry. Hence, SSM comes under the general mining umbrella without a separate identity. In India, as per the National Institute of Small Mines (NISM) Kolkata (MMSD, 2001), if the quantum of mineral production is 0.1 million tonnes per year (MTPY) maximum, it is termed a *Small-Scale Mine*. All other mines from 0.1 to 0.5 MTPY and above 0.5 MTPY fall in the category of medium and large/big mines category respectively. This criterion has been adopted without inducting any other parameters such as the value of production, degree of mechanization, and employment. Moreover, artisanal mining is generally considered an unauthorized or illegal mining operation but it is not always the case. These mines are highly labour-intensive and no mining equipment, except simple tools, is used in their operation. However, the contribution of Small-Scale Mining (SSM) is quite significant in developing countries and third-world countries. In the race for higher production and better productivity, SSM adopts even higher rung, follows mechanized practice partially, and deploys a skilled workforce, quite often. SSM activities provide sustenance to the local people and people living below the poverty line and are thus socially preferred. Even the government is also supportive as it provides regular employment to many of the locals.

Having explained the quarry and small-scale mine, now let us discuss the productivity measurement concept and its suitability concerning the mine type and size including its relation (Figure 2.3). For a quarry or SSM, productivity is a measure that relates to profitability. The productivity concept for the underground and opencast (coal mines) can be explained by the old terminology, Output Per Man Shift (OMS). In very simple terms, it is the price recovery factor that has a relationship with company profits and economic welfare. Thus, OMS is calculated as follows:

For underground coal mines

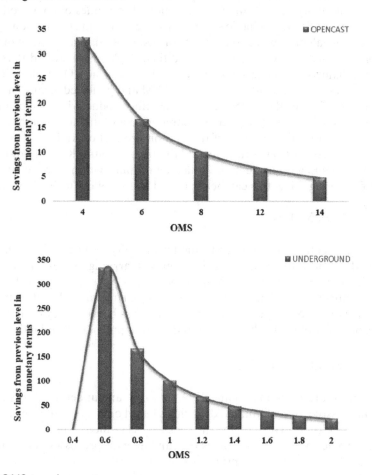

Figure 2.3 OMS trend vs savings.

OMS = Tonnage of coal or mineral produced / (Manpower × shifts)

For opencast coal mines

OMS=P+1.4 Q/M (1+1.4 R)

where P is the production of coal in tonnes, Q is the overburden removed in m^3, R is stripping ratio in m^3 of the overburden removed per ton of coal, M is the manpower, and 1.4 is taken as the assumed average specific gravity of coal.

A trend of OMS vs savings in monetary terms (Figure 2.3) shows that considerable saving is possible in the case of underground mines (a rise of OMS by 0.2 leads to the cost-saving from Rs. 166 to Rs. 67), whereas in the case of open mines with OMS improvement by 2, the cost-saving is only marginal (Mathur, 1999).

Essentially, OMS is a partial measure of total productivity only. The primary reason for measuring productivity in terms of OMS is that it has the apparent merit of simplicity; however, the OMS suitability is often questioned. This is because of the problems with the OMS measurements, uniformity, compatibility between different countries, and even the difficulty in ensuring its calculation. The tonnes of coal produced from mines are based on ROM production, while in some mines, the production figures may be based on equivalent saleable tonnage of washed coal. Similarly, the manpower (the number of men used in OMS calculation and their deployment level) differs. With the advent of mechanization and larger man–technology interface, OMS alone is not suitable for bigger mines, but it is still prevalent for SSM and preferred means for measuring productivity. The higher the OMS, better is the mine productivity. The opencast mine productivity index is generally higher compared to that of underground mines.

On analysing different schools of thought on production and productivity, it was found that productivity (P) is a function of labour (L), capital (K), manpower (M), and energy (E) as inputs and production (T) as the output, within a specified timeframe (t=1–2). Concerning mine, this can be expressed by the following basic equation:

$$P = 1\int 2 \ (L, K, M, E)/T \tag{2.1}$$

Thus, mine productivity is an index to the profitability and economic welfare of the miners in which they are engaged in the mining/quarrying operation irrespective of the mine site or quarry dimension. While productivity per se is vital for the mining industry, there are major difficulties in measuring it, particularly when it comes to fixing the weightage of various input and output parameters because they do not have common units of measurement. This has led to the development of 'Productivity Indices'.

2.6   New mineral deposit

Mineral deposits are found in so many different forms and under so many varying conditions. The classification of mineral deposits is based on major Earth-process systems that include rock types. Universally, the rocks are classified as igneous, sedimentary, and metamorphic rocks, which express the fundamental processes active in the crust of the Earth. In the twentieth century, many classifications of mineral deposits were based on the types of rocks hosting the ore or on the geometry of the deposit and its relation to the host rocks. Since a perfect classification is utopic, in summary, creating a simple classification of mineral deposits linking deposit types directly to ore-forming processes and genesis is certainly the preferred way to classify the mineral deposit forms (Bustillo, 2018). Accordingly, many distinguished classifications/categorizations came into foray, e.g. massive ore deposits, vein-like deposits, layered deposits, stratified deposits, and disseminated ore deposits.

Further, there are two broad categories of a new mineral deposit, namely:

a. A **virgin deposit** that is yet to be opened for mining.
b. A **new or recently opened** deposit that is where mining has recently started.

Criteria used to classify mineral deposits vary widely and there are many forms of deposits. For simple understanding, they are classified depending on their geological formations, genesis, tectonic history, use, size, topography, and mineral category types. The following are the different classification categories of mineral deposit forms:

Geological and Genitival

- Magmatic deposit
- Sedimentary deposit
- Metamorphic/metamorphosed deposits and
- Hydrothermal deposit

Uses

- Industrial mineral deposit
- Non-industrial deposit
- Strategic mineral deposits of industrial and non-industrial category
- Minor

Topography

- Hill deposit
- Plain deposit
- Coastal deposit

Size

- Major deposit
- Minor deposit

Mineral Category

- Coal and Non-coal deposit
- Metallic and non-metallic deposits
- Placer deposit

Each of them has several types and sub-types for practical exploration and exploitation.

Mineral deposits (ore/coal) are also named and known by their shape, size, and location, being the most representative indicative words for their description. Sometimes, the name of a place, region, or city is used (e.g. Raniganj coal, Himalayan type, Alpine type, Sudbury type, Cyprus type, Mississippi valley type). Other times, the deposits are known using their acronyms, e.g. BHQ, SEDEX, BIF, and MVY. In addition, the deposits may be called according to the rock type, such as pegmatite (large crystals), porphyry copper (disseminated stockwork linked to plutonic intrusive), and skarn (a calcium-silicate rock).

Some deposit types, as defined below, are according to different criteria:

- **Massive deposit**: A deposit with mineralization comprising more than 50% of the host rock is said as massive.
- **Vein-type deposit**: A vein-type deposit is one in which the mineralization occurs in the fractures, filling, or fissures of the underlying rock. Its orientation and size are non-restrictive and may be in any direction or shape. Vein-type deposits are the major source of metals in contrast to the layered rock formations.
- **Pipe-type deposit**: In this deposit type, the mineralization body has the form of a carrot or pipe-like shape hence called a pipe-type deposit, e.g. diamond deposits are this type of deposit typically.
- **Lens-type deposit**: The mineralization body is much thicker in the centre than around the edges, and it may be flat-lying, dipping, or vertical.
- **Tabular deposit**: An ore zone that is extensive in two dimensions but has restricted development in its third dimension is referred to as 'tabular'.
- **Epithermal, mesothermal, and hypothermal deposits**: Epithermal deposits are those formed at less than 1,500 m depth and formed as a result of temperatures between 50°C and 200°C. Mesothermal deposits originated at intermediate depths (1,500–4,500 m) and formed as a result of temperatures between 200°C and 400°C. The hypothermal deposits are those formed at a greater depth of more than 4,500 m and temperature range between 400°C and 600°C.
- **Stratiform deposit**: A mineral deposit related to concrete stratigraphic bedding.
- **Layered deposit**: A mineral deposit in a layer or bedded form is called a layered deposit, e.g. coal seams are formed in layers.
- **Strata-bound deposit**: A mineral deposit limited to a determined part of the stratigraphic column.

A deposit, either virgin or recently opened, as a result of the mining, gets converted into a full-fledged mine (surface or underground). The operational mine slowly reaches either the partially developed or fully developed stage. When a fully developed deposit reaches the end of its life on exhaustion of ore/coal, it turns into an 'abandoned deposit' which means all ore/coal reserves have been exploited. In this way, the difference between *a deposit* and *a mine* can be understood.

Here, BHQ is the banded haematite quartzite, BIF is the banded iron ore formations, MVT is the Mississippi valley type lead-zinc ores, and SEDEX is the sedimentary exhalative ore

2.7 Earth system – coexistence with mining

While dealing with mining, the Earth system always comes into the picture. The coexistence of mining with geology is beyond doubt and interlinkage with the Earth system is proven because the formation of mineral deposits took place in Mother Earth. A geological, tectonic, lithological, or geochemical feature of the Earth system is believed to have played a significant role in the formation of ore deposits and the concentration of one or more metal elements in it. Hence, it is clear that minerals and metals contained in the Earth co-exist.

The mining, i.e. the exploitation part of the mineral resources, is related to the depth and geometry of the ore, its content type, and genesis. Therefore, it is necessary to understand how mineral/ore formation took place on Earth. Being interdisciplinary, the theoretical aspects of mineral deposit formation and related geological terms have been explained as they are a part of the Earth system.

2.7.1 Theoretical aspects of mineral deposit formation in the Earth system

Agricola (1556) formulated the first reasonable theory of ore genesis. In his book, *De Re Metallica,* he showed that lodes originated by deposition of minerals in fissures for circulating underground waters, largely of surface origin, that had become heated within the Earth and had dissolved the minerals from the rocks. Agricola made a clear distinction between homogeneous minerals (rocks) and heterogeneous minerals (rocks). Little progress was made in the study of ore genesis from the time of Agricola until the middle of the 18th century. By the 1700s, more remarkable progress was made in Germany in the Erzgebirge mining district. At the end of the eighteenth century, the polarized views of either *plutonist* or *neptunist theories* were developed. In the middle of the nineteenth century, Von Cotta affirmed judiciously the various theories of mineral genesis and correctly concluded that no one theory was applicable to all ore deposits. At the end of the 1800s and starting of the 1900s, different authors created a new controversy related to the ore formation theories. In 1913, a classification of mineral deposits based on their origin, whether they were products of mechanical or chemical concentration, and if chemical, whether they were deposited from surface waters, from magmas, or inside rock bodies, has been proposed. Other theories about the origin of mineral deposits resulting from the injection and rapid freezing of highly concentrated magmatic residues were brought forward in 1923 (Spurr, 1923). A metallurgical interpretation of the ore deposits was also proposed: During the former molten stage of the Earth, the metallic minerals sank into deep zones due to their specific gravity, and they were later brought to the surface. According to this model, the upper layers first and the lower layers later moved upward in the form of vapours from which the metals and minerals were deposited. Simultaneous to this exotic theory, Bateman (1951) suggested that the formation of mineral deposits is complex, and eight diverse processes can account for their formation: *magmatic concentration, sublimation, contact metamorphism, hydrothermal action, sedimentation, weathering, metamorphism,* and *hydrology.* The advent of plate tectonics improved considerably the understanding of the lithotectonics of rocks and ore occurrences. Because mineral deposit systems require a conjunction of processes to produce exceptional metal enrichment over background terrestrial concentrations that result in ore deposits, they can form only under specific conditions in particular tectonic environments. As an example of modern theories on mineral deposit genesis, a classification based on the different geological processes that form mineral deposits can be outlined (Kesler, 1994). These ore-forming processes can be surface processes, including weathering, physical sedimentation, chemical sedimentation and organic sedimentation, and subsurface processes, involving water or magma. This broad expression of ore-forming processes is actually the most frequently used (Bustillo, 2018).

Ore-forming processes can be classified into four main categories: *Internal, hydrothermal, metamorphic,* and *surficial processes.* The former three processes are related to the subsurface phenomena, while the last one covers those processes occurring at the Earth's surface. The hydrothermal process has been further subdivided into magmatic, metamorphic, diagenetic, and surface to refine the nature of the hydrothermal process (Bustillo, 2018).

Thus, it is clear that some mineral deposits are formed by magmatic processes, while other mineral deposits are produced by sedimentation or surface weathering. Probably, the main difference between both is the secondary importance of metamorphism in the enumeration of the substantial ore-forming process compared to its fundamental role in generating rocks. Another major difference is the essential function of hydrothermal fluids (hot aqueous fluids) in the genesis of ore deposits. The circulation of aqueous hot fluid in the crust is usually cited as a factor that modifies locally the composition and texture of rocks.

Coal and its formation on the Earth: The formation of coal is different compared to that of metallic minerals (ore). Coal is primarily formed as a consequence of temperature and pressure build-up underneath the Earth's surface. It is used as a fuel mineral and is known from most geologic periods. Nearly, 90% of all coal beds were deposited in the 'Carboniferous and Permian Periods', which represent just 2% of the Earth's geologic history.

Coal is composed of macerals,[2] minerals, and water, and fossils were found in coal formations quite commonly. At various times in the geologic past, the Earth had dense forests in low-lying wetland areas. In these wetlands, the process of coalification (conversion of dead vegetation into coal is called coalification) began when dead plant matter was protected from biodegradation and oxidation, usually by mud or acidic water, and was converted into peat. This trapped the carbon in immense peat bogs that were eventually deeply buried by sediments. Then, over millions of years, the heat and pressure of deep burial caused the loss of water, methane, and carbon dioxide and increased the proportion of carbon. The grade of coal produced depended on the maximum pressure and temperature reached, with lignite (also called brown coal) being produced under relatively mild conditions and sub-bituminous coal, bituminous coal, or anthracite (also called hard coal or black coal) being produced in turn with increasing temperature and pressure. Of the factors involved in coalification, the temperature is much more important than either pressure or time of burial. Sub-bituminous coal can form at temperatures as low as 35°C–80°C, while anthracite requires a temperature of at least 180°C–245°C (Wikipedia).

2.7.2   *Some geological terms*

Related to mineralization in the Earth system, words like *Metallogenic Province* and *Metallogenic Epoch* are often referred to. Since these words are associated with the mineral/ore formations within the Earth, their meaning should be known to a mining engineer.

- **Metallogenic Province**: A metallogenic province may be defined as a mineralized area or region containing mineral deposits of a specific type or a group of deposits that possess features (e.g. morphology, style of mineralization, or composition) suggesting a genetic relationship.
- **Metallogenic Epoch**: A metallogenic epoch is a geological time interval of pronounced formation of one or more kinds of mineral deposits.

    The size of a metallogenic province can be as large as the Superior Province (Canadian Shield), and a metallogenic epoch can be as broad as the entire Proterozoic.

2.8   Commencement of mining operation

Actual mining operation starts when all necessary clearances, i.e. mining clearances, environment clearances, forest clearances etc., required for a mine are in place. The work of obtaining the surface rights from landowners is first completed by the mining company to make claims that the land is owned by the company. The lease allotment work, which is a part of the mining clearances, commences from the concerned government department, either state or central, as per the applicable rules of the concerned ministry or government agency. Before the commencement of mining operation, a notice of intimation for opening/reopening a mine is given in an appropriate proforma to the local and statuary agency concerned (district authorities and bureau of mines) indicating the name of mine and company, lease registration details, number, etc., including the owner's address and particulars. A lease allotment letter, mining plan as approved by the state/central government concerned and a firm date of opening is a pre-requisite to obtaining the *consent to operate a mine*. In addition, the name and address of the previous owner (if any), present owner, and particulars of owner, agent, manager, geologist, and mining engineers to be employed in the mine shall also be furnished. Once these formalities are completed, the mine can be started in actuality and the mining operation can be commenced legally and authoritatively.

2.9   Opening a mine: By surface methods and underground methods

At once when it is confirmed through geological investigations that the mineral deposit is located at shallow horizons or a depth (more than a shallow horizon), it is decided whether the deposit will be exploited by a surface mine or an underground mine, e.g. a coal deposit lies at a shallow depth compared to a metallic ore formation, namely, gold. Considering this fact, the access to a mineral deposit is decided accordingly and two principal approaches may be applied to opening a mine. A surface mine is opened through a 'box cut', whereas an underground mine is opened by an incline/decline or a shaft as the case permit. In the following paragraphs, its technical description has been given.

2.9.1   *Surface mine*

In a flat surface topography, the opencast mines are opened up by an initial open cut technically termed a 'box cut'. The box cut slopes down at a suitable gradient to reach the desired mine bench level. The opening cut can be either internal or external to the pit/mine. External openings are preferred for shallow pits, whereas internal siting on berms at one side of the mine pit is preferable for deeper open pits. The internal opening cut saves space and caters for a large number of benches simultaneously but blocks the ore underneath. It should be excavated at a location following certain criteria (Box 2.1) to conserve space, be safe, and have minimal excavation cost. For shallow pits and deep pits, the length of the box cut could be short or long. From the box cut end, the first working bench of the mine is developed.

In the case of a hill deposit, the opening up of the deposit is much simpler and occurs along the contour. The first bench in the hill deposit is opened from a *central trench* cut across the top level or from one side following the hill contour over a face length sufficient to give the desired production of ore/mineral. Such an opening up usually gives a pair of long straight mine working faces, which, if located, in the mineralized zone can give a quick return too. Conventional blasting of toe holes from the hillside gives access to the mine inside. The central trench facilitates easier transportation for the mine, faster opening up, and space to accommodate the mining equipment and machinery because space is a major constraint in the hills.

How to excavate a box cut is the first question that comes to mind for opening a surface mine. The box cut is started from the surface down to the floor level of the first bench in a conventional manner using excavating machinery such as a shovel, dumper, and draglines which are to be subsequently used in the pit for minerals and waste

---

Box 2.1: Siting Criteria for the Box Cut Location

1. The box cut opening area shall be free from geological disturbances and stable.
2. The box cut should be designed to bring quick returns.
3. Maximum safety and economy through minimum cost of transportation of waste or mineral, as the case may be, from the pit to the waste dump or at the boundary of the mineral deposit.
4. The box cut should be located where the overburden:ore ratio is the least for minimum volume of waste excavation (mineral outcrops are preferable).
5. Where no shifting is needed and where you have command over the maximum possible mining areas.
6. The box cut should be at a level which is high enough, well drained, and preferably not flooded (above highest flood level).

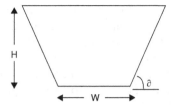

Figure 2.4 An opening trench (box cut for a surface mine).

excavation. In this open-cut excavation, the floor of the box cut is first deepened slowly and excavated material is dumped away from the box cut, or transported using trucks or dumpers. When the required depth is reached, the box cut floor is levelled off. The deployed excavator may be located either near the excavation point on the surface or the floor, depending on the type of equipment used. Dragline excavators are also used in place of shovels/excavators/diggers and are more suitable for making a deeper and wider box cut standing on the surface. In general, a power shovel is suitable for a limited depth standing on the box cut floor itself.

The box cut shape is trapezoidal (Figure 2.4) and has a suitable slope on both sides of the cut at 50°–60°, and the floor gradient ranges from 2% to 12%. The gradient shall be such that the safety and the vehicular movement, in favour and against the load, are adequately maintained. Since the opening box cut has to stand for a long time, it is safe to have their sidewalls sloping at the angle of repose of the rocks encountered during excavation.

The width of the box cut or trench depends on the size of the excavator and the transport system. The minimum width should be twice the turning radius of the excavator deployed. Depending on the transport system the minimum width ranges from 5 to 20 m. It is also required that the width of the box cut should be sufficient to accommodate approach roads to all benches of the pit and shall be maintained properly for the safe movement of man and machinery.

2.9.2 Underground mine

Briefly, two alternatives exist for opening an underground mine and they are opening by shaft and opening by incline/decline. A mineral bed can be easily accessed through an incline/decline or a shaft, serving as an initial opening or access for the underground mine. When compared, the incline/decline is a cheaper access method compared to the shaft because the shaft sinking/shaft drivages are a skilled and intricately planned operation requiring more capital investment as well. Sitting criteria for the underground access structures, i.e. incline or shaft, are more or less similar to those used for the box cut location for surface mines (Box 2.1). To excavate these accesses, the mining is done by conventional mining methods only. If the depth of the mineral bed is not more or very shallow, opening an underground mine by an adit or incline/decline is the most preferred way and should be planned. If the mineral deposits have a higher stripping ratio and lie at a greater depth, the access and approach to the excavation site should be made by vertical shaft for mineral production. An adit or a cross-measuring drift connects the incline or the shaft with the actual working faces from where the mineral is extracted. There may be a case of mineral beds at different depths, e.g. coal seams occurring at different horizons. In such cases, the deposits are also entered and opened by an incline/decline or a shaft depending on the depth criteria.

An incline/decline is a sloping road like a ramp, driven from the surface up to the deposit or very near to the deposit from where horizontal access is feasible. The incline/decline is driven at a gradient of 1 in 4 to 1 in 5 to provide access and movement for both the man and the material.

The normal practice of incline/decline drivages for access is along the true dip of the deposit. They could have a rectangular, a square, or a horse-shoe shape like a tunnel. A rectangular incline/decline is usually 4.2 m wide and 2 m high for access to a mine. The size may be decided depending on the production planned. A conventional method of excavation is the practice for the incline drivages, and the opening remains supported permanently to serve as an entry for the mine life throughout.

Shafts serve as a better and only means of access to deep-located mineral deposits which are excavated by underground mining (Figure 2.5). They are excavated from the surface ground level up to the deepest level at which the mineral is found or many times more than one shaft is also excavated in a mine. They are circular shaped in general with varying diameters. The size, shape, and number of shafts for access will depend on the production size of the underground mine. Since the shaft is a permanent means of access throughout the life of the underground mine, the *shaft pillar* should be kept on all sides of the shaft and only essential roads should be driven through it. It should also be understood that more technical intricacies are involved in opening an underground mine through a shaft compared to the incline/decline access or open cut of a surface mine.

2.10   Merits and demerits of the basic mining methods

As explained in the rudiment of mining in Chapter 1 (Section 9), there are two basic methods, viz. surface methods and underground methods. Each of them has its advantages and disadvantages (Tables 2.1 and 2.2).

Besides basic conventional mining methods, i.e. surface and underground, other special mining methods that are also practised in the industry at noted and specific mining sites, e.g. hydraulic mining, marble mining, sand mining/dredging, and marine mining, offer many merits and demerits, separate for each special method. They will be described in Section 2.13 along with the special method.

2.11   Stages of mining

The mineral excavation process (mining) involves various stages. The following are the four broad and important stages in mining (Figure 2.6):

a. Prospecting and exploration (Exploratory drilling for ore assessment)
b. Development of a mine
c. Exploitation (mining for mineral extraction) and milling/ore beneficiation
d. Decommissioning (Closure through Reclamation, rehabilitation, and revegetation processes as applicable in mining)

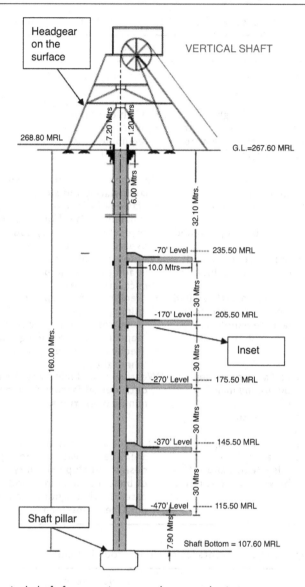

Figure 2.5 A vertical shaft for opening an underground mine.

A mining engineering student must become familiar with these mining stages first and remain aware of their technicalities to know about the mine details which is engaged with the mineral extraction. *Exploration*, the first stage of mining activities, encompasses all actions in the field that include initial reconnaissance, exploratory drilling, trenching/pitting, geophysical surveys, geochemical assessment, and reserve estimation work so that a deposit can be declared economically viable for coal/ore extraction. Exploration activities also include determining the location, form, size, shape, position, and value of the mineral deposit using various methods of prospecting and exploration.

Table 2.1 Advantages and Disadvantages of the Surface Method of Mining

Advantages	Disadvantages
High productivity and labour conservation. Good output per man shift for both ore and waste	Limited by depth factor and stripping ratio
Degree of mechanization – A higher degree of mechanization is possible to yield high production output. Remote and safe operation is feasible in hazardous places	• High capital investment • More dependency on machine and equipment use (with large-scale mechanization)
Lowest cost per ton mining cost for coal/ ore/ or any other mineral	Detrimental to surface environment, i.e. land, water, and air
• Early production (less gestation period) • Mine development can be programmed better • Suitable for large equipment use and high production as well as productivity	Surface damage may require reclamation, rehabilitation (afforestation), and revegetation thereby adding to the overall production cost Maintenance of surface damage cost in addition to the per ton mineral excavation cost
Low requirement of labour, mostly skilled labours are required as key operators and heavy earth machine handlers, e.g. drill, shovel, dragline, and dumpers	• Requires large deposits to realize the mining • Requires large reserves and massive deposits for the mining operation • Requires sufficiently high grade of ore/ mineral/coal
• Mine development and access are simple compared to an underground mine • Relatively flexible • Can vary output easily if there is a change in demand	Weather and climate are the major detrimental factors to impede open mining operations
Superior over underground mining in terms of risk and safety (No underground hazards; Health and safety of workmen and the workplace is not a cause of great concern)	Requires an essential provision of dump area (heavy waste disposal); proper slope design and slope stability; maintenance of benches including good drainage is essential throughout the mine life and must be maintained

Table 2.2 Advantages and Disadvantages of Underground Mining

Advantages	Disadvantages
Low productivity compared to the surface method of mining	Technological limitations, lower safety, and higher risk are the bottlenecks of underground mines
Not limited by depth but limited by the stripping ratio	Higher cost per ton compared with opencast mining
Environment friendly compared to the surface counterpart	High capital investment
Better high-grade ore, located at depth and restricted for mining due to the depth factor being easily extractable. Good recovery is possible through underground mining	Less surface damage and disturbances hence easily acceptable to the society

(Continued)

Table 2.2 (Continued) Advantages and Disadvantages of Underground Mining

Advantages	Disadvantages
Weather and climate are no deterrents to impede underground mining operations	More labour requirements (both skilled * and unskilled)
Very large deposit/reserve to realize, slope stability, large equipment, waste disposal, and space for dump area are some key advantages over its surface counterpart	The higher gestation period (late production starts because of slow development in the underground
Fairly good for selective mining	Relatively less flexible
Do not require reclamation and restoration of land which has added advantages to the production cost	

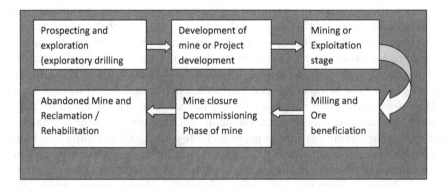

Figure 2.6 Stages or life cycle of mining.

The exploration stage, i.e. surveys and mapping drilling, generally produces environmental impacts in terms of clearing of trees and vegetation, displacement and disturbances to the wildlife, and bringing landform changes temporarily. These impacts are localized and disappear when the exploration camp is dismantled or work is over. At the end of this stage, the feasibility of the mineral deposit is declared and it is decided whether the deposit is commercially exploitable or not.

In the second stage, the *development of the mine* (or project) is started and this consists of several principal activities right from determining the project feasibility in detail to conducting financial analysis to determine how to develop the mineral property. Acquiring surface rights and mining rights up to the consent for an actual mining operation is done at this stage. A mine deposit is then developed as a mine for mineral production and its consequent sale in the market. The development phase includes activities such as opening up of deposit for mining, overburden stripping and placing, mine approach road and haul road preparation, equipment and machinery procurement, drilling and blasting for mineral excavation, and creation of various infrastructures for the mine project such as transportation roads, building, power line, plants, water supply, drainages, and sewerages for constructed houses, buildings, and different facilities as amenities. This developmental phase of the mine could cause environmental damage and should be kept contained through the managerial efforts of the mine management. Consequent to this stage, the mine site becomes ready for mineral production.

The third stage of *mining and milling/ore beneficiation* starts together so that the excavated ore is converted into the form of metal for its subsequent use by humans. The mining process, either conventional or mechanized, is geared up at the mine and the 'ore processing plant' is planned for erection within the project boundary or outside. The smelting and refining phase, which will be executed in the mine mill for ore beneficiation, is a metallurgical activity involving a variety of online and offline processes depending on the ore to be handled, e.g. gravity separation, magnetic separation, froth flotation, heat treatment, electrical/optical sorting, agglomeration, ore leaching, and several types of electrochemical processes. On-site metallurgical processes are meant to get the metal or alloys extracted from the excavated ore. All associated activities, i.e. transportation of ore & minerals, storage, preparation of ore concentrate, and the making of metal ingots, are a part of this third stage. Indeed, the sale and marketing part of the extracted metal or coal is the ultimate aim of the ensuing mining operation, which is done as a part of this stage by the mining company that runs the mine. The potential environmental impacts of this mining phase revolve around the land water air, noise, and socio-economic pollution in addition to heavy metal pollution and toxicity. High energy consumption also needs attention because both the mine and the ore processing plants are being developed and made operational at this stage of the mine.

Mine decommissioning is the fourth and last stage of the mine and its life cycle. It usually reaches and occurs when the economic recovery of minerals has ceased. The overall mine decommissioning process is integrated with the mine operation and planning process. In other words, the mine should be designed and operated with a continual focus on its closure altogether. Depletion of ore reserves that can be mined, changes in the market condition, the financial viability of the company mining the ore, or the adverse unavoidable environmental, social, and political conditions contribute to the mine closure. The mine closure phase includes the end-use of the mine for miscellaneous purposes and reclamation, rehabilitation, and revegetation activities so that the mine activities can be brought back to their original shape from which the mine started its mining journey. The mine closure plan and closure funds, required at the last stage of the mine, are the essential components of this last mining stage. The mine closure phase thus includes such activities as the re-contouring of mine pit slopes and waste dumps, handling of tailing dumps and decommissioning of mine roads, and dismantling of buildings apart from the reclamation work which restores the greenery of the mine area.

Reclamation/revegetation is an important component of this last phase referred to as mine closure or decommissioning. Hence, issues to consider in developing a *reclamation plan* include the long-term stability of impoundments, slopes, and surface materials; safety issues relating to open pits, shafts, subsidence; toxic or radiological hazards, e.g. the physical characteristics, nutrient status, and inherent toxicity of tailings or waste rock, which may constrain revegetation. The costs of ongoing and post-decommissioning rehabilitation should be managed from the closure fund collected during the operation phase of the mine. Thus, mine decommissioning ultimately determines what is left behind as a benefit or legacy for future generations. If mine closure operations are not undertaken in a planned and effective manner, chances are that the site will continue to be hazardous and a source of pollution for many years to come. The overall objective of mine closure is to prevent or minimize adverse long-term environmental, social, and economic impacts, as well as to create a stable landform suitable for future land use.

2.12 Abandoned mine: Physical, environmental, financial, and socio-economic considerations

Generally, abandoned mines are those sites where advanced exploration, mining, or mine production ceased without rehabilitation. On cessation of the mining operation for a long period, the mine turns into an 'abandoned mine'. Many thousands of former mining sites became abandoned and continue to pose a real or potential threat to human safety and cause health and/or environmental damage in many areas. An 'abandoned mine' is a part of decommissioning stage of the mining cycle. In general, the presence of abandoned mines negatively influences public perceptions of the industry. Governments are usually and ultimately responsible for the cost of dealing with the abandonment of a mine. Therefore, the physical, environmental, financial, and socio-economic aspects of such abandoned sites become important, particularly where the existence and economic survival of large communities may depend on a mine or a group of mines, e.g. the potential for acid mine drainage from abandoned pits and shafts (as a consequence of oxidation of sulphides contained in the ore or wastes), groundwater contamination from tailings and waste rock dumps, and the potential for methane emissions from coal mines are considered a negative legacy of the mining industry and extremely important for the society. In the absence of care and inadequate adherence to regulations, the past mining practice will be turned into a hazard. Therefore, planning and handling abandoned mines can produce significant benefits both during and at the end of operations. The site's progressive and effective improvement not only benefits the mining organization but also produces a positive impression in favour of the mining company. Society and abandoned mines are at loggerheads, and if the essential requirements of such mines have been complied with, they prove a boon for society and are accepted easily.

Mining, by its nature, has environmental impacts in all the phases of a mining project including the last phase of the mine. Some abandoned mines present only physical concerns, but in our opinion, environmental and socio-economic considerations are of prime importance. The most important socio-economic concern caused by abandoned mines is the loss of employment and business activities in the community due to the closure. The other socio-economic considerations of abandoned mines mostly arise include the safety hazards. The physical impacts of abandoned mines, including contamination of the soil by acid mine drainage, metal released from the waste piles (toxicity), and the risk associated with slope instability, most of these cause the loss of productive land. Therefore, abandonment of mine needs aftercare. Most abandoned mines require funds for the rehabilitation of the degraded sites. The questions when dealing with abandoned mines include who will provide these funds, what mechanisms exist, and who is ultimately responsible for the rehabilitation work and the long-term care of such sites. Two schools of thought exist regarding these financial considerations: (a) The government should pay for rehabilitation. (b) The mine owners or heirs of mine should be held responsible for the clean-up actions based on the polluter-pays principle. In many countries, mine closure legislation can enable the regulating authority to control this part, but in its absence, either government or mine owner takes up the responsibility of rehabilitating abandoned mines, either jointly or individually. The incurred cost to rehabilitate abandoned mine lands is very uncertain because these are affected by the lack of agreed criteria as to what conditions need to be mediated and what the goals of rehabilitation should be.

2.13 Conventional vs special methods

The art/science of mining involves excavations of the type which is not just limited to mineral extraction from Mother Earth. Shaft sinking (excavation, deepening and widening, shaft lining, support of shaft, etc.); raising and winzing; tunnelling; cavern excavations; underground penstocks for hydel projects, waste repositories; and sub-ways for civic needs, etc. are various other application areas of mining where different aspects of excavation and engineering are usually applicable.

Besides conventional methods of mining, i.e. surface and underground methods, there are hydraulic mining (aqueous extraction method); mining for placer deposits (dredging); deep-sea (Marine) mining; marble mining and dimensional stone mining; uranium mining (strategic minerals); and sand mining to recover minerals from the Earth. Each of these special methods is mastered by a few companies globally and expertise extended, e.g. shaft sinking is mastered by *DMC Mining Services*, whereas expertise in hydraulic mining is available to other companies.

When it is asked to describe the novel methods and technologies in mining, both conventional and special methods must be included. Besides the manual method, which is no longer economical, the mechanical method of extraction, i.e. mining by machine (continuous miners), is common these days for mineral extraction. Some noted special methods are described below in brief along with the suggested readings:

2.13.1 *Hydraulic mining*

Hydraulic mining, also referred to as *hydraulicking*, is an aqueous method of mining wherein water is used to extract minerals. This method makes use of a powerful water jet (under high pressure) or liquid solvent to dislodge mineral/ore from the host rocks. It is useful for excavating unconsolidated rocks/rock material, e.g. placer deposits, alluvium, laterites (soil rich in iron oxides), and *Saprolites* (soil rich in clay), as well as the mine tailings, which still contain valuable mineral residues even after mineral processing.

Hydraulic mining has two variants, firstly for the placer deposit extraction involving *dredging*, and secondly, solution mining through solvent extraction (leaching). A brief description of mining placer deposits is given in this sub-section below, and for solution mining, a case study of Canada (through audio-visual presentation) is given in this book in Chapter 6.

This special method employs a sequence of development and a cycle of operations, however, finds restrictive and limited uses in mining.

2.13.2 *Shaft sinking*

Though the *shaft* is a means of access for those mineral deposits which are lying at depth, excavation of the shaft called 'shaft sinking' requires specialized mining skill, be it for its digging or supporting. Hence, shaft sinking is described here as a 'special method of mining' (Figure 2.7).

Figure 2.7 Shaft-sinking work at a location in India for going to deeper depths for mining.
Courtesy: CSIR-CIMFR, Nagpur.

Selection of site: Several factors are considered for the selection of a site for the shaft to be excavated. They are surface considerations, geological considerations, and rock/rock mass properties. A suitable/not suitable site is mainly decided based on these factors.

The mineral deposit (flat/incline deposit) and its position (centre of the property), HFL (highest flood level) of the area, road and railway access for the dispatch of coal or mineral including supply of material, if the shaft is a production shaft, the availability of sufficient space to provide infrastructural facilities and public utility services, and the availability for the dumping of waste during shaft excavation are the main surface considerations. The strength of the rock/strata, rock density, and rock swell factor through which the shaft has to pass shall be determined. The running strata, unstable ground, and water-bearing ground shall be avoided and it should be checked that the site is free from geological disturbances like dyke, faults, and sills. Other factors that matter are the cost of sinking; capital available; social, economical, and political considerations.

Shaft-sinking methods: The following are the shaft-sinking methods:

1. Pilling method
2. Open caisson method (Sinking drum method)
3. Forced drop shaft method
4. Pneumatic caisson method
5. Freezing method
6. Cementation method

Applicability of these methods varies as per the encountered ground conditions. Each method has its own merits and demerits which are considered while selecting the method and its suitability for a particular site. Cost considerations are equally important for the shaft and its excavation. However, the support system (temporary and permanent) and ventilation are vital for the shaft, both during excavation and in the post-excavation stage because it is a permanent structure serving the whole mine life.

Shaft sinking through stable strata, hard rocks, unstable porous rocks (loose-ground), firm and fissured strata, running sand, and water-bearing strata, when the in-rush of water is controlled/uncontrolled (heavy), is different. Depending on the ground type, the sinking method and all its parameters are finalized.

Support in sinking shaft: Usually, the most common shape of the shaft is *circular* even though the square, rectangular, or elliptical-shaped shafts are also excavated when the requirement arises. From the stability angle, the circular-shaped shafts are the most preferred ones. Their support (lining) and excavation are easy. The supports provided in the shaft could be either temporary or permanent. Being a permanent structure of importance, either for the mine access, production, or underground ventilation to the mine workings, the shafts are temporarily supported using bolts, and shotcrete or a permanent lining of concrete or MS steel girders can be provided. These supports serve the purpose for the whole life of the shaft to seal off the water by pre-grouting, protect the shaft sidewalls from collapse, restrict the running of sand, gravel, alluvium, or other loose material into the shaft, and reduce the effect of weathering and minimize the resistance of ventilating air.

Ventilation in sinking shaft: In a shaft exceeding nearly 25 m in depth, ventilation during sinking is produced by a mechanical ventilator which is commonly a forcing fan of suitable capacity. *Axial flow fans* are commonly used in mines or shaft workings. Besides fan ventilation, natural ventilation (without mechanical means) is also used for the shafts having less depth. Normal fan ventilation is a forcing-type system; however, reverse or exhaust ventilation is also used.

The shaft sinking of a new shaft and the deepening of the existing shaft require continuous innovations and development. New approaches, modern techniques, and mechanization have been used in Poland for the Leon IV shaft (Pawel, 2021). Typical experiences in handling water hazards during mine shaft sinking can enrich the reader with new and latest practices (Pior Czaja et al., 2021).

2.13.3 Marble and dimensional stone mining

Marble ($CaCo_3$) is a stratified geological deposit which is mined intact in slab form without breakages. Therefore, the *mining of marble* is different from other mining methods. The marble industry, which is prominent in Italy, India, China, Iran, and other parts of the world, needs its mention under the category of a special method.

Marble is a type of *dimensional stone* like granite and Kotah stone, but it's not that hard. Its mining resembles dimensional stone mining meaning that both these rocks/stones are to be mined out intact with no cracks. The marble mining industry involves mining (diamond wire-saw, jet piercing), cutting (gang saws and slab cutters), and tiling units for the production of slabs, tiles, and strips to be used for miscellaneous purposes in buildings, floors, and articles. During wire-saw cutting of in situ marble slabs and during its processing and polishing, huge amount of water is used (as a cooling agent for the cutting blades). The extracted marble stone after cutting and polishing is sold in the market by the size (per m^2 rates) and not by the weight per ton rate. So, the selling price increases with the size of the slabs. All the operations including mining and processing are targeted to get slabs of bigger sizes as far as possible (Figure 2.8).

Marble slurry is a by-product of *marble mining*. The marble mining industry produces slurry at every step. The marble slurry consists of fine particles generated in the cutting, grinding, and polishing process of marble and it has dissolved water in it. This suspension of marble fines is becoming a major threat to the environment where

Figure 2.8 Marble mining: a view of the Morwad mine of M/s R.K. Marble, India.

marble-related activities are concentrated. Marble waste is harmful to vegetation, the common man, as well as animals and creates aesthetic problems. The top fertile soil becomes unfertile due to marble dust. The polluted slurry water affects the water sources, and when it flows with rainwater into the rivers and other water bodies, such water sources get polluted. Pollution of air is also caused by the marble dust produced in large quantities. Thus, the bad effects of the marble slurry on the environment are well recognized and established. Though categorized as waste, many beneficial uses are discovered for such marble mining slurry (Soni, 2008, 2010).

Dimensional stone/decorative stone mining: Mining of decorative/dimensional stones is similar to marble mining where mining operations require extra consideration of the volume and size of the slabs being produced. All over the world, dimensional stones found ample uses and their marketable blocks should satisfy not only stone quality requirements and dimensional standards but also conform to certain risk factors (Yari et al., 2020). Hence, mining sites for dimensional stones must be carefully selected.

2.13.4   Highwall mining (a clean coal technology for improved safety and coal recovery)

Highwall mining is a surface coal mining method being applied in the mines/collieries of the USA, Australia, Indonesia, and India. The highwall mining technology originated in the 1970s in the USA and later spread wherever it was found feasible to apply.

Highwall mining is generally adopted where an open pit reaches its pit limit and further expansion is not viable. In those mines where coal seams are flat and moderately flat, roof condition is good and the seam is free from geological disturbances or intermediate dirt/stone bands. Highwall mining is suitable for old abandoned mines to recover locked-in coal, selective mining of high-value coal seams, and in conditions where the transition from surface mining to underground mining is probable.

A highwall mining operation involves driving a series of parallel entries (web entries) using remotely operated continuous miners with an attached conveying system into a coal seam exposed at the highwall. These web entries are separated by web pillars of pre-designed width and remain unsupported, unmanned, and unventilated throughout their life. Highwall mining systems are of two types:

i.   Auger Mining System
ii. Continuous Highwall Miner (CHM) System

The first version of the highwall mining system was of auger type which was later modified to a CHM system with a fully automated navigation control system. These machines/Highwall miners can produce 0.5–1.0 million tonnes of coal in a year depending on the geology and site conditions.

CHM produces coal by cutting from the rectangular entries (2.9–3.5 m width), while the auger mining system circularly cuts the coal seam with a diameter from 1.35 to 1.8 m (Figure 2.9). The cutter head could drive in individual circular or twin circular excavations. The maximum penetration depth that was achieved ranged from 60 to 100 m. The continuous miner that cuts the coal loads it on the ADD CARS. ADD CAR

Figure 2.9 A Highwall mining system for clean coal extraction.

is a belt conveyor car attached to the system for coal transportation. The fragmented coal conveying system for CHM, i.e. ADD CAR, each of these cars is 12 m long, individually powered, and has self-mobility for flexibility (John et al., 2017).

A glance at the various coalfields, extracting coal of both low value and high value, from open as well as underground mines, provides an opportunity for highwall mining because these coal-bearing areas have multiple extractable seams of both thick and thin contiguous categories. The highwall mining system involves low capital and low development cost, which makes it much cheaper compared to the longwall mining underground. In some cases where the locked-up coal recovery is planned, the highwall system proves to be a profit-making proposition. Since men and machinery of highwall mining are operated remotely and exposure to hazardous conditions is minimal, highwall system mining is a safe system. By and large, the highwall mining system is a successful, safe, and special modern system for coal mining.

2.13.5 *Seabed mining/Deep-sea mining/Ocean (marine) mining*

Oceans, which cover 70% of the globe, remain a key part of our life. About 95% of the deep ocean remains unexplored though around 15%–20% of the world's population lives in coastal areas. The economies of many countries, e.g. those supporting fisheries and aquaculture, tourism, livelihoods, and blue trade, are dependent on the sea surrounding, and the ocean is a major economic factor. Oceans are a storehouse of minerals too besides the food, energy, medicines, and modulator of weather and climate that underpin life on Earth. As demand for minerals is intensified, humans are turning to the seabed for new resource deposits.

Polymetallic nodules are potato-shaped, porous nodules found on the deep sea bed in abundance carpeting the sea floor of the World's oceans. These nodules are also known as manganese nodules and are found in the deep sea only. *Polymetallic nodules* found on the deep sea bed contain valuable minerals like manganese, iron, nickel, copper, cobalt, lead, molybdenum, cadmium, vanadium, and titanium. Minerals like cobalt, antimony, and nickel are available on the Earth in limited quantity, but their availability in the form of nodules on the ocean floor has triggered the scope for sea-bed mining. Thus, *'seabed mining'*, *'deep-sea mining'*, *'ocean mining'*, and *'marine mining'* are synonyms for the extraction of minerals from the polymetallic nodules. These polymetallic nodules are rocks scattered on the seabed containing iron, manganese, nickel, and cobalt.

Deep-sea mining includes three different kinds of resources – polymetallic nodules, massive sulphides, and cobalt-rich manganese crusts. Hence, in this century, the development of technologies for *Deep-Sea Mining* and *Manned Submersibles* is underway. A manned submersible is a tool that carries people to a depth of sea >6,000 m in the ocean with a suite of scientific sensors and various tools required for ocean mining. An 'Integrated Mining System' works for mining polymetallic nodules at those depths in the ocean where a normal approach is not feasible.

However, deep-sea mining has been a contentious issue in terms of the marine environment and its degradation due to less known information about the deep sea. Deep-Sea Mining and Polymetallic nodules (nodule mining) could be crucial, but their future is still uncertain because deep-sea habitats of living marine species and cultural heritage in the area may be adversely affected (https://oceanfdn.org/seabed-mining/). However, the exploration studies of minerals will pave the way for commercial exploitation shortly. As and when international seabed authorities and agencies of the United Nations (UN) organization will evolve safe procedures for their commercial exploitation, the deep-sea polymetallic nodule extraction will gear up. This may be noted that the nodules are collected by a mining device and pumped or lifted to a mining support vessel. The nodules may be separated from the sediment and processed to extract minerals for human consumption. The operational vessels (on-board) for R&D, seafloor exploration and exploitation, are available for deep-sea mining works.

2.13.6 *Hill mining (strip mining and mountaintop removal)*

Surface mining is a method of extracting minerals near the surface of the Earth. The three most common types of surface mining are open pit mining, strip mining, and quarrying. Hill mining is a type of opencast stripping to be done along the hill contours. Mostly, hills are mined by removing the top hill cover and constructing berms or benches to reach the deposit (Strip mining). Thus, strip mining – the method applicable for the hills and mountains – is worked from the top downward. Contour mining progresses in a narrow zone following the outcrop of a mineral body or coal seam in mountainous terrain.

Strip mining destroys the landscape and beauty of the hill, which generates a large amount of dust at the source and disturbs the villagers and nearby human settlements during ongoing excavation operation or related activities. To eliminate this, eco-friendly hill mining is a viable and possible solution for the hills (Dwivedi and Soni, 2021). By adopting the best mining practices for the hills, the country rocks/waste rock/waste material, excavated together with minerals, can be managed conveniently despite the space constraints of the hills.

In the past, strip-mined mineral deposits that became exhausted or uneconomical to mine often were simply abandoned, but now, the *'conventional hill mining by stripping'* can be turned into eco-friendly hill mining through technological interventions.

Several options for different unit operations in hill mining are available for their practical applicability. The extraction of mineral deposits is possible without disturbing the
serene hill environment.

2.13.7 *Uranium mining (extraction of minerals of strategic importance)*

Uranium is a radioactive mineral of strategic importance. Its military uses and uneven distribution globally raise its importance significantly. Uranium and its ore are
found in various minable forms which are economically extractable. A case study
of the Tummalapalle uranium mine (Case Study 6.6, Chapter 6) located in Andhra
Pradesh, India has been presented in this book to describe how minerals of strategic
importance are mined.

It is significant to note that some minerals of the rare earth group, which are extractable as placer deposits, are also minerals of strategic importance because these
minerals found their uses and application in high technology devices and defence, e.g.
Zircon, Lanthanum, Rubidium, Strontium, and Rutile.

2.13.8 *Mining for placer deposits (dredging)*

'Placer' deposits are formed by surface weathering/wind weathering action or by ocean
or river action, resulting in the concentration of some valuable heavy-metallic minerals of economic importance. The placer deposits can be an accumulation of valuable
mineral quantities, formed by gravity separation during sedimentary processes. The
types of placer deposits are, namely:

- Alluvial (transported by a river),
- Colluvial (transported by gravity action),
- Eluvial (material still at or near its point of formation),
- Beach placers (coarse sand deposited along the edge of large water bodies), and
- Paleo-placers (ancient buried and converted rocks from an original loose mass of
  sediment).

The most common placer deposits are those of the gold & platinum group, gemstones,
pyrite, magnetite, tin/cassiterite, wolframite, rutile, monazite, and zircon (rare earth
mineral group).

Dredging is usually done to remove the placer deposits (an ore). This special method
of mining, namely, dredging, is useful for gold or tin extraction from the water-sedi-
ment slurry through sluice boxes. Similarly, 'Panning' is yet another manual age-old
technique for the mining of placer deposits. It was used by miners in the 19th century
or even earlier and continues in African countries to extract gold and heavy minerals.
Thus, placer deposits are the deposits of the coastline, river line, or marine formations
and their mining is called *placer mining*. To excavate, transport, concentrate, and recover heavy minerals from the placer deposits, *placer mining* is practised widely since
ancient times. Mineral processing is done thereafter to extract metals (gold/rare earth

minerals) from the placers. Gold panning, Gravel mining, sand mining, etc., all belong to the same group of mining.

Sand mining: The sand, used widely as a building material, is extracted by 'dredging' which is not a conventional method. Sand mining is usually done on the river beds using a machine such as scrapers and dredgers. Dredger and dredging operation is done using a bucket-ladder dredger, paddock dredger, or other mechanical means. It is an important method of mining for placer deposits and is one of the most applied for sand mining in this century.

2.13.9 Raising/winzing for mining of metallic deposits

A metallic ore deposit, available in the form of *veins* or *lodes* and lying at a depth from the surface, needs good mining skills. There are several stoping methods, some of which are for special conditions. In each of these methods of stoping, raise/winze are common to connect the levels from where the actual ore is extracted. Hence, equipment that is common in most of the stoping operations to increase productivity and performance is specially covered here under a special category.

Alimak Raise Borer (Raise Climber) is a practically tested underground mining equipment for metal mines operations. A fleet of mechanized raise climbers and LHDs/loaders can give good productivity irrespective of the stoping method either special or conventional.

The stoping method, whatever is selected for the ore production, has to be safe and economical too and equipment must add to the overall productivity of the mine as well as machinery. *Alimak Raising* enables mining approaches that cannot easily be carried out using other methods and is a viable, faster alternative for practical situations (Figure 2.10). The *raise boring* is considered for excavating raises/winzes than the traditional drilling and blasting methods. This equipment is a continuous, mechanical method of excavating vertical or near-vertical holes between two levels of an underground metal mine.

Specialized methods unlike conventional methods continue to be important, e.g. in the coal industry, thick coal seam mining is a big challenge. At places, thick coal seams (>4.5 m thick) are encountered for excavation which is to be worked by underground mining methods, principally by conventional Bord & Pillar/Room-and-Pillar method. Such seams pose difficulties of dirt bands in between, spontaneous heating, premature collapse, working on the soft floor, and caving-related problems during de-pillaring operation. Thick coal seams are usually worked in lifts from bottom to top and with or without caving depending on the site conditions. Treating a thick seam as multiple contiguous seams, the *Slice and Rib method* or *Blasting Gallery method* with or without support is generally applied for mining. Such methods are special as they are applicable in limited areas, e.g. Wardha valley coalfields, India, where more than 10-m-thick coal seams with multiple bands are economically extractable for mining. In one mine of Singerani Collieries Company Limited (SCCL), India, a thick seam is being extracted in two sections simultaneously, keeping the top section one slice ahead of the bottom section. The top section (0.3 m) is worked manually and the bottom section (0.3–4.5 m) is worked by LHDs using the conventional *slice* and rib method with 3 m parting in between (Mathur, 1999).

Figure 2.10 Alimak raise mining.
Source: DMC mining services; https://dmcmining.com/vertical-raising/

Practically, it is learnt that mining is that form of applied engineering which require tailor-made solutions several times, though standard methods remain in vogue. The suitability and feasibility factors together with the economics of mining (cost factor), which varies from one site to another, have tremendous importance in mineral excavation.

2.14 Mine plan/mining plan

The mine plan/mining plan is the backbone of a mine and the most essential part of an operational mine. The mining plan in accordance with the existing rules and regulations is required by all mines whether small or big. In India, the preparation of a mining plan/modified mining plan or the review of the mining plan comes under the Mineral (Other than Atomic and Hydro Carbons Energy Minerals) Concession Rules, 2016. The preparation of the mining plan and system of certification has been allowed as statuary compliance under the Act or rule framed by the government. In India, the Indian Bureau of Mines (IBM) has been mandated to exercise the power stipulated under clause(b) of sub-section(2) of section 5 of the MMDR Act 1957 and as per Rules 15,

16, and 17 of the Minerals (Other than Atomic and Hydro Carbons Energy Minerals) Concession Rules, 2016 for processing and approval of *Mine/Mining Plan.* The Government of India for ease of doing business has developed a new simplified format for Mining Plan (preparation and submission) which is available online and in force. The facility of online submission of the mine plan also provides data updation for the new mining plan, its review and modification, which is applicable for all the Indian mines extracting different minerals. After payment of requisite fees, these statuary documents can be submitted digitally. Downloadable templates and web links for these plans and their review are made available to the mine owners, officials, and concerned agencies, namely, the ministry and statuary organizations (IBM). An exclusive website is under preparation by the Govt. of India for Indian mines that will provide for the submission and disposal of the Mining Plan separately.

To elaborate further, a mining plan is a plan that contains several plans and sections (drawings) with site-specific features (Table 2.3) that delineate all planning about the mineral excavation (mining) process. It depicts and describes general information about the mine project, geology and exploration of the deposit, operations planned at the mine, and sustainable mining practices. The impact of mining on the environment based on the baseline information forms a part of a mine plan, though not in detail, a preliminary overview is a must. If mineral processing, crushing, and grinding are involved, its detail should be mentioned in the mine plan document. The mine plan should have all operational parameters of the mine, production planning and scheduling, material handling including transportation, dump and stacks, and tailing details, along with the progressive closure plan and financial assurance from the company to statuary authorities when the mineral is exhausted. Manpower deployment at the mine (managerial, supervisory, skilled, and semi-skilled workers) which is given in the plan gives an idea about the number of persons engaged at the mine in a day or per annum.

Lease, land ownership, exploration, and statuary compliance details in terms of forest clearance, environment clearances, and pollution control board approvals make it a comprehensive document for referral throughout the mine life. The future exploration proposal (geological mapping) not only makes the mine plan a comprehensive one but also advantageous. Another important feature of the mine plan is the CSR initiative taken by the company at the mine and the resettlement and rehabilitation of the affected people called PAP (project-affected person/people) under the R&R policy (resettlement and rehabilitation) of the mining company. In this way, an area covered with mining, topsoil stacking, overburden/waste dumping, mineral storage/stacking, backfilling, tailing pond, effluent treatment plant, mineral separation plant, and infrastructure such as workshop, offices, administrative building roads/railways, and township area can be specified at that mine. Anticipated impacts of mining on the environment (air, water land, noise, vibration, flyrock, etc.) and mitigative measures are extremely helpful in improving the socio-economic profile of the mine and raising of quality of life in the core zone and buffer zone of the mining area consisting of 5–10 km radius area. The mine plan also gives a piece of information about the health profile of the mine area population in the core and buffer zone. Historically, culturally and ecologically important places in and around the mine project may take benefit from its details.

Every mining plan shall be prepared by a person having the requisite qualification and experience, i.e. a degree in mining engineering or a post-graduate degree in geology granted by a university established or incorporated by or under a Central Act,

Table 2.3 The Mining Plan Should Contain These...

1	Lease Cadastral Plan (scanned image)	The scanned copy of the drawing shall be of the original lease map issued by State Government along with other details
2	Surface cum Reclamation and Rehabilitation Plan (.Kml/.kmz/.shp format)	The Plan should be submitted showing different colour codes for (1) active pits and excavation area, (2) excavated area reclaimed & rehabilitated, (3) active dumps, (4) stabilized and rehabilitated dump area, (5) green belt, (6) mineral stacks, (7) utilities such as plant and buildings, and (8) lease boundary
3	Surface Geological Plan of the Lease (.Kml/.kmz/.shp format)	The Plan should be submitted showing different colour codes for (1) lithological/geological Occurrence; (2) area under G1, G2, G3, and G4 geological reserve grades; (3) active pits and excavation area; (4) dump area; (5) mineral stacks; and (6) lease boundary
4	Surface Geological Sections (in .pdf format)	Year-wise geological sections with different colour coding to depict all the features shown in Surface Geological Plan
5	Five-year Production and Development Plan (.Kml/.kmz/.shp format)	This plan should be submitted periodically showing different colour coding for (1) active pit and excavation area, (2) year-wise excavation proposal for the year I to V, (3) active dump and year-wise dump proposal for the year I to V, (4) year-wise dump working proposal for the year I to V, and (6) lease boundary (with ref. to chapter/section 4)
6	Five-year Production and Development Sections (in .pdf format)	Year-wise excavation and dumping proposals with different colour coding depicting all the features as shown in the Five-year Production and development plan
7	Progressive Mine Closure Plan (.Kml/.kmz/.shp format)	The plan should be submitted showing different colour coding for (1) year-wise excavated area reclaimed & rehabilitated for the year I to V, (2) year-wise dump area to be stabilized and dump area to be rehabilitated for the year I to V, (3) year-wise green area proposed from the year I to V, (4) any other reclamation and rehabilitation measures proposed, and (5) lease boundary (with ref. to chapter/section 6)
8	Progressive mine Closer sections (in .pdf format)	Year-wise progressive mine closure sections show all the year-wise details of reclamation and rehabilitation proposals as depicted in the progressive mine closure plan
9	Conceptual Plan and Section	The plan should depict the status of the lease area as envisaged at the end of the mine life. The status of land use shall be depicted by different colour coding

A Provincial Act, or a State Act including any institutions recognized by the government or any equivalent qualification granted by any university or institute recognized within and outside the country. Professional experience in a responsible capacity in the field of mining after obtaining the degree is desired for the preparation of a mining plan. Modification and review of a mining plan shall be carried out by the qualified person, as specified above.

Progressive improvement, which differs from country to country regarding updation periodically for mine plan preparation/review/modification, keeps on changing.

Quoting examples of India for the reader's reference, it should be mentioned that the earlier norms of 'Registered Qualified Person (RQP)' recognized by IBM (India) for the mining plan preparation have been withdrawn, and for this purpose, requisite qualification and experience has been modified and reframed as described in the paragraph above. Similarly, in India, the word 'mining schemes' has been replaced and substituted with new words and is now referred to as the 'Mining Plan Review'. These new and latest changes have improved the mining plan approval process and provided convenience to the mine professional.

2.15   Closure plan

'Closure plan' is a plan, depicting decommissioning phase of the mine but is required to be prepared since the inception of the mining operation, i.e. at the exploitation stage itself. In India, the *Final Mine Closure Plan (FMCP)* is prepared as per the various provisions and clauses of MCDR 2017 [Rule 21(4); 22 (3); 24, 25, 26][3] currently in force, especially for metal mines in India. Every holder of a mining lease shall take steps to prepare mine closure plans as per the guidelines and format given by the Indian Bureau of Mines (IBM) from time to time. The holder of a mining lease shall not abandon a mine unless an FMCP duly approved by the competent authority is implemented, and for this purpose, the lessee shall be required to obtain a certificate from the authorized officer, as the case may be, to the effect that protective, reclamation, and rehabilitation work as per the FMCP or with such modifications as approved by the competent authority have been carried out before abandonment of mine. FMCP shall be submitted 2 years before the proposed closure of the mine and approval from IBM shall be obtained. As enumerated in Rule 26 of MCDR 2017, it is the responsibility of the holder of a mining lease to ensure that the protective measures including reclamation and rehabilitation works have been carried out under the approved mine closure plan or with such modifications as approved by the competent authority.

As per the guideline issued by the government, the *FMCP* contains introductory details about the name of the lease area/mine, the lessee, the location and extent of the lease area, the type of land within the lease area (forest or non-forest), geology, topography and rock types along with the mining, and mineral beneficiation details that may be given as per the approved mining plan document for two years closure period. An updated geological reserve/resource (Annexure 2.1) for in situ as well as mineralized dumps shall be reported. An account of all mineralized dumps/mineral stacks indicating their quantity, grade, size classification/distribution shall be shown on all relevant plans mentioning its dimensions (length, breadth, and height or area and height) with an appropriate number of sections. The closure plan also contains particulars about the status of land in the core zone of the mine, i.e. total area degraded, the total mined-out area reclaimed and rehabilitated, and other areas reclaimed and rehabilitated. Backfilling area, the quantity of waste, fill material available at the site (in cubic m), availability of topsoil for spreading (m^3), and spread area in the square metre have been given in the closure plan.

The closure plan has baseline information as given in the mining plan, mined-out land details, air quality and water quality management, waste management, topsoil management, management of mineral rejects (stacks) available in the lease area and tailing dam management. How the infrastructure will be utilized in the future?

How the disposal of mining machinery or utilization should be done shall be evaluated in FMCP. Disaster management, risk assessment, economic repercussions of closure of mine and manpower retrenchments, and compensation (if any) are also addressed in the closure plan. The details of the schedule of all closure operations and financial assurance form part of the FMCP.

The 'mine closure plan' is processed online for approval, when prepared and is considered as a part of the mining plan (Figure 2.11). As the mining progresses, alteration in the closure plan seems inevitable because mining is a dynamic process. FMCP has to be kept ready since the inception of the mine when actual mining is going on.

FMCP for the coal mines and metal mines in India has a slightly different module/procedure for implementation, but the basic intent of FMCP is the same which indicates the preparedness of the mine management to answer what will happen when the mine is closed. A real-time case record of coal mine closure is described in Annexure 2.2 for assessing the difference and gaining a deeper understanding.

To monitor the FMCP, a report on the status of the implementation of closure plans on a half-yearly basis is required from the company or mine lessee. FMCP shall be checked through periodic inspection preferably at an interval of 1–2 months and a certificate be issued by the authorized officer. Modification in the approved FMCP is permissible.

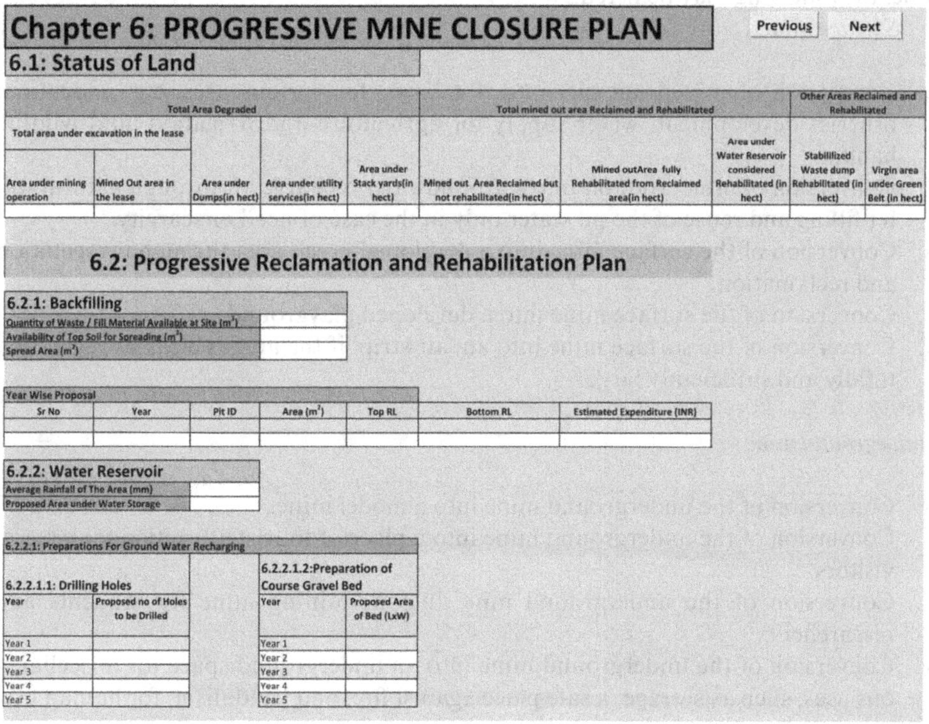

Figure 2.11 A closure plan view (online) details.

As enumerated earlier, FMCP is executed at the end phase of life when the mine is closed or about to be closed. The advantageous feature of the closure plan is the 'end-use of mine area', i.e. extracting value from the mine when it reached its end life. It is researched and found that a *surface mining pit* can be converted into a water body for pisciculture use and recreation, and at the same time, the pit act as a groundwater recharge structure for the nearby areas of the mine. The end-use of an underground mine could be in the form of a tourist attraction to the common public. More value addition, depending on case-to-case basis, is feasible thereby extracting value. Hence, it is evident that a 'closure plan' and 'effective planning and implementation of closure' is likely to create a win-win situation for all who are beneficiaries.

2.15.1   *End-use of mine and value addition*

To enhance the return from the mine to the mining enterprise, an ample scope exist even at the decommissioning phase of the mine when the mine has stopped its normal production either for technical reasons (mineral/ore exhaustion, higher stripping ratio) or for legal reasons, i.e. authorized permission from the competent authorities. At this stage, it is possible to make use of the mined-out area for miscellaneous purposes. Depending on the topography and local site conditions, the feasibility and purpose of end-use may be decided. Among the many possibilities that exist for the end-use of a mine, which is divided as per the mine type, i.e. the surface mine or the underground mine, the following may be listed:

Surface mine

   i. Conversion of mined-out pits into 'Pit lakes' for various uses, e.g. recreation, fisheries development, water supply for agriculture and irrigation, and wildlife habitat.
  ii. Conversion of mined-out pit into a groundwater recharge pit through natural water filling and reuse of the pit water only in the case of need or scarcity.
 iii. Conversion of the surface mine into a developed green area through revegetation and reclamation.
  iv. Conversion of the surface mine into a developed playground.
   v. Conversion of the surface mine into an 'air strip' if the area is clear of vegetation totally and sufficiently large.

Underground mine

   i. Conversion of the underground mine into a model mine.
  ii. Conversion of the underground mine into a place of tourist attraction for general visitors.
 iii. Conversion of the underground mine into a training mine for students and researchers.
  iv. Conversion of the underground mine into an underground space for miscellaneous uses, such as storage, a safe place against fire, and vandalism, for human use.

It is anticipated that mining 'pit lakes' formed at the closure of mining may have environmental issues. Internationally, pit lakes as self-sustaining aquatic ecosystems have been developed in the past, e.g. Alberta Pit Lake in Canada, Sleeper pit lake in Nevada USA, and Westfield pit lake in Scotland. Since pit lakes are created water bodies formed by artificial flooding, or allowing the pits to fill naturally through normal hydrological processes, namely, rainfall or groundwater infiltration, they are to be in tune with the existing ecosystem. Since pit lakes have long-term benefits as a 'water resource' for industrial or other activities, the quality of water contained in them is important and their environmental reclamation is usually required to keep them biologically active. Excavated mine pit of metallic ores (like Cu, Zn, and Mn) and their water environment is economically manageable by long-term measures and the scene can be set for the pit lake development. It is the responsibility of implementing mining organization to prepare such pit lakes at the decommissioning operation of the mine as per the mine closure plan of the company.

Value addition: Mine waste (overburden and tailings), mine water, and mine gases (methane from gassy coal mines) all have a practical value when they are handled scientifically. All or most of them can be reused, recycled, and put into use for various applications, e.g. slate mine waste and marble mining waste and slurries can be developed as bricks, tiles, construction blocks for application in the construction/building industry.

Mine water has the utility to develop cost-effective technologies and value-added products as depicted in Box 2.2. Such applications have tremendous utility for mining companies (MDOs) who are operating mines for mineral extraction.

A packaged drinking water plant of 'Coal Neer' at Patansaongi mine of Western Coalfields Ltd.(WCL), near Nagpur (India), is yet another example of extracting value from the mine water by the coal mining company. This plant makes use of mine water as raw water and using the reverse osmosis (RO) process, converts it to drinking water.

Box 2.2: Mine Water for Value Addition

- Mine water for coal washing for quality improvement of coal.
- Mine water for haul road wetting in mines and cement factories and other campuses to suppress dust.
- Mine water commercial sale for irrigation and agriculture use in both treated and non-treated forms.
- Mine water for plantation and greenbelt development.
- Mine water for construction-related civil works in industry, mines and plants.
- Mine water for non-potable uses to industrial purposes.
- Mine water for miscellaneous industrial applications outside the mine.
- Mine water for large colonies, societies in gardening, and other uses for miscellaneous applications.
- Mine water as an alternative to fresh water and cost-saving in terms of taxes on freshwater and overexploitation of freshwater resources.
- Mine water for overcoming water crisis in the case of necessity.
- Mine water for industrial operations, satellite colony supply, etc., thus reducing freshwater consumption and considerable value addition.
- Mine water for the development of 'salt pans' to make 'salt' in the coastal areas.

Mine overburden for sand production: During opencast mining of coal, the strata lying above the coal seam, known as 'overburden', comprising sandstone, shale, clay, and alluvium rocks with rich silica content is mined together with coal. The overburden is removed to expose and extract coal from beneath. After completion of coal extraction, the overburden is either dumped in waste dumps or used for backfilling to reclaim the land to its original shape. This mined overburden can be easily converted to sand by crushing, sieving, and cleaning. A 'sand plant' near the pit head is needed for this. This signifies that mines are factories of sand too (Figure 2.12a and b).

Figure 2.12a Sand plant with OB in the background that serves as a raw material source for the plant (Gondegaon Opencast coal mine, India).

Figure 2.12b A close view of OB dump used for sand production.

In this way, the demand for s*and*, a major building material for construction uses, which is in short supply, can be easily met in urban areas. This may be noted that the demand for river sand is very high in urban localities. and restrictions on *riverbed mining* are a major cause of the shrinking sand supply.

Mine dust from iron ore mines, limestone mines, and coal mines is useful to make agglomerate, pellets, and briquettes (Soni, 2021).

Coal bed methane (CBM) found in gassy coal mines is locked methane and could be a factory for methane generation which is a greenhouse gas. Thus, coal mines are a major sink that can reduce the carbon footprints across the coal mining industry globally (Soni, 2021).

Thus, the mine has its value as a resource even when the mining operation is discontinued. As an enterprise, the productivity of a mine can be enhanced till the end following the end-use of mine and value addition philosophy.

2.16 Statutory compliances

A legal framework pre-defined by the government and within which any organization must function is termed 'statutory compliance'. It is expected that any industrial company, be it a mine or another organization like a factory, should perform its assigned tasks as per the various central and state laws framed for the purpose. The employer and employees both must treat and facilitate in such a way that adherence to rules and regulations is made compulsory to follow. This synergy is helpful in the implementation of the law of land meant for the betterment of the company/organization.

In the case of a mine operation, the need for statutory compliance is felt at many operational levels by many functionaries be it a mine owner or manager, supervisor, or worker, e.g. a blaster has to follow the explosive handling rules for safety, whereas the mine manager has to comply with rules for their workers' safety. Mine plans, environment plans, closure plans, rescue plans, and water danger plans, all are prepared to fulfil the statuary compliance of that particular mine. If strictly adhered to, the production of coal/mineral will be uninterrupted and smooth. Especially for workplace and mine safety, the rules follow-up is very advantageous because it is planned and designed considering the highest safety factor. Similarly for the deployed equipment or heavy earth moving machine's safety, obeyance of rule makes them efficient and breakdown free. Periodical refinement of statuary provision increases the contribution of mining to the national GDP (Gross Domestic Product), e.g. in 2021, the Indian government brought amendments to the existing laws leading to a 1% increase in the GDP contribution from the mining sector. In addition, statutory compliance brings transparency in the mine functioning process, improves employment rates, and also helps in attracting domestic and foreign investment for the company because it gets the recognition of the disciplined industry. Incorporation of safe, effective, and sustainable technologies, including industry friendliness, is some added benefits of statutory compliance.

2.17 Policies vis-à-vis compliance and productivity

A modern mining company engaged in coal/ore production as raw material feed for the industry and related diversified activities must ensure both growth and productivity.

To shine and maintain its reputation, the company should have strong foundations and a base that has good services, delivery, and welfare schemes. Both at the micro and macro level, the approach of policy should be inclusive meaning that equality and participation for its employees and citizens have to be balanced. The organization's policy should address the economic, social, and environmental challenges with strict compliance. An operative mine or a mining company should have a social licence to operate particularly for its smooth functioning. The policy efforts (local and corporate) should be such that they understand, felicitate, and help the government, new developments, and concerns for the company's betterment. Since the policy/policies are instrumental in overall profitability, they should help enhance mine production as well as be in tune with the current trends. Response of corporate governance at the forefront must enhance the productivity of the mine and mining company from an overall perspective. Sectoral improvement, when added, will lead to improved results in production and productivity. It is evident that the company's domestic and international needs and performance both can be fulfilled through forward-looking approaches and robust policies of international standards. The diversification and globalization trend of the present time, leading to prosperity, is possible only with productive and compliance-friendly policies.

2.18   Role and responsibilities of different organizations

For a mine, from exploration to mining and till closure, organizations concerned and different divisions/departments of the company who matters in the management of the mine have their separate roles and responsibilities. But, everyone involved is directed towards a common goal to produce minerals with safety. We are not defining here the role of every unit or division but covering it broadly in the context of the entire framework/organization that any mine or mining company may have (Table 2.4).

2.19   How to modernize a mine?

Most of the mines, particularly in developing countries, are still using old and primitive techniques for several reasons either technical or financial. Every manager of the mine wants that the mine/quarry should be a modern mine with all infrastructure facilities, be it a surface pit mine, a hilly mine, or an underground mine. The mining operation shall be carried out with new and innovative techniques, waste and fines are properly recycled/utilized, and a clean environment should exist. No more old practices are followed and they should be shelved in the bygone era.

The development of a mining facility on a modern line requires significant planning and upfront investment. The process of developing a new mine from a virgin deposit is quite lengthy and complex and has a long gestation period with risk and uncertainty at its fore. But here, we will be describing the modernization aspects concerning an 'existing mine', which is operational. Compared with a new (virgin) deposit, some general practical requisites to modernize a mine are:

Table 2.4 Role, Responsibilities, and Core Functions in the Organization Framework Context

S. No.	Name of different organizations	Role, responsibilities, and core functions
1.	Government Organizations	**Federal** Facilitation and regulation of exploration and exploitation (mining) activities, making provision for the development of infrastructure, mining and environmental clearances, tax collection, and statuary compliance through involved government agencies **State** Legal permit and permissions, tax collection
2.	Mining Company and its divisions	• Corporate governance: Policy implementation • Production department: Mining and production • Geology department: Exploration, reserve estimation, and geological assessment • Planning and survey department: Mine planning, reconnaissance, and survey • Electrical and mechanical (E&M) division: Electrical and mechanical maintenance of equipment and heavy earthmoving machinery (HEMM) including their operation, running, and handling • IT division: Digitization role, software use, communication tasks, and role/responsibility for the computer maintenance of the organization • Environment division: Reclamation, revegetation, and rehabilitation of mine and mining areas and related ancillary roles for environment protection • Mine safety division: Safety and rescue • Ore processing division: Mine mill, process plant, and metallurgical processes including the maintenance • Maintenance Division: General civil and infrastructure
3.	Industrial Liaison	• Inside (local/domestic) and outside (national/ international) The organization
4.	Support Functions	• Administration • Accounts • Stores and Purchases • Transportation
5.	Welfare, Legal, and Vigilance	• From corporate to government • From miners, employees, local communities and other stakeholders to trade, e.g. CSR
6.	Business and Marketing	Promote sales of minerals and carry out marketing of minerals and other by-products produced by the company

Technical

• Most of the mine functions, wherever feasible, shall be computer-driven including reporting practices. Whether for production, processing, or any machine maintenance, adopt '*best mining practices* for the various sub-operations and unit operations in the mine.

• To convert conventional mining operation into a digitized operation or process-controlled mine, make use of the internet of things (IOT) and IIOT for operation, control, and communication.

- Whenever the choice of selection comes for the plant, equipment, machinery, or method and process, always go for the electronic-based latest systems as they are compact and easy to use. Deploy HEMM that has the GPS-based remote operation latest technology and systems.
- Safety and Rescue operations of the mine shall be computer-based, modern, and remotely controlled. Equip miners and employees with modern personal appliances and devices having state-of-art facilities in them.
- Make use of only environmentally friendly techniques and clean technologies in mining and allied operations.
- Various technical issues (Problems and solutions) should be solved through an interdisciplinary approach only as each area is an expert area that needs the best solutions through the expertise of that area.
- Modernize the statuary compliance mechanism and make it transparent. It will give a fillip to the modernization efforts.
- Decommissioning (closure of a mine) shall be on modern principles only. Wherever needed, modern equipment and method shall be used and adapted for reclamation and revegetation practices.
- Develop a skilled and modern task force for disaster management in mines.

Managerial

- Adopt and implement only those practices at the mining facility, be it an HRD cell or an office functioning for routine mine operation, which is new, professional, and modern as per the current trend.
- Resource use through new material management tools shall be ensured by the mine management.
- Labour and personnel management, to the satisfaction of the maximum, is a must for smooth mine operation and modernization. The highest degree of trustworthiness between employer and employee should be the management approach. Consider this philosophy to modernize a mine.
- Be sound and clear in the financial management of the mine. Export/import and sale of ore, raw material, or products required for running a mining facility shall be transparent and on the modern lines only.
- Mine should be equipped with the latest gazettes and devices in the daily affairs of management and supervision. Make the office function smart with good housekeeping practices in place and a new type of architectural look.
- Adopt ethical practices only by amalgamating the new and old practices applicable in the mine and company.
- Try to use skilled manpower in mine management and supervision for modernization. Remain tactful, alert, and up to date with the adopted interdisciplinary approach and techniques.
- Different forms of publicity and propaganda shall be used for dissemination, production practices, an employee credential, or a company affair.

Rhythmic conversion of a conventional mine to a modern mine is advisable as many practical problems may be inevitable. Hence, implement different technical and managerial tasks in a phased manner, considering the requisites as pointed out above. This will slowly and steadily convert the conventional mine into a modern mine.

2.20   Tutorials

Tutorials play a crucial role in learning the mining operation quite closely because mining engineering is a field-oriented subject of importance. It is plausible to fill the learning gaps between theory and practical aspects with modern tools, namely, video techniques/audio-visual aids, computers, powerpoint presentations, mobile cameras (scenes captured with a moving camera), and software-based tools. We as authors have made use of it in the case study chapter of the book, refer case study on solution mining.

In this section, we are depicting some tutorials to know the practical aspects of production and productivity in mining. Their numbers could be many more depending on the topic and the need; however, we have made the tutorials very simple and interesting to learn despite knowing the fact that these days many of the world's mining operations and projects are done through modern *software in different subjects areas, e.g. mine planning, ventilation planning, and blasting.*

PROBLEM 2.1

How can the productivity of the mines be improved through blasting? What needs to be done for productivity improvement?

Solution

It is well known that mining is an industrial operation for the extraction of minerals from the Earth. From mine to the consumer, i.e. excavation of ore/coal up to the dispatch, there are systems, sub-systems, and various unit operations (at least seven) involved as shown in Figure 2.13. This whole chain is referred to as the *Mine-Mill Fragmentation System (MMFS)*. Blasting is one of the essential unit operations of mining. Through the blasting process, we convert an explosive's chemical energy in a few milliseconds to mechanical energy for breaking rocks. In this way, the sudden release of the energy cause rock fragmentation and rocks are thrown to a distance from the blasting site called flyrocks.

Fragmentation is the prime variable in the determination of productivity performance and hence the economics of the mining operations (Figure 2.14, left). A conflicting situation in the cost relationships of the unit operations exists as evident from Figure 2.14 (left). There is an impact of both undersized and oversized blasted rock materials on the productivity and the value of ore produced by blasting (Figure 2.14, bottom). This is the reason why MMFS is important and needs optimization to achieve better production in a mine. Thus, in any mine or mining operation where blasting is involved, fragmentation has a close linkage with productivity. Consequently, *fragmentation and its measurement* became an essential and integral part of mining systems worldwide. In addition to fragmentation, proper heaving of blasted material and control of unwanted outcomes are also essential to achieve the objectives of blasting. Several scientific methods can be applied for evolving proper blast design for a mine. However, we are still stuck with the old concept called the '*Powder Factor'*.

The fragmentation analysis of *blasted muck* is now done using software, namely, Fragalyst-4.0 or higher version which is designed and developed by CSIR-CIMFR, Nagpur and is commercially available in the open market. Several other imaging

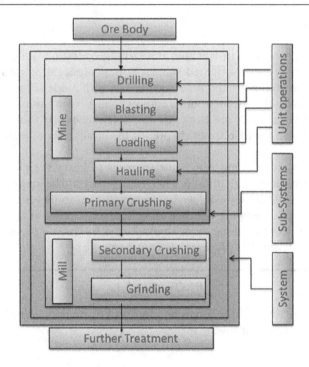

Figure 2.13 Mine-Mill Fragmentation System (MMFS).

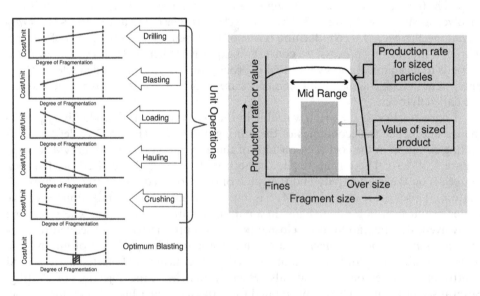

Figure 2.14 Relation of the degree of fragmentation, fragment size on mining operations, and production rate/productivity. (Left): Effect of degree of fragmentation (size) on cost per unit of unit operations. (Bottom): Fragment size vs. production rate and value of the sized product.

software is also available for the computerized fragmentation analysis to improve mine productivity. The better the fragmentation, good is productivity.

   Rock fragmentation in a blast (assessment and monitoring both) is far more important in determining the MMFS performance than any other factor. Though *Powder Factor* (a scientific term for a specific charge) does not tell anything about rock–explosive interaction, its simplicity and ease of use make it very popular. The powder factor just represents a fraction of the MMFS and not the total impact. Let us consider an example to demonstrate the same:
Example:

1. Total cost on the MMFS=100 (say)
2. Total cost on drilling and blasting=25 (@ Save 10%)=2.5
3. Total cost on loading and hauling=75 (@ Lose 10–15%)=7.5 (taking minimum value)
4. Percentage loss=7.5 – 2.5=5.0

The example above assumes that explosive and rock mass have been matched to a good degree, and unless the mine-mill fragmentation system is optimized, the losses in loading and hauling will continue to happen. This points to the fact that it is better to save on loading and hauling cost (75% segment) compared to drilling and blasting. The loading/hauling optimization is feasible through MMFS and is the best way for productivity enhancement.

   For sustained, safe, and economic production of minerals from mines, the production patterns for different mine benches are different, and it varies from the type of mineral excavated and from one mine to another mine as well.

   Concerning an operative mine, the following are some measures that need to be taken for productivity improvement:

1. Measurement/Instrument monitoring and Auditing

   a. Regulatory (current and futuristic)
   b. Performance evaluation

   • Bench profiling
   • Explosive characterization
   • Rock mass characterization for establishing blast ability
   • Fragmentation analysis using software

2. Blast Auditing: This involves the evaluation of all the variables entering into the blasting domain and analysis of such variables for identification of the operations resulting in financial loss. Some of these are Zonation of mines (rock mass assessment), drilling variables, blast design variables, explosive quantity and quality and transportation variables, fragmentation studies, performance analysis of shovel and dumper cycle time analysis, etc., crusher and other equipment studies, and identification of mining vs human-related issues.
3. ANN and Multivariate Modelling of different processes.

Figure 2.15 shows the parameters (OPTIBLAST) to be touched for productivity enhancement. The expected output of production enhancement is possible through blast design, optimization, monitoring, performance analysis, and training of the personnel.

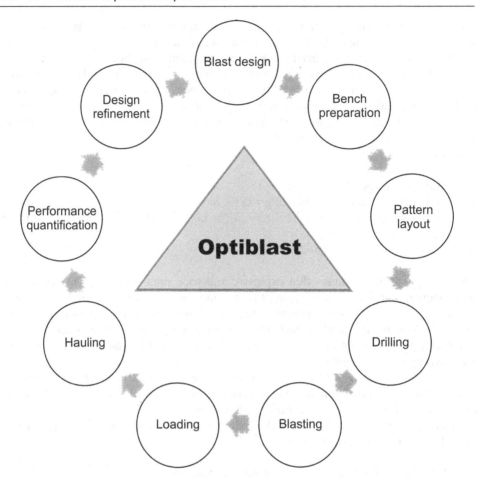

Figure 2.15 OPTIBLAST – parameters to be touched for productivity enhancement.

PROBLEM 2.2

A D-shaped tunnel of dimension shown in Figure 2.16 is planned to be excavated in limestone. SemiGel Explosive of a reputed brand will be used for loading the 32 mm diameter blast holes. The density of the explosive and rock is 1.3 and 2.6 g/cm^3, respectively. Calculate the basic blast design parameters, namely, burden, spacing, stemming length, sub-grade drilling, hole depth, and holes per row for a scientific blast design which can excavate a smooth tunnel surface.

Solution

The D-shaped tunnel dimension shown in the figure is 15.2 m × 15.2 m (width × height). As the tunnel height (at the centre) is large, scientific excavation is possible with the heading and benching method. This conventional method is a combination of underground face blast and surface mine bench blast. The top heading is driven ahead of the lower bench. The pattern of the blast can be planned with various types of cuts and

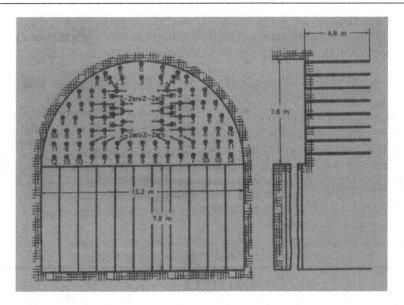

Figure 2.16 A D-shaped tunnel with blast design.

charged appropriately to achieve the desired fragmentation with fewer vibrations and flyrocks. If established norms are followed, the production from the blast will be good.

Now, we will be estimating various parameters using simple math formulas available in mining literature. The V-cut pattern is to be followed for this tunnel blast.

a. **Burden (B)**

Formula $B = (2 \times SG_{explosive} / 2 \times SG_{rock}) + 1.5)$ dia of explosive
$B = 0.012 \, [(2 \times 1.3) / 2.6 + 1.5] \, 32 = 0.96$

Hence, Burden **B=0.96 m**

b. **Spacing (S)**

If delays are used in the initiation of the blast, then bench height divided by burden or $\dfrac{H}{B}$ is used to calculate the ratio. To calculate spacing (S) for a bench blast with V-cut, the formula used is $S = 1.4\,B$. Thus,

$H/B = 7.6/0.96 = 7.9$

$S = 1.4\,B = 1.4 \times 0.96 = 1.34$ m

c. **Sub-Grade drilling (Sgd)**

Formula $Sgd = 0.3B$
$= 0.3 \times 0.96 = 0.29$ m

d. **Stemming (St)**

Formula $St = 0.7 \times B = 0.7 \times 0.96 = 0.67$ m

e. Hole depth (D)

The height of the bench is 7.6 m and sub-grade drilling is 0.29 m. Hence, the hole depth will be the addition of two for a stemming length of 0.67 m.

$$D = H + Sgd = 7.6 + 0.29 = 7.89 \text{ m}$$

f. No. of rows and Holes per row

The number of rows is normally 3–5 depending on the availability of delays in blast initiation and vibrations to be achieved at the project site. The number of holes per row is calculated by the width of the tunnel divided by spacing plus one. Here, the width of the tunnel is 15.2 m (given) and the spacing is 1.34 m (calculated):

$$(15.2/1.34) + 1 = 11.34 + 1 = 12 \text{ (Twelve)}$$

Therefore, 12 holes will be used per row.

PROBLEM 2.3

a. Calculate the optimum capacity of a mine if workable reserves available are 50 MT.
b. Calculate optimum reserves allocation to a mine for an annual production capacity of 1.2 MT per annum given that

- Cost function for amortization cost, $KA = 962.7 \ q^{0.7727}$
- Cost function for operating cost, $KB = 217.7 \ q^{0.9167} L^{0.2789}$

where q is the annual production capacity of mine, MT/annum, L is the weighted average transportation distance, and R = 139.8.

Solution

This problem in mine planning is based on various cost functions such as operating and amortization costs. The formula for this is:

$$\text{Amortization cost } KA = aq^{\mu} \tag{2.2}$$

$$\text{Operating cost } KB = b \times q^{n} \times L^{W} \tag{2.3}$$

For the above cost functions, the following are the expression to estimate 'q' and 'Q':

$$\text{Annual production capacity of mine} \ q = \left(\frac{R(1-v)}{a\mu} \times Q^{\frac{2+w}{2}} \right)^{\left(\frac{1}{\mu-v+1}\right)} \tag{2.4}$$

$$\text{Workable reserves } Q = \left(\frac{2aq^{\mu-v+1}}{bQ\mu} \right)^{\left(\frac{2}{w+2}\right)} \tag{2.5}$$

The constant R in Equation (2.4) is a coefficient determined by the regression analysis as per the optimization equation of mine planning (Zambo, 1968). R will vary with changes in KA and KB values. For this problem, given value is 139.8.

The input value given for **a, b, μ, v,** and **W** is 962.7, 217.7, 0.7727, 0.9167, and 0.2789, respectively. As per the cost function (Equations (2.2) and (2.3)), substituting these values q and Q can be calculated as

$$q = \left(\frac{139.8(1-0.9167)}{962.7 \times 0.7727} \times 50^{1.304} \right) \frac{1}{(0.7727 - 0.9167 + 1)}$$

ANNUAL PRODUCTION CAPACITY = **1.42 MT/annum.**

Here, the target capacity (q) is given as 1.2 MT/annum. Therefore, workable reserve Q is calculated as

$$Q = \left(\frac{2 \times 962.7 \times 1.2^{0.856}}{139.8 \times 0.2789} \right)^{0.08755}$$

WORKABLE RESERVE = 34.84 MT

PROBLEM 2.4

Calculate the number of pillars of a bord and pillar panel, given that

  i. Annual production = 0.6 Mt
 ii. No. of working days in a year = 300
iii. Thickness of coal seam = 2.8 m
 iv. Coefficient of recovery = 0.80
  v. Coal density = 1.3 t/m^3
 vi. Incubation period = 9 months
vii. Pillar size, centre to centre = 45 m × 45 m
viii. Width of gallery = 4.8 m

To estimate, assume the other required conditions.

Solution

The size of the panel should be such that it can be worked before the incubation period.
  By pillar extraction

$$\text{Monthly production} = \frac{0.6 \times 10^6}{12} = 50,000 \text{ t}$$

Total production within Incubation period = 50,000 × 9 = 4,50,000 t

$$\text{Coal locked up in pillars} = \frac{\text{Production}}{\text{coefficient of recovery}} = \frac{450,000}{0.80}$$
$$= 5,62,500 \text{ t}$$

$$\text{Size of pillar (m)} = 45.0 - 4.8$$
$$= 40.2 \text{ m}$$

$$\text{Coal locked in one pillar} = 40.2 \times 40.2 \times 2.8 \times 1.3$$
$$= 5,882 / -$$

$$\text{No. of Pillars} = \frac{562,500}{5,882} = 96 = \text{Ninety Six Pillars}$$

PROBLEM 2.5

Calculate the ultimate depth up to which an open pit mine can be worked (Figure 2.17), given that

  i.   Value of one=INR (Rs.) 6,780 per ton
 ii.   Production cost=INR (Rs.) 4,210 per ton
iii.   Overburden removal cost=INR (Rs.) 850/m³
 iv.   Dip of one body=65°
  v.   Ultimate pit slope =52°
 vi.   Thickness of deposit =70 m
vii.   Tonnage factor=0.4 ton per m³

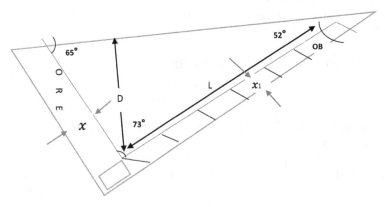

x = Width of the Strip of Ore Body ; X1 = X * Sin 73.
$x1$=Width of the strip of OB Strip; OB =Overburden

Figure 2.17 Ore and OB width of a strip in a mine.

Solution

$$\text{Break ever stripping retro (BESR)}=\frac{\text{Sale price of ore/-production cost/-}}{\text{overburden removal cost-Rs/m}^3}$$
$$=\frac{6,780-4,210}{850}$$
$$=\ 3.02$$
$$\cong 3$$

Let the thickness of the strip at the bottom-most point be x_m. At depth D, for removing a strip of ore of x metre, a length L of overburden with thickness x metre shall be removed. Considering 1 m along the strike

$$\text{BESR}=\frac{\text{Volume of overburden}}{\text{weight of ore}}$$

$$3=\frac{X_1L.1}{x_x\text{thickness of ore}\ \times\ 1/\text{tonnage factor}}$$

$$3=\frac{X\text{-}L\text{-}Sin73}{X\text{-}70/0.4}$$

$$3=\frac{L\ \times\ 0.4\ \times\ 0.9563}{70}$$

$$L=\frac{70\ \times\ 3}{0.9563\ \times\ 0.4}=548.99\cong 549\text{ m (say)}_{=\text{(say)}}$$

Ultimate depth $D=L\times Sin\,55$
$$=549\ \times sin\,55=449.71\cong 450\text{ m}$$
$$D=450\text{ m}$$

PROBLEM 2.6

For a shovel with a bucket capacity of 11 m³, calculate the size and number of dumpers (trucks) required. Assume bucket factor as 90% and cycle time t as 15.0 min. Given that the density of rock=1.7 t/m³ and the daily production of mine is 10,000 tons, 85-ton dumpers are deployed at the mine for both ore and overburden transportation.

Solution

Bucket capacity=11 × 0.9 × 1.7
= 16.83 ton
Find truck size for 4–6 swings/load:
4 swings ×16.83 = 67.32
5 swings ×16.83 = 84.15
6 swings ×16.83 = 100.98

$$\text{No. of trips} = \frac{7 \text{ hour} \times 50 \text{ min/h}}{15.0}$$
$$= 23.3 \cong 23 \text{ trips per dumper or truck}$$

Output/shifts = $85 \times 23 = 1{,}995$ tons per dumper/truck

$$\text{Number of dumper/trucks} = \frac{10{,}000}{1995}$$
$$= 5.12 \text{ or} \cong 5 \text{ (Say)}$$

Therefore, dumper/trucks required = five (5) of 85-ton capacity each.

References

Agricola, G. (1556), De Re Metallica. *The Mining Magazine*, London (Translated from the first Latin edition by Herbert Clark and Lou Henry), p. 1912.

Bateman, A. M. (1951), *The Formation of Mineral Deposits*, New York: Wiley, 371 p.

BullockRichard, L. (2011), Introduction to Underground Mine Planning, Chapter 12.1, In *Society of Mining Engineers (SME) Handbook*, Two Volumes, pp. 1134–1141.

Bustillo Revuelta, M. (2018), Mineral Deposits: Types and Geology. In: *Mineral Resources*, Cham: Springer Textbooks in Earth Sciences, Geography and Environment, pp. 49–119. DOI: 10.1007/978-3-319-58760-8_2.

Deshmukh, D. J. (2001), *Elements of Mining Technology*, Vol. 1, Nagpur, India: Central Techno Publication, p. 423.

Dwivedi, R. D. and Soni, A. K. (2021), Eco-friendly Hill Mining by Tunnelling Methods, In: *Mining Technique -Past, Present and Future*, edited by Abhay Soni, London: IntechOpen. DOI: 10.5772/intechopen.95918.

Goel, S. C. (1991), Geotechnical Aspects of Vertical Crater Mining Method in a Deep Mine, In: *African Mining '91*, Dordrecht: Springer, 45–53. DOI: 10.1007/978-94-011-3656-3_5.

Jairo, N. and Chileshe, P. R. K. (2020), Global Mine Productivity Issues: A Review, *International Journal of Engineering Research & Technology (IJERT)*, Vol. 9, Issue 5. DOI : 10.17577/IJERTV9IS050071.

JohnLoui, P., Roy, P. P., Baotang, Shen, and Karekal, S. (2017), *Highwall Mining: Applicability, Design and Safety*, London, UK: CRC Press (Taylor & Francis), p. 323.

Konya, Calvin J. (1995), *Blast Design*, Montville, OH: Intercontinental Dev, p. 230.

Kesler, S. E. (1994), *Mineral Resources: Economics and the Environment*, New York: Macmillan, 400 p.

Mathur, S. P. (1999), *Coal Mining in India*, Bilaspur, India: M.S Enterprises Publication, p. 472.

MMSD. (2001), Artisinal and Small–scale Mining in India, Country Paper No. 78, Prepared for Mining, Minerals and Sustainable Development (MMSD) project by S.L. Chakraborty, A project of IIED & WBCSD (International Institute for Environment and Development-IIED, London and World Business Council for Sustainable Development—WBCSD); published as United Nations Documents relating to Mining Industry by UNEP and UNCTAD; October, p. 81.

Pawel, Kaminski. (2021), Polish Experiences in Shaft Deepening and Mining Shaft Elongation, In: *Mining Technique – Past, Present and Future*, edited by Abhay Soni, London: IntechOpen, pp. 79–96.

Pior, Czaja, Pawel, Kaminski, and Artur, Dyczko. (2021), Polish Experiences in Handling Water Hazards during Mine Shaft Sinking, In: *Mining Technique – Past, Present and Future*, edited by Abhay Soni, London: IntechOpen, pp. 97–107.

RaiSheo, Shankar. (2020), Digital Transformation for Improving Productivity of Mining – An Approach, *MGMI News Journal,* Vol. 46, Issue 3, pp. 34–41.

Soni, A. K. (2008), Utilization of Marble Slurry and other Stone Wastes: Some Tips to Prepare Road Map and Action Plans, In: *Proceedings of Workshop on Gainful Utilization of Marble Slurry and Other Stone Waste*, January, Jaipur, India: Centre for Development of Stones (CDOS), pp. 1–4.

Soni, A. K. (2010), Quarrying and Processing Waste for Development of Value Added Products, In: *Proceedings of CII Conference on "Technology for Sustainable Livelihood and Community Development" (AP- TEC -2010)*, Hyderabad, 9–10 December, p. 6.

Spurr, J. E. (1923), *The Ore Magmas*, New York: McGraw-Hill, 915 p.

Wikipedia, the free encyclopedia. https://en.wikipedia.org

Yari, M., Bagherpour, R., Khoshouei, M., and Pedram, H. (2020), Investigating a Comprehensive Model for Evaluating Occupational and Environmental Risks of Dimensional Stone Mining, *The Mining-Geology-Petroleum Engineering Bulletin*, Vol. 35, Issue 1, pp. 101–109.

Zambo, J. (1968), *Optimum Location of Mining Facilities*, 1st Edition, Budapest: Akademiai Kiado, p. 112.

Annexure 2.1: Resources and reserves

Definitions of resources (Table A2)

i. **Inferred Mineral Resource** is that part of a *mineral resource* for which quantity and grade, or quality can be estimated based on geological evidence and limited sampling; and reasonably assumed, but not verified, geological and grade continuity.

Table A2 UNFC Classification Table

UNFC classifications	Geological reserves	Economic viability (EV)	Mineral resource	UNFC classifications
The UNFC classification consists of a three-dimensional system with three axes, namely Geological Assessment, Feasibility Assessment and Economic Viability.	Proved Mineral Reserves (111)	---add EV---	Measured Mineral Resource (331) Indicated Mineral Resource (332)	In this three-digit code system, the first digit represents economic viability, the
	Probable Mineral Reserves (121 & 122)	---add EV---	Inferred Mineral Resource (333) Reconnaissance Mineral Resource (334) Prefeasibility Mineral Resource (221 and 222) Feasibility Mineral Resource (211)	second digit shows the feasibility, and the third letter shows geologic position

Note: Economic Viability = 1,2,3 in descending order with 1 being highest and 3 being lowest. Feasibility Assessment = 1,2,3 in descending order with 1 being highest and 3 being lowest. Stages of geological assessment : 1,2,3,4 (1= detailed exploration; 2 general exploration; 3= prospecting; and 4 = reconnaissance)

[1] Mineral occurrence: A mineral occurrence is an indication of mineralization that is worthy of further investigation. The term mineral occurrence does not imply any measure of volume/ tonnage or grade/quality and is thus not part of a UNFC classification as a mineral resource.

[2] Uneconomic Occurrence: Minerals that are too low in grade and potentially not economical to extract are said as the 'Uneconomic Occurrence'. They are not part of UNFC mineral resource categorization.

The estimate is based on limited information and sampling gathered through appropriate techniques from locations such as outcrops, trenches, pits, workings, and drill holes.

ii.   **Indicated Mineral Resource** is that part of a *mineral resource* for which quantity, grade or quality, densities, shape, and physical characteristics can be estimated with a level of confidence sufficient to allow the appropriate application of technical and economic parameters, to support mine planning and evaluation of the economic viability of the deposit. The estimate is based on detailed and reliable exploration and testing information gathered through appropriate techniques from locations such as outcrops, trenches, pits, workings, and drill holes that are spaced closely enough for geological and grade continuity to be reasonably assumed.

iii. **A Measured Mineral Resource** is that part of a *mineral resource* for which quantity, grade or quality, densities, shape, and physical characteristics are so well established that they can be estimated with confidence sufficient to allow the appropriate application of technical and economic parameters, to support production planning and evaluation of the economic viability of the deposit. The estimate is based on detailed and reliable exploration, sampling and testing information gathered through appropriate techniques from locations such as outcrops, trenches, pits, workings, and drill holes that are spaced closely enough to confirm both geological and grade continuity.

Definitions of reserves (Table A2)

i.   **Mineral Reserve** is the economically mineable part of a measured or indicated mineral resource demonstrated by at least a preliminary feasibility study. This study must include adequate information on mining, processing, metallurgical, economic, and other relevant factors that demonstrate (at the time of reporting) that economic extraction can be justified. A mineral reserve includes diluting materials and allowances for losses that may occur when the material is mined.

ii.   **Probable Mineral Reserve** is the economically mineable part of an *indicated*, and in some circumstances, *measured mineral resource* demonstrated by at least a preliminary feasibility study. This study must include adequate information on mining, processing, metallurgical, economic, and other relevant factors that demonstrate that economic extraction can be justified.

iii   **Proven Mineral Reserve** is the economically mineable part of a *measured mineral resource* demonstrated by at least a preliminary feasibility study. This study must include adequate information on mining, processing, metallurgical, economic, and other relevant factors that demonstrate that economic extraction can be justified.

Annexure 2.2: Coal mine closure

A case record of Makardhokra – I. Opencast Coal Mine of Western Coalfields Limited, India

T.N. Suryavanshi

About the mine and region

The Makardhokra-I OC mine (MKD-1) of 3.5 MT per year planned production capacity is a completed project of M/s Western Coalfields Limited (WCL), a coal mining company in India. The mine is located 10 km N-W of Umrer town of Nagpur district, Maharashtra. The nearest major city Nagpur is 53 km away (road distance) and the Umrer town is about 43 km from Nagpur town in the S-E direction. The Umrer town is well connected by road and rail, and the mine area is approachable with an all-weather metalled road. State Highway No. 92 passes through the town. The Umrer-Butibori railway line connects to the northern boundary of the mine.

Makardhokra-I OC mine (MKD-1) is located in the existing Makardhokra block (MKD block). The MKD block is subdivided into the Makardhokra-I, II, and III, all of which are collieries, either surface or underground. The easternmost of the MKD block, adjacent to *Amb River*, was named Makardhokra-II block located east of the MKD-I OC mine and an underground mine of WCL is also operating nearby. Another nearest operating mine is Makardhokra-III (Dinesh OC) on the N-W northwestern side of the project. Makardhokra-II OC mine located east of Makardhokra-I OC mine is a mine where coal is exhausted. Umrer OC mine which is near exhaustion is located further east of Makardhokra-II OC and the Amb river separates the two mines. At this mine (MKD-1 mine), the coal is exhausted and the mine is in the decommissioning phase.

The region is characterized by ample coal reserves of the Gondwana formations. Coal deposits occurrence is spread over a strike length of 11 km in the Makardhokra block on the western side of Umrer beyond the 'Amb River' – a major river of the area that was established through exploration in 1984.

Closure planning: Makardhokra – I. A surface coal mine

Mine closure planning has been carried out from the start of the mine with periodical reviewing and revision during its life cycle to cope with the technical constraints, safety economic risks, social, and environmental challenges. For the mine closure activities, the guidelines mentioned above and a corpus fund were created by opening an Escrow account with the CCO (Coal Controller Organization) in a nationalized bank. Financial details and its calculation for the MKD closure planning are given in Table A2(i).

The annual mine closure cost with a 5% escalation in subsequent years is tabulated below. The total mine closure corpus for the next 5 years (post-closure period), i.e. 2021–2026, works out as Rs. 6,521.669/- [Table A2(ii)]. The annual production

Table A2 (i) Financial Details for the MKD Closure(up to 2020–2021)

S. No.	Parameters	Details
1	Escrow Account Fund (for MKD mine)[a]	2,340.67 Lakhs
2	WPI – April 2019 (for all commodities)	121.1
	WPI – July 2021 (for all commodities)	135.0
3	Ratio of WPI = (135/121.11)	1.114781
4	Total land area requirement for the project (in hectares)	650.02 ha
5	Mine closure cost @ Rs. 9 lacs/Ha (Rs. in lakhs)	5,850.18 Lakhs
6	Mine closure cost after indexing from April 2019 to July 2021 = (5 × 3) (Rs. in lakhs) =	6,521.669 Lakhs
7	Net mine closure cost as of 2021–2022 = (6–1) (Rs. in lakhs)	4,180.999 Lakhs
8	Life of mine (in years starting from first year and in 2021–2022)	5
9	Annual contribution to Escrow fund as of 2021–2022 = (7/8) (Rs. in lakhs)	836.19 Lakhs

[a] This amount is already deposited with CCO for the Makardhokra-I OC mine up to 31.03.2021 during its operative life (Rs. In lakhs).

Table A2 (ii) Corpus Fund Details for the MKD Closure

	Year	Corpus fund (Rs. in lakhs)
1	2021–2022	836.19
2	2022–2023	878.00
3	2023–2024	921.90
4	2024–2025	968.00
5	2025–2026	1,016.40
Total		**4,620.49**

(2021–2022) has reached up to 4.9 million tonnes per year (mtpy) beyond the targeted capacity of 3.5 mtpy.

Besides financial provisions kept for the closure planning, the implementation part of the closure included the preparation of the Progressive or Concurrent Mine Closure Plan (MCP) and Final Mine Closure Plan (FMCP) following the guidelines framed for the purpose.

These plans would include various land use activities like mined-out land/filling (stowing)/backfilling of void areas, their technical and biological restoration plan, biological reclamation of left-out overburden dump, water quality management, plantation, rehabilitation measures, revegetation, landscaping, and cleaning of sites. Environment monitoring, supervision, power cost, protection, and rehabilitation measures including their maintenance and monitoring should be done for the specified closure period. Activity-wise break-up of closure at MKD coal mine including the cost [Table A2(i)] is summarized below. For the MKD mine whose case record is described here, e.g. presently, 7.30 Mm^3 overburden (OB) has been externally dumped (External dump: D-3). This dump has been covered with a green cover and plantation is done in the 12.50 ha land area of the MKD block [Figure A2 (i)]. Apart from these, the effluent water treatment plant for water purification during closure and water discharge outside the mine premises also forms a part of the planning [Figure A2 (i)]. In addition, the infrastructure dismantling, fencing all around the working areas cleaning of sites, and rehabilitation and disposal of the mining machinery are also included in the closure. Some details of which are given in Table A2(iii).

Note: This case record is written and developed by the contributor as per the 'Revised guidelines of 2020' (Ministry of Coal, Govt. of India). See below for excerpts of the guideline ▼The Mining Plan of MKD-I is prepared by CMPDIL (2021) and the mine data are real data arranged from the coal mining company, i.e. Western Coalfields Limited (WCL), Nagpur, India.

Figure A2 (i): Dump plantation and effluent water treatment plant at MKD- I mine.

Table A2(iii) Activity-Wise Break-Up of Closure at MKD Coal Mine Including the Cost

S. No.	Activity	Weighted % of mine closure cost			Mine closure amount (Rs. in lakhs)
		Progressive	*Final*	*Average*	
A.	**Dismantling of structure**				
	Service Building	0.00	8.50	4.25	295.85
	Residential Building				
	Industrial Structure				
B.	**Safety & Security**				
	Random rubble masonry/ concrete wall				
	Toe wall around dump/Gabion wall	6.50	3.20	4.85	337.62
	Barbed wire fencing				
	Fencing/Boundary wall, fencing around the water body				
	Garland drains				
C.	**OB Dumping Reclamation & Plantation**				
C.1	**Technical Reclamation**				
	Re-handling of OB				
	Levelling by Dozer	60.50	60.50	60.50	4,211.50
	Grading				
	Levelling and Grading of highwall slopes and OB dump				

(Continued)

Table A2(iii) (Continued) Activity-Wise Break-Up of Closure at MKD Coal Mine Including the Cost

S. No.	Activity	Weighted % of mine closure cost			Mine closure amount (Rs. in lakhs)
		Progressive	Final	Average	
C.2	**Biological Reclamation & Plantation** Top Soil Management Grassing of OB Dump Plantation around the virgin area, safety zone, green belt over the external dump, and internal reclamation area Plantation post-care (incl. manpower) Plantation over cleared area obtained after dismantling	15.00	11.70	13.35	929.31
D.	Landscaping of the open space in the leasehold area for improving its aesthetic. Drain, pipelines, peripheral roads, gates, viewpoints, and cement steps on the river bank Development of Agriculture Land	4.00	5.50	4.75	33.65
E.	**Environment Mitigation and Manpower** Air quality (Water tanker, Sprinkler, and other control measures) Water Quality (ETP & STP operating cost) Manpower cost and Supervision	12.00	1.50	6.75	469.88
F.	**Post-Closure Monitoring** Air Quality Water Quality Power Cost Manpower cost and supervision	0.00	3.20	1.60	111.38
G.	Entrepreneurship Development (Vocational/ Skill development training for sustainable income of affected people)	1.00	0.50	0.75	52.21
H.	Miscellaneous and other measures like Golden Handshake, one-time financial grants, alternative jobs, and other services	1.00	5.40	3.20	222.76
	Total	**100**	**100**	**100**	**6,961.16**

Coal mine closure: some guideline details
[*Based on the "Revised guidelines for preparation, formulation, submission, processing, scrutiny, approval and revision of the mining plan for the coal & lignite mines/Blocks'*]

1. For all coal and lignite mines in India, the Mine Closure Plan (MCP) and Final Mine Closure Plan (FMCP) are an integral part of the *Mining Plan*. No separate approval of MCP/FMCP is required. The closure plan and its implementation shall be enforced into practice as per the guidelines of 2020 listed as a reference of this annexure.

2. The mine closure report/information must be submitted at least 180 days before the expiry of 5 years, or the date of execution of the duly executed mining lease deed, whichever is later. The Mineral Concession (Amendment) Rules, 2020 will be applicable for execution and implementation.

3. The closure Plan (FMCP/MCP) duly approved by the respective company board shall be submitted along with the last mining plan when it is decided to close the mine. The project proponent shall submit related documents to the Coal Controller (CCO), Delhi. The details of the Final Mine Closure Plan along with the details of the updated cost estimate for various mine closure activities and the escrow account already set up shall be submitted and required at the time of final mine closure.

   Implementation of the approved mining and closure plans shall be the sole responsibility of the mine owner. It is the responsibility of the mine owner to ensure that the protective measures contained in the mine closure plan, including reclamation and rehabilitation works, have been carried out as per the approved MCP/FMCP.

4. *Mine Closure Plans* will have two components, viz., (a) progressive or concurrent mine closure activities and (b) final closure or cessation plan. Progressive Mine Closure Plan would include various land use activities to be done continuously and sequentially during the entire period of the mining operations, whereas the Final mine closure activities would start towards the end of mine life and may continue even after the reserves are exhausted and/or mining is discontinued till the mining area is restored to an acceptable level. The mine closure details of the mining plan should be oriented towards the restoration of land back to its original form as far as practicable or further improved condition.

5. The 'Progressive Mine Closure Plan' shall be prepared for a period of every 5 years from the beginning of the mining operations. These plans would be examined periodically (every 5 years) and subjected to third-party monitoring by the identified agencies approved by the Central Government.

6. The total cost for carrying out closure activities shall be estimated in advance to know the financial implications of abandonment. This cost of the mine involves progressive and final mine closure activities of all types, e.g. fencing of the working area, dismantling of structures/demolition, cleaning of sites, reclamation/rehabilitation/revegetation of site, mining machinery rehabilitation, plantation, landscaping, dump reclamation, overburden and de-coaled void management, post-environmental monitoring, supervision, power cost, and protection and rehabilitation measures including their maintenance and monitoring,. for the specified post-closure period.

7. The Mining Company/Mine Owner as a part of financial assurance will open an **'Escrow Account'** (a kind of fixed deposit) with the Coal Controller Organization on behalf of the Central Government as the exclusive beneficiary, before the commencement of any mineral getting activities that include mining as well, on the land/project area of the mine and shall submit the same to Coal Controller Organization (CCO) before the permission is granted for opening the mine. The mining company shall cause the payment to be deposited regularly without any notice/perusal at the selected scheduled bank and inform the same to CCO.

8. **Escrow Account Calculation**: In August 2009, it was estimated that typically closure cost for an opencast mine was around rupees six lakhs per hectare of the total project area and rupees one lakh per hectare for the underground project area at the the-then price level. Periodical modifications in these rates based on the *Wholesale Price Index* (WPI) as notified by the Government of India have been done to plan mine closure activities. The escalated rates are Rupees 09 Lakh/ha for opencast and Rupees 1.50 lakh/ha for underground mines in 2019 (based on the WPI of 01.04.2019). These rates will be considered as 'Ease Rates' and subjected to changes periodically.

   Annual closure cost is to be computed considering the total project area of the mine multiplied by escalated rate (at the above-mentioned rates) and dividing the same by the balance life of the mine in years. An amount equal to the annual cost is to be deposited each year throughout the mine life compounded @5% annually, e.g. if the annual cost works out as Rs. 100, then in the first year, the amount to be deposited will be Rs. 100, in the second year $100 \times (1+ 5\%)^1$, and in the third year $100 \times (1+ 5\%)^1$. Further, in the case of the mine, where the Escrow account is already open, the annual closure cost is to be computed considering the total project area at the above-mentioned rates minus the amount already deposited and dividing the same by the balance life of the mine in years and annual cost as arrived should be compounded @5% annually.

   The Coal Controller shall get the WPI updated, which is used for closure cost calculation and escalation, at the time of formulation of the mining plan itself. The mine owner/company including all public/private sector companies shall keep an updated record of the Escrow account deposit as well as the WPI from the mine inception to the mine closure.

9. **Time Scheduling for abandonment:** The Action plan for carrying out all abandonment operations (progressive and final mine closure) should be furnished in the form of a bar chart for the period of life of the mine plus the post-closure period. Post-closure period shall be taken as 3 years for Underground mines and Opencast mines having a stripping ratio lesser than 6 and 5 years for mines having a stripping ratio more than 6.

10. The mine owner shall be required to obtain a final mine closure certificate from the Coal Controller of the effect that the protective, reclamation, and rehabilitation work following the approved mining plan, covering final mine closure provisions/activities, has been carried out by the mine owner. The balance amount at the end of the final mine closure shall be released to the mine owner in compliance with all statuary provisions of the Closure Plan and surrendering of the reclaimed land shall be done to the State government.

11. The mine owner is required to submit to the Coal Controller a yearly progress report before 1st July of every year setting forth the extent of protective and rehabilitative works carried out as envisaged in the approved mine closure plans (Progressive and Final Closure Plans).

Notes

1 This plan contains all information about mine planning and scheduling based on the geological grade, ore contents, and ore availability at the mining faces.
2 A coal petrology word meaning an organic component related to the origin of coal or oil shale. The term 'maceral' about coal is analogous to the use of the term 'mineral' about igneous or metamorphic rocks, e.g., inertinite, vitrinite, and liptinite.
3 Means rules as applicable in India.

References

CMPDIL. (2021), Mining Plan for Makardhokra -I Expansion OCM (4.90 MTY), October, pp. 1–8.
MOC. (2020), Revised Guidelines for Preparation, Formulation, Submission, Processing, Scrutiny, Approval and Revision of Mining Plan for the Coal and Lignite Blocks, Govt. of India, Ministry of Coal (MOC), vide OM No. 34011/28/2019-CPAM dtd. 29th May, p. 39.

Chapter 3

Mine system engineering

> Live a lifestyle that matches your vision and have enough surrender to see the truth as it is.
>
> (Bhagwad Gita)

A system is a collection of different elements that together produce results not obtainable by the elements alone. Simply stated, a system is an integrated composite of people, products, and processes that provide the capability to satisfy a stated need or objective. For any operation and activity, it is essential to improve performance and optimize the cost in complex situations. To reduce the complex problem into a simpler one, the systems approach is used by dividing the complex problem into parts and establishing a relationship among them to achieve the desired goal.

A system consists of three related sets namely:

i. Set of elements
ii. Set of interactions between the elements
iii. A set of boundary conditions

Different elements: People, Hardware, Software, Machines, Facilities, Policies, and Documents.

All these things are required to produce systems-level results.

Boundary conditions: The boundary conditions are interactions of elements with outside or surrounding objects in which the system is operating. It may include:

a. Input
b. Output
c. Environmental conditions

Interactions between the elements take place in a flow in a logical way and in functional sequences (Maier and Rechtin, 2000).

DOI: 10.1201/9781003274346-3

3.1   System engineering: Concept

Systems engineering (SyE) is the carring of planning, designing, operating, and engineering of any system of interacting elements so that the objective of the system is automatically optimized. Thus, systems engineering normally is concerned with systems not yet activated. Systems analysis may be described as a scientific method of making decisions based on a quantitative or other objective evaluation of all action alternatives.

Operations research (*OR*), which is a part of system engineering, can be more specifically defined as the application of mathematical models to the problem of optimizing the objective of any predefined system. The system engineering is having the following contents which make the system more comprehensive (Figure 3.1).

Both SyE and OR are identical but not similar in terms of engineering concept but usually are not considered to be restricted to problems of the design before use. These two sub-disciplines are somewhat vague and can encompass a great variety of engineering functions.

The system engineer is instrumental in the design of a system having a resemblance with the real-world or real system itself, but this engineer is required to have in-depth knowledge of all engineering disciplines which are having the application while designing the system. This is evident as shown in Figure 3.2.

3.2   Systems engineering process

A systems engineering process provides a structure for solving design problems and tracking requirements flow through the design efforts. It includes the following:

* Development of a total system design solution (that balances cost, schedule, performance, and risk).
* Development and tracking of technical information (needed for decision-making).

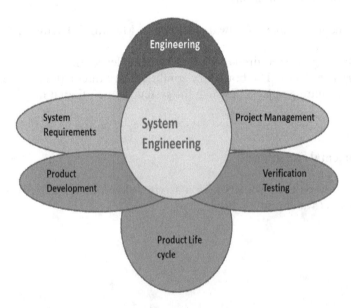

Figure 3.1 Components of system engineering.

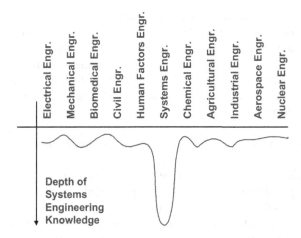

Figure 3.2 The depth of knowledge necessary for a system engineer.

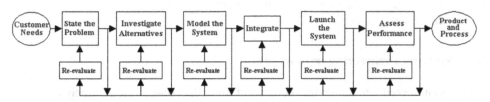

Figure 3.3 The system engineering process.
Source: Bahill and Gissing (1998)

- Verification (that satisfies customer requirements for their technical solutions).
- Development of a system (that is reproducible economically and supported throughout the life cycle).
- Development of compatibility and monitoring (internal and external interface of the system and sub-systems using an open systems approach).
- Establishment of baselines and configuration control.

Systems engineering is an interdisciplinary process that ensures that the end user's needs are satisfied throughout the life cycle of the entire system and given the summarized acronym as **SIMILAR** (Figure 3.3; Bahill and Gissing, 1998; Bahill and Dean, 1999). This process is comprised of seven tasks principally and they are

- **State the problem**: Stating the problem is the most important systems engineering task. It entails identifying customers, understanding customer needs, establishing the need for change, discovering requirements, and defining system functions.
- **Investigate alternatives**: Alternatives are investigated and evaluated based on their performance, cost, and risk.
- **Model the system**: Running models clarifies requirements, reveals bottlenecks and fragmented activities, reduces cost, and exposes duplication of efforts.
- **Integrate**: Integration means designing interfaces and bringing system elements together so they work as a whole. This requires extensive communication and coordination.

- **Launch the system**: Launching the system means running the system and producing outputs, and making the system do what it was intended to do.
- **Assess performance**: Performance is assessed using evaluation criteria, technical performance measures, and measures – measurement is the key. If you cannot measure it, you cannot control it. If you cannot control it, you cannot improve it.
- **Re-evaluation**: Re-evaluation should be a continual and iterative process with many parallel loops.

Most systems engineers accept the following basic core concepts:

a. Understand the whole problem before you try to solve it.
b. Translate the problem into measurable requirements.
c. Examine all feasible alternatives before selecting a solution.
d. Make sure you consider the total system life cycle. The birth to death concept extends to maintenance, replacement, and decommission. If these are not considered in the other tasks, major life cycle costs can be ignored.
e. Make sure to test the total system before delivering it.
f. Document everything.

3.3   Systems engineering management

Systems engineering management consists of three different phases, namely the development phase, the System engineering process, and life cycle integration (Figure 3.4).

As the name envisages, the development phasing is that which controls the design process and provides a baseline to coordinate design efforts. The management process, if phased out, is completed through the *concept level, system level,* and *sub-system/component levels.*

The concept level, which produces a system concept description (usually described in a concept study), is the first one. The system level, which produces a system description in performance requirement terms, is at the intermediate in the hierarchy

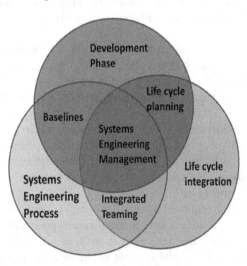

Figure 3.4 System engineering management.

supported by the sub-system or the component level, which produces first a set of sub-system and component product performance descriptions. A set of corresponding detailed descriptions of the products' essential characteristics for their production comes as a final phase of system engineering management.

3.4 Mine system engineering

Having got the introduction to System Engineering – a comparatively new subject for structured-project uses – to provide a holistic, flexible yet solid foundation, we will take you through these key concepts to the mining engineering applications. Mine is a system meant to produce minerals. The development of a mine (a complicated system) requires a synthesized approach and key concepts, which are then reinforced with methods, problems, and practical exercises, to strengthen human learning about mining. Before we look at the various systems engineering activities in more detail in forthcoming paragraphs, the needs and requirements for a system must be fully understood first. The systems designers, business managers, and principals of the business operations (MDOs) will examine the needs and requirements and develop them for future implementation/applications. Considering the views of various stakeholders, one will enter into the real practices of systems engineering.

Mine system engineering consists of basic mining engineering problems together with management and administration. As illustrated in Figure 3.5, a combination of engineering, management, and administration develops into a complete system engineering for an actual
mine, either surface, underground, or any other mine type of special category. In mining engineering, the following are the main systems:

- Winning system
- Ventilation system
- Support system
- Transport system
- Drainage system
- Pumping system
- Safety system

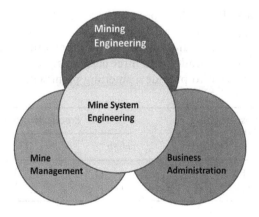

Figure 3.5 Scope of mine system engineering.

The techniques of systems engineering most widely used in practice in the mining industry are linear regression models, linear programming, models and simulation, network analysis/CPM (critical path method)/PERT (project evaluation and review techniques), and transportation problems and decision methods. They are described below in brief.

3.4.1   Linear regression models

This has become one of the most widely used statistical tools for analysing multi-factor data. It is appealing because it provides a conceptually simple method for investigating functional relationships among variables. The standard approach in regression analysis is to use sample data to compute an estimate of the proposed relationship and then evaluate the fit using statistics such as t, F, and R^2 (Chatterjee et al., 1977). For example, in mining, productivity depends to some degree on the dimensions of the mining plan, the seam thickness, top and bottom conditions, equipment performance characteristics, and possibly other physical factors. If equations could be developed to predict productivities in advance of mining based on these simple measurements, the problems of mine planning, design, and economic justification would be greatly simplified. Thus, the solution would be proposed as a linear Equation 3.1 of the form:

$$\text{Productivity} = A_1 + A_2 \times \text{THICK} + A_3 \times \text{TOP} + A_4 \times \text{BOTTOM} \tag{3.1}$$

where A_1 is a constant and A_2, A_3, etc., are the coefficients determined from the regression analysis; THICK, TOP, and BOTTOM are the values for the thickness, top condition, and bottom condition of seam respectively for the mine being evaluated. Necessary data would then be collected on these factors and the resultant *productivity* from as many production operations as possible is calculated. Regression analysis to find the linear coefficients that best fit the data is able to minimize the sum of squares error.

3.4.2   Linear programming (LP)

A linear programming (LP) problem for a mine can be described as given in the next paragraph.

Problem: The mining company owns two different mines that produce an ore which, after being crushed, is graded into three classes: high, medium, and low grades. The company has contracted to provide a smelting plant with 12 tons of high-grade,

Mine	Cost per day ('000')	Production (tons/day)		
		High	Medium	Low
X	180	6	3	4
Y	160	1	1	6

8 tons of medium-grade, and 24 tons of low-grade ore per week. The two mines have different operating characteristics as detailed below:

How many days per week should each mine be operated to fulfil the smelting plant contract?

Solution: *This is a very simple example to start with and progress to a more complicated level.*

In these two mines' problems, we have data for cost and production. What we need to do is to translate that data into an *equivalent* mathematical description. In dealing with problems of this kind, we often do best to consider them in the order:

- variables
- constraints
- objective

We do this below and note here that this process is often called *formulating* the problem or more strictly formulating a mathematical representation of the problem.

Variables: These represent the *unknowns* (the decisions that have to be made). Here, the variables are as follows:

x = number of days per week mine X is operated
y = number of days per week mine Y is operated

Note that here $x \geq 0$ and $y \geq 0$.

Constraints: It is best to first put each constraint into words and then express it in a mathematical form, e.g. ore production constraints – balance the amount produced with the quantity required under the smelting plant contract.

Ore:

High $6x + 1y \geq 12$
Medium $3x + 1y \geq 8$
Low $4x + 6y \geq 24$

Here, one would note that we have inequality rather than equality. This implies that we may produce more of some grade of ore than we need. We have the general rule: given a choice between equality and inequality, choose the inequality.

For example: If we choose equality for the ore production constraints, we have the three equations $6x+y=12$, $3x+y=8$ and $4x+6y=24$, and there are nil values of x and y which satisfy all three equations (the problem is therefore said to be *'over-constrained'*). For example, the values of x and y which satisfy $6x+y=12$ and $3x+y=8$ are $x=4/3$ and $y=4$, but these values do not satisfy $4x+6y=24$.

The reason for this general rule is that choosing an inequality rather than equality gives us more flexibility in optimizing (both maximum and minimum) the objective, i.e. the deciding values for the decision variables that optimize the goal. In this problem, days per week is a constraint – we cannot work more than a certain maximum number of days a week, e.g. for a 5-day week, we have $x \leq 5$ and $y \leq 5$. Constraints of this type are often called *implicit constraints* because they are implicit in the definition of the variables.

Objective: Our objective is (presumably) to minimize the cost which is given by $180x + 160y$

Hence, we have the complete mathematical representation of the problem as:
minimize $180x + 160y$......subject to

$$6x + y \geq 12$$
$$3x + y \geq 8$$
$$4x + 6y \geq 24$$
$$x \leq 5$$
$$y \leq 5$$
$$x, y \geq 0$$

Since there are only two variables in this LP problem, we have the graphical representation of the LP given below with the feasible region (Figure 3.6) outlined.

To draw the diagram above, we turn all inequality constraints into equalities and draw the corresponding lines on the graph (e.g. the constraint $6x+y \geq 12$ becomes the line $6x+y=12$ on the graph). Once a line has been drawn, then it is a simple matter to work out which side of the line corresponds to *all* feasible solutions to the original inequality constraint (e.g. all feasible solutions to $6x+y \geq 12$ lie to the right of the line $6x+y=12$).

We determine the optimal solution to the LP by plotting $(180x+160y)=K$ (K constant) for varying K values (iso-profit lines). One such line $(180x+160y=180)$ is shown dotted on the diagram. The smallest value of K (remember we are considering a minimization problem) such that $180x+160y=K$ goes through a point in the feasible region is the value of the optimal solution to the LP (and the corresponding point gives the optimal values of the variables).

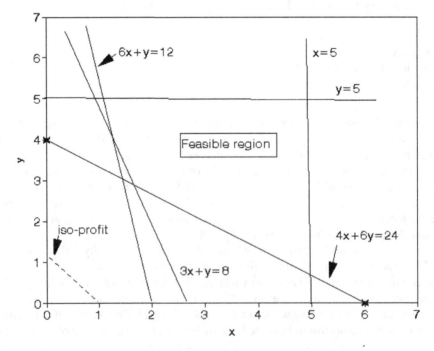

Figure 3.6 Mine system engineering problem using linear programming.

Hence, we can see that the optimal solution to the LP occurs at the vertex of the feasible region formed by the intersection of $3x+y=8$ and $4x+6y=24$. Note here that it is *inaccurate* to attempt to read the values of x and y off the graph and instead we solve the simultaneous equations:

- $3x + y = 8$
- $4x + 6y = 24$

to get $x=12/7=1.71$ and $y=20/7=2.86$, and hence, the value of the objective function is given by $180x+160y=180(12/7)+160(20/7)=765.71$

Hence, the optimal solution has cost 765.71

From the explanation given above, it is clear that the graphical approach to solving LPs can be used for LPs with two variables, but most LPs have more than two variables. This brings us to the *simplex* algorithm for solving LPs. Note that in the example considered above, the optimal solution to the LP occurred at a vertex (corner) of the feasible region. It is true that for any LP (not just the one considered above), the optimal solution occurs at a vertex of the feasible region. This fact is the key to the simplex algorithm for solving LPs. Essentially, the simplex algorithm starts at one vertex of the feasible region and moves (at each iteration) to another (adjacent) vertex, improving (or leaving unchanged) the objective function as it does so until it reaches the vertex corresponding to the optimal LP solution.

3.4.3 Models and simulations

Models: A model is defined as a representation of a system (operation, situation) to study the system. For most studies, it is not necessary to consider all the details of a system; thus, a model is not only a substitute for a system; it is also a simplified form of the system. Models cannot have all the attributes of the real system; they tend to be abstract, general, and often have very moderate details. The model contains only those components that are relevant to the study at hand. The building of a model is a major goal of the system engineer.

3.4.3.1 Type of models

Models can be classified, according to their degree of abstraction, into the following groups: (a) physical models, (b) deterministic (exact) and probabilistic (stochastic) models, and (c) symbolic (mathematical) models.

A physical (or iconic) model looks like the part of reality that it represents. A scale model of a building is a physical representation of a real building. Likewise, model aeroplanes and trains are physical replicas of real objects. This physical modelling approach is rarely used in practice. Iconic models are easy to observe, build, and describe, but are very difficult to manipulate and not very useful for prediction. Commonly, these models represent a static event.

Analogue models (diagrammatic) differ from iconic models in that their physical appearance is different from that of the object they represent. A slide rule is an analogue model because it substitutes distances for numerical quantities. Many graphical representations are analogue models in that they represent numerical data pictorially; for example, a map where different colours represent water, desert, or mountains. Likewise, a thermometer is an analogue model because distances (mercury level) on the thermometer are used to represent degrees of temperature. Analogue models are more abstract than iconic models. These models are easier to manipulate, can represent dynamic situations, and are generally more useful than the iconic models.

Deterministic (exact) models are used at a time chance plays no role and when the effect of a given action will be reasonably closely determined. A deterministic model can be used, for example, in long-range production scheduling problems in the case of known or committed demand. These models are sufficiently accurate over the long-range run planned and actual production will be reasonably close.

A probabilistic (stochastic) model, on the other hand, gives recognition to uncertainty. Such models are of great use in the analysis of problems concerning Advertisements and investments in the stock problems where the unpredictability of consumers lays a big role. Symbolic models represent a real situation by symbols such as words or mathematical relationships. These models are built easily and are more economically compared to the physical models.

Mathematical models are the most commonly used and are considered to be the backbone of engineering. Both abstract and symbolic models can be developed to do away with the client's solution. Moreover, with the availability of high-speed computers, it has become relatively easier to handle mathematical models of great complexity. These models are more precise yet complex. Items, such as common and complex business solutions, income statements, and the company's organization chart, all can be represented with the help of a mathematical model.

3.4.3.2   Reasons to use mathematical models in projects

Some of the major reasons for employing the mathematical models for project solutions are:

i.   The use of mathematical models enables the identification of a very large, sometimes almost infinite number of possible situations.
ii. Models enable the compression of time. Years of operations can be simulated in minutes or even seconds of computer time.
iii. Manipulation of the model (changing variables) is much easier than manipulating the real system; therefore, experimentation is easier.
iv. The cost of making mistakes during a trial-and-error experiment is much smaller when done on the model.
v.   The cost of modelling analysis is much smaller than it would be if a similar process was conducted with the real system.
vi. Models enhance and reinforce the learning process.

3.4.3.3   Characteristics of a good model

The characteristics of a good model are as follows:

-    The model must be an abstraction.

- The model should contain enough details to allow the decision-maker to accomplish his objectives.
- The model should be simple. Simplicity may be obtained by eliminating certain factors from the model. As the model is simplified, the solution process is simplified too.
- If a model is large enough, it should be broken down into smaller segments (i.e. sub model).
- A model should be as simple as possible and still represent reality as well enough to be useful to the decision-maker.

To conclude, the models are not true or false; rather, they are useful and appropriate for the analysis. Model building is at least as much an art as a science, requiring a balance between realism and simplicity. It is the art/act that makes this field so challenging.

3.5   Monte Carlo simulation: A modelling tool

Monte Carlo methods were originally developed for the Manhattan Project during World War II. The term 'Monte Carlo' comes from the name of a city in Monaco. The city's main attractions were casinos, which run games such as roulette wheels, dice, and slot machines. These games provide entertainment by exploiting the random behaviour of each game. However, they are now applied to a wide range of problems – nuclear reactor design, econometrics, stellar evolution, stock market forecasting, etc.

Monte Carlo simulation is a stochastic technique used to solve mathematical problems. The word 'stochastic' means that it uses random numbers and probability statistics to obtain an answer.

Monte Carlo methods randomly select values to create scenarios of a problem. These values are taken from within a fixed range and selected to fit a probability distribution (e.g. bell curve and linear distribution). This is like rolling a dice (Figure 3.7). The outcome is always within the range of 1–6, and it follows a linear distribution – there is an equal opportunity for any number to be the outcome.

In Monte Carlo simulation, the random selection process is repeated many times to create multiple scenarios. Each time a value is randomly selected, it forms one possible scenario and solution to the problem. Together, these scenarios give a range of possible solutions, some of which are more probable and some less probable.

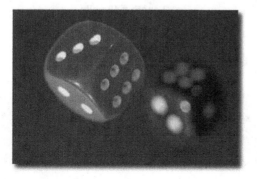

Figure 3.7 Monte Carlo: based on probability and chance.

When repeated for many scenarios [10,000 or more], the average solution will give an approximate answer to the problem. The accuracy of this answer can be improved by simulating more scenarios. The accuracy of a Monte Carlo simulation is proportional to the square root of the number of scenarios used.

3.5.1   What is Monte Carlo simulation good for?

Monte Carlo simulation is advantageous because it is a 'brute force' approach that can solve problems for which no other solutions exist. Unfortunately, this also means that it is computer intensive and best avoided if simpler solutions are possible. The most appropriate situation to use the Monte Carlo method is when other solutions are too complex or difficult to use.

Example: To further understand Monte Carlo simulation, let us examine a simple problem. Below is a rectangle for which we know the length [10 units] and height [4 units]. It is split into two sections which are identified using different colours. What is the shaded area (Figure 3.8)?

Due to the irregular way in which the rectangle is split, this problem is not easily solved using analytical methods. However, we can use Monte Carlo simulation to easily find an approximate answer. The procedure is as follows:

Step 1: Randomly select a location within the rectangle.
Step 2: If it is within the shaded area, record this instance as a hit.
Step 3: Generate a new location and repeat it 10,000 times.

After using the Monte Carlo simulation to test 10,000 random scenarios, we will have a pretty good average of how often the randomly selected location falls within the shaded area. We also know from basic mathematics that the area of the rectangle is 40 square units [length × height]. Thus, the shaded area can now be calculated by:

$$\text{shaded area} = \frac{\text{Number of shaded hits}}{10,000 \text{ scenarios}} \times 40 \text{ unit}^2$$

Naturally, the number of random scenarios used by the Monte Carlo simulation does not have to be 10,000. If more scenarios are used, an even more accurate approximation for the shaded area can be obtained.

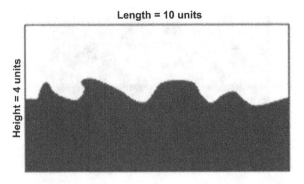

Figure 3.8 What is the shaded area?

The expression 'Monte Carlo method' (MC) is very general and the MC methods are based on the use of random numbers and probability statistics to investigate problems. You can find MC methods used in everything from economics to nuclear physics to regulating the flow of traffic. Of course, the way they are applied varies widely from field to field, and there are dozens of subsets of MC. But, strictly speaking, to call something an MC experiment, all you need to do is use random numbers to examine some problem.

The use of MC methods to model physical problems allows us to examine more complex systems than we otherwise can. Solving equations which describe the interactions between two atoms is fairly simple; solving the same equations for hundreds or thousands of atoms is impossible. With MC methods, a large system can be sampled in several random configurations, and that data can be used to describe the system as a whole. 'Hit and miss' integration is the simplest type of MC method to understand, and it is the type of experiment used in this lab to determine the HCl/DCl energy level population distribution. Before discussing about the lab, however, we will begin with a simple geometric MC experiment which calculates the value of pi based on a 'hit and miss' integration.

By using random inputs, you are essentially turning the deterministic model into a stochastic model. Let us use simple *uniform random numbers* as the inputs to the model. However, a uniform distribution is not the only way to represent uncertainty. Before describing the steps of the general MC simulation in detail, a little word about uncertainty propagation is necessary.

The Monte Carlo method is just one of many methods for analysing *uncertainty propagation*, where the goal is to determine how random variation, lack of knowledge, or error affects the sensitivity, performance, or reliability of the system that is being modelled. Monte Carlo simulation is categorized as a *sampling method* because the inputs are randomly generated from probability distributions to simulate the process of sampling from an actual population. So, we try to choose a distribution for the inputs that most closely matches data we already have or best represents our current state of knowledge. The data generated from the simulation can be represented as probability distributions (or histograms) or converted to error bars, reliability predictions, tolerance zones, and confidence intervals (Figure 3.9).

The steps in Monte Carlo simulation corresponding to the uncertainty propagation shown in the above figure are fairly simple and can be easily implemented in Excel for simple models. All we need to do is follow the five simple steps listed below:

Step 1: Create a parametric model, $y = f(x_1, x_2, ..., x_q)$.
Step 2: Generate a set of random inputs, $x_{i1}, x_{i2}, ..., x_{iq}$.
Step 3: Evaluate the model and store the results as y_i.
Step 4: Repeat steps 2 and 3 for $i = 1$ to n.
Step 5: Analyse the results using histograms, summary statistics, confidence intervals, etc.

Monte Carlo simulation furnishes the decision-maker with a range of possible outcomes and the probabilities they will occur for any choice of action. It shows the extreme possibilities – the outcomes of going for broke and for the most conservative decision – along with all possible consequences for middle-of-the-road decisions.

The technique was first used by scientists working on the atom bomb; it was named for Monte Carlo after the Monaco resort town renowned for its casinos. Since its introduction in World War II, Monte Carlo simulation has been used to model a variety of physical and conceptual systems.

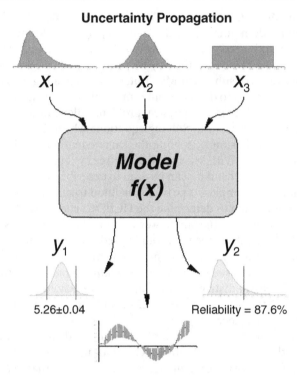

Figure 3.9 Schematic diagram showing the principal of stochastic uncertainty propagation.

3.6   Load haul dump (LHD) repair system: A simulation model of mining

General description of Model: This model will give an in-depth understanding of the concept of simulation in probabilistic modelling. The application is selected for mining application Simulation is often used to observe the behaviour of a system under varying conditions or to gain an understanding of the relationship among components of the system. This simulation model is useful to get an idea for selecting the repairman/mechanic for maintenance of the load haul dump (LHD) machines, and at the same time, to render the minimum cost of breakdowns.

In a Bord & Pillar mining system, a set of four to five LHD is deployed in a district for loading coal from the face and unloading it into the mine car chain conveyor. Some parameters of LHD operations are derived based on continuous monitoring of loading operations about the occurrence and type of breakdowns. Depending upon the frequency of the breakdown and time lost, the appropriate number of mechanics has to be selected. This model will help in this selection.

Objectives of model simulation: The objective of this model study is to determine the number of repairmen (mechanics) to be employed to minimize total cost and work time.

Probabilistic assumptions: Here, to solve the problem, we have made the following assumption:

- It is equally likely that either none or one breakdown occurs per hour.
- Half of the breakdowns occurring are of 'X' types and the other half are 'Y' types; X types require 1.5 hours to repair and Y types require 2.25 hours for repairing.
- For each minute the machine is out of service, the company loses Rs. 2.
- Repairman/Mechanic for these machines is each paid Rs. 10 per hour whether or not the mechanics are working.
- Only one repairman/mechanic is required to repair a given machine at a time.
- Rs. is Indian Currency INR representative only.

The Experiment: A simulation of this machine–repairman system experimented for 10–12 hours from the beginning of a mine production shift. Initially, a simulation with one repairman is performed and the resulting total cost is determined. Then, two repairmen are considered and the total cost is calculated. This can be continued for more repairmen also.

The equation for calculating the total cost (TC) of various numbers of repairmen based on simulation time is

$$TC = RT \times ST \times (K) + MD \times (DT) \tag{3.2}$$

where RT is the repairman payment per hour, ST is the simulation time, K is the number of repairmen, MD is the machine down cost per minute, DT is the number of minutes of machine delay to attend, i.e. waiting time.

The value of total cost (TC) is obtained for several different number of repairmen. Then, select the policy that results in the lowest cost. Cost is indicative and can be higher based on the price escalation.

Procedure: The simulation procedure is simple. Consider coin no. 1 and coin no. 2. Coin no. 1 represents whether a breakdown occurs or not and coin no. 2 determines the breakdown that occurs – whether 'X' type or 'Y' type – if one does. The coin no. 1 is tossed, and if it comes up tail, it is assumed that "no" breakdown occurs in LHD machine. If it comes up head, then it is assumed that one machine broke down. If the machine breakdown occurs, the coin no. 2 is tossed to determine whether it is a 'X' type breakdown if tail comes (1.5 hours repair time) or a 'Y' type break down if head comes (2.25 hours repair time).

Looking into the simulated results (Table 3.1), notice that one repairman should be able to handle the job. Since, on an average, only one machine is breaking down in each 2 hours and the average repair time is only 1.875 hours. With only one repairman, however, a chance is taken such that machine will break down in a short period of time or that several will require a long repair time causing avoidable waiting time and increasing the cost.

The results of the experiment are mentioned in the table. From the results, one can see that with one repairman, there are 1.5 machine hours delay (90 min) which is avoidable waiting time when two repairmen are engaged:

One repairman engages: $TC = (10) \times (10) \times (1) + (2) \times (90) = Rs. 280$

Two repairman are engaged: $TC = (10) \times (10) \times (2) \text{ (no delay in attending)} = Rs. 200$

Table 3.1 Simulated Results for Two Repairmen (a and b are Two Repairmen)

Hour of operation of a shift	COIN-1	No. of break down	COIN-2	Repair time (hours)	1 Repair man system			2 Repairmen System			Remarks
					Start repair time	End repair time	Waiting time (hours)	Start repair time	End repair time	Waiting time (hours)	
1	T	0	T								End time = starting time+ repair time
2	H	1	T	1.5(X)	2	3.5	0	2(a)	3.5(a)	0	(2 + 1.5) = 3.5
3	H	1	H	2.25(Y)	3.5	5.75	0.5	3(b)	5.25(b)	0	No wait for 2 men
4	T	0	T								
5	H	1	T	1.5(X)	5.75	7.25	0.75	5(a)	6.5(a)	0	Repair started late for 1 man
6	T	0	T								
7	T	0	T								
8	H	1	H	2.25(Y)	8	10.25		8(b)	10.25(b)	0	
9	T	0	T								
10	H	1	H	2.25(Y)	10.25	12.5	0.25	10(a)	12.25(a)	0	
Total Delay Time							1.5			0	

Note: Monetary values expressed are in Indian rupees (Rs.)
TC = (Rs. 10) × (10 hours) × (K) + (Rs. 2) × (DT)
where K is the Number of repairmen and DT is the number of minutes of machine delay to attend, i.e. waiting time.

With two repairmen, there is no waiting time in our simulated example; however, the cost for two repairmen is Rs. 200. Although three repairman were not considered, their repair cost would have been Rs. 300. The results indicate that we should hire the services from two repairmen for rendering cost to minimum.

One repairman engagement: $TC = (10) \times (10) \times (1) + (2) \times (90)$
$$= Rs.\ 280\ 1.5\ hours\ delay\ time = 90\ minutes$$

Two repairmen (a and b) are engaged:

$$TC = (10) \times (10) \times (2)\ (no\ delay\ in\ attending) = Rs.\ 200$$

Result/Conclusion: It is advised to select *two mechanic systems* based on the simulation results.

3.7 PERT/CPM for project scheduling & management

Basically, CPM (Critical Path Method) and PERT (Programme Evaluation Review Technique) are project management techniques, which have been crafted out of the need of industrial and military establishments to plan, schedule, and control complex projects. This network analysis technique (CPM/PERT) has a history of its own (**Box 3.1**), mostly developed along two parallel streams, one industrial and the other military.

3.7.1 *Planning, scheduling, and control*

Planning, scheduling (or organizing), and control are considered to be basic Managerial functions, and CPM/PERT has been rightfully accorded due importance in the literature on Operations Research and Quantitative Analysis.

Far more than the technical benefits, it was found that PERT/CPM provided a focus around which managers could brainstorm and put their ideas together. It proved to be a great communication medium by which thinkers and planners at one level could communicate their ideas, their doubts, and fears to another level. Most importantly, it became a useful tool for evaluating the performance of individuals and teams.

Box 3.1: CPM/PERT History

CPM was the discovery of M.R. Walker of E.I. Du Pont de Nemours & Co. and J.E. Kelly of Remington Rand, circa 1957. The computation was designed for the UNIVAC-I computer. The first test was made in 1958, when CPM was applied to the construction of a new chemical plant. In March 1959, the method was applied to a maintenance shutdown at the Du Pont works in Louisville, Kentucky. Unproductive time was reduced from 125 to 93 hours. Similarly, PERT was devised in 1958 for the POLARIS missile programme by the Program Evaluation Branch of the Special Projects office of the U.S. Navy, helped by the Lockheed Missile Systems division and the Consultant firm of Booz-Allen & Hamilton. The calculations were so arranged that they could be carried out on the IBM Naval Ordinance Research Computer (NORC) at Dahlgren, Virginia.

There are many variations of CPM/PERT which have been useful in planning costs, scheduling manpower, and machine time. CPM/PERT can answer the following important questions:

How long will the entire project take to be completed? What are the risks involved?

Which are the critical activities or tasks in the project which could delay the entire project if they were not completed on time?

Is the project on schedule, behind schedule, or ahead of schedule?

If the project has to be finished earlier than planned, what is the best way to do this at the least cost?

3.7.2   The major differences and similarities between PERT and CPM

PERT and CPM are very similar in their approach; however, two distinctions are usually made. The first relates to the way in which activity duration is estimated. In PERT, three estimates are used to form a weighted average of the expected completion time, based on a probability distribution of completion times. Therefore, PERT is considered a probabilistic tool. In CPM, there is only one estimate of duration; that is, CPM is a deterministic tool. The second difference is that CPM allows an explicit estimate of costs in addition to time. Thus, while PERT is basically a tool for planning and control of time, CPM can be used to control both the time and the cost of the project. Extensions of both PERT and CPM allow the user to manage other resources in addition to time and money, to trade off resources, to analyse different types of schedules, and to balance the use of resources. The PERT and CPM allow the user to manage other resources in addition to time and money, to trade off resources, to analyse different types of schedules, and to balance the use of resources (Wiest et al., 1974; Freund, 1979).

3.7.3   The framework

Essentially, there are six steps which are common to both the PERT and CPM techniques. The procedure is described below:

a.   Define the Project and all of its significant activities or tasks. The Project (made up of several tasks) should have only a single start activity and a single finish activity.
b.   Develop the relationships among the activities. Decide which activities must precede and which must follow others.
c.   Draw the 'Network' connecting all the activities. Each activity should have unique event numbers. Dummy arrows are used wherever required to avoid giving the same numbering to two activities.
d.   Assign time and/or cost estimates to each activity.
e.   Compute the longest time path through the network. This is called the critical path.
f.   Use the Network to help plan, schedule, monitor, and control the project.

The key concept used by CPM/PERT is that a small set of activities that make up the longest path through the activity network control the entire project. If these 'critical' activities could be identified and assigned to responsible persons, management resources could be optimally used by concentrating on the few activities which determine the fate of the entire project.

Non-critical activities can be replanned and rescheduled, and resources for them can be reallocated flexibly without affecting the whole project. Five useful questions to ask when preparing an activity network are:

- Is this a start activity?
- Is this a finish activity?
- What activity precedes this?
- What activity follows this?
- What activity is concurrent with this?

Some activities are serially linked. The second activity can begin only after the first activity is completed. In certain cases, the activities are concurrent, because they are independent of each other and can start simultaneously. This is especially the case in organizations that have supervisory resources that allow work to be delegated to various departments, each of which will be responsible for the activities and their completion as planned. When work is delegated like this, the need for constant feedback and coordination becomes an important senior management pre-occupation.

3.7.4 Drawing the CPM/PERT network

Each activity (or sub-project) in a PERT/CPM Network is represented by an arrow symbol. Each activity is preceded and succeeded by an event, represented as a circle and numbered (Figure 3.10).

At Event 3, we have to evaluate two predecessor activities – Activity 1–3 and Activity 2–3, both of which are predecessor activities. Activity 1–3 gives us an Earliest Start of 3 weeks at Event 3. However, Activity 2–3 also has to be completed before Event 3 can begin. Along this route, the Earliest Start would be 4+0=4. The rule is to take the longer (bigger) of the two Earliest Starts. So, the Earliest Start at Event 3 is 4. Similarly, at Event 4, we find we have to evaluate two predecessor activities – Activity 2–4 and Activity 3–4. Along Activity 2–4, the Earliest Start at Event 4 would be 10 wks, but

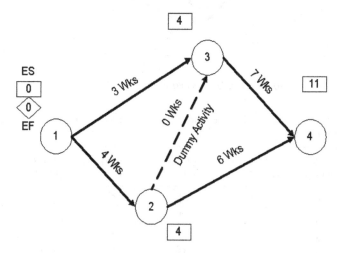

Figure 3.10 The PERT/CPM network.

along Activity 3–4, the Earliest Start at Event 4 would be 11 wks. Since 11 wks is larger than 10 wks, we select it as the Earliest Start at Event 4. We have now found the longest path through the network. It will take 11 weeks along with activities 1–2, 2–3, and 3–4. This is the critical path.

3.7.5   The backward pass – latest finish time rule

To make the Backward Pass, we begin at the sink or the final event and work backwards to the first event (Figure 3.11).

At Event 3, there is only one activity, Activity 3–4 in the backward pass, and we find that the value is 11 − 7=4 weeks. However, at Event 2, we have to evaluate 2 activities, 2–3 and 2–4. We find that the backward pass through 2–4 gives us a value of 11–6=5, while 2–3 gives us 4–0=4. We take the smaller value of 4 on the backward pass.

Tabulation and analysis of activities: One can tabulate the various events and calculate the Earliest and Latest Start and Finish times along with the Total Float (or Slack) which is defined as the difference between the Latest Start and Earliest Start and computed (Table 3.2).

- The Earliest Start is the value in the rectangle near the tail of each activity.
- The Earliest Finish=Earliest Start+Duration.

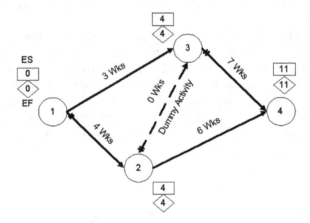

Figure 3.11 A diagram showing the backward pass.

Table 3.2 Computed Values of the Earliest and Latest Start and Finish Times and a Total Float

Event	Duration (weeks)	Earliest start	Earliest finish	Latest start	Latest finish	Total float
1–2	4	0	4	0	4	0
2–3	0	4	4	4	4	0
3–4	7	4	11	4	11	0
1–3	3	0	3	1	4	1
2–4	6	4	10	5	11	1

- The Latest Finish is the value in the diamond at the head of each activity.
- The Latest Start = Latest Finish – Duration.

There are two important types of Float or Slack. These are *Total Float* and *Free Float*.

   Total Float is the spare time available when all preceding activities occur at the earliest possible times and all succeeding activities occur at the latest possible times.

- Total Float = Latest Start – Earliest Start

Activities with zero Total float are on the critical path.

   Free Float is the spare time available when all preceding activities occur at the earliest possible times and all succeeding activities occur at the earliest possible times.

   When an activity has zero Total float, Free float will also be zero.

   There are various other types of float (Independent, Early Free, Early Interfering, Late Free, Late Interfering), and float can also be negative. We shall not go into these situations at present for the sake of simplicity and be concerned only with Total Float for the time being.

   Having computed the various parameters of each activity, we are now ready to go into the scheduling phase, using a type of bar chart known as the Gantt Chart.

   There are various other types of float (Independent, Early Free, Early Interfering, Late Free, Late Interfering), and float can also be negative. We shall not go into these situations at present for the sake of simplicity and be concerned only with Total Float for the time being. Having computed the various parameters of each activity, we are now ready to go into the scheduling phase using a type of bar chart known as the Gantt Chart.

3.8 Gantt Chart: Scheduling of activities

Once the activities are laid out along a Gantt Chart (see Figure 3.12), the concepts of Earliest Start & Finish, Latest Start & Finish, and Float will become very obvious.

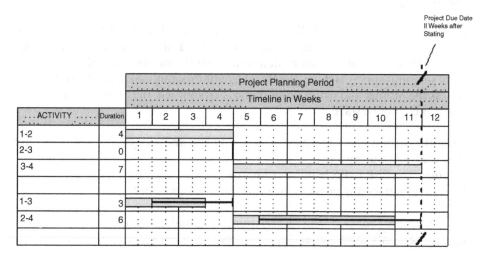

Figure 3.12 The Gantt Chart.

Activities 1–3 and 2–4 have a total float of 1 week each, represented by the solid timeline which begins at the latest start and ends at the latest finish. The difference is the float, which gives us the flexibility to schedule the activity.

For example, we might send the staff on leave during that one week or give them some other work to do. Or we may choose to start the activity slightly later than planned, knowing that we have a week float in hand. We might even break the activity in the middle (if this is permitted) for a week and divert the staff for some other work, or declare a National or Festival holiday as required under the National and Festival Holidays Act.

These are some of the examples of the use of float to schedule an activity. Once all the activities that can be scheduled are scheduled to the convenience of the project, normally reflecting resource optimization measures, we can say that the project has been scheduled.

3.9   PERT calculations and the social project exercise

A *Social Project Manager* around the mine is faced with a project with the following activities:

Activity-ID	Activity – Description	Duration
1–2	Social Work Team to live in Village	5 Weeks
1–3	Social Research Team to do the survey	12 Weeks
3–4	Analyse the results of the survey	5 Weeks
2–4	Establish Mother & Child Health Program	14 Weeks
3–5	Establish Rural Credit Programme	15 Weeks
4–5	Carry out Immunization of Under-Fives	4 Weeks

Draw the arrow diagram, using the helpful numbering of the activities, which suggests the following logic:

- Unless the social work team lives in the village, the Mother & Child health programme cannot be started due to the ignorance and superstition of the villagers.
- The analysis of the survey can be done only after the survey is complete.
- Until the rural survey is done, the Rural Credit Programme cannot be started.
- Unless the Mother and Child Programme is established, the immunization of under-fives cannot be started.

Calculate the earliest and latest event times, tabulate and analyse the activities, and schedule the project using a Gantt Chart.

Solution: The PERT (Probabilistic) Approach

In mining engineering projects, two situations comes across

a.   High certainty (cause–effect logic/relationship is well known);
b.   High uncertainty (cause–effect logic/relationship is not so well established).

However, in Research & Development projects, or in Social Projects which are defined as "Process Projects", where learning is an important outcome, the cause–effect relationship is not so well established. In such situations, the PERT approach is useful, because it can accommodate the variation in event completion times based on an expert or an expert committee's estimates.

For each activity, three-time estimates are taken:

- The Most Optimistic
- The Most Likely
- The Most Pessimistic

The Duration of activity is calculated using the following formula:

$$t_e = \frac{t_0 + 4t_m + t_p}{6}$$

where t_e is the expected time, t_o is the optimistic time, t_m is the most probable activity time, and t_p is the pessimistic time.

It is not necessary to go into the theory behind the formula. It is enough to know that the weights are based on an approximation of the Beta distribution.

The standard deviation, which is a good measure of the variability of each activity, is calculated by rather simplified formula:

$$S_1 = \frac{t_p - t_0}{6}$$

The Variance is the Square of the Standard Deviation.

3.10 PERT calculations and the social project

In our social project, the Project Manager is now not so certain that each activity will be completed based on the single estimate he gave. There are many assumptions involved in each estimate, and these assumptions are illustrated in the three-time estimate he would prefer to give to each activity.

In Activity 1–3, the time estimates are 3, 12, and 21 (Figure 3.13). Using the PERT formula, we get:

$$t_e = \frac{3 + (4 \times 12) + 21}{6} = 72/6 = 12$$

$$S_1 = \frac{(21 - 3)}{6} = 18/6 = 3$$

The standard deviation (SD) for this activity is also calculated using the PERT formula (Table 3.3). The PERT event times for each activity and other details are calculated as below:

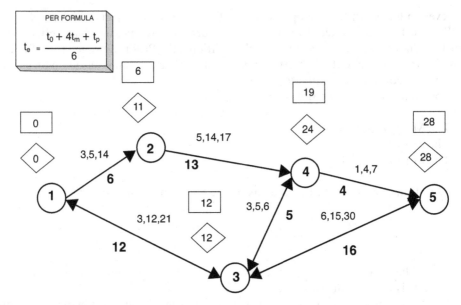

Figure 3.13 PERT calculation diagram for the social project.

Table 3.3 PERT and the SD

Event	t_o	t_m	t_p	t_e	ES	EF	LS	LF	TF	S.D.	Var.
1–3	3	12	21	12	0	12	0	12	0	3	9
3–5	6	15	30	16	12	28	12	28	0	4	16
1–2	2	5	14	6	0	6	5	11	5	2	4
2–4	5	14	17	13	6	19	11	24	5	2	4
3–4	2	5	8	5	12	17	19	24	7	1	1
4–5	1	4	7	4	19	23	24	28	5	1	1

3.10.1   *Estimating risk*

Having calculated the SD and the variance, we can do some risk analysis. Before that, we should be aware of two of the most important assumptions made by PERT:

- The Beta distribution is appropriate for the calculation of activity durations.
- Activities are independent, and the time required to complete one activity has no bearing on the completion times of its successor activities in the network. The validity of this assumption is questionable when we consider that, in practice, many activities have dependencies.

3.10.2   *Expected length of a project*

PERT assumes that the expected length of a project (or a sequence of independent activities) is simply the sum of their separate expected lengths. Thus, the summation of all the t_e's along the critical path gives us the length of the project. Similarly, the variance of a sum of independent activity times is equal to the sum of their variances.

In our example, the sum of the variance of the activity times along the critical path, VT, is found to be equal to (9+16)=25. The square root VT gives us the standard deviation of the project length. Thus, ST= 25=5. The higher the standard deviation, the greater the uncertainty that the project will be completed on the due date. Although t_e's are randomly distributed, the average or expected project length T_e approximately follows a normal distribution. Since we have a lot of information about normal distribution, one can make several statistically significant conclusions from these calculations. A random variable drawn from a normal distribution has a 0.68 probability of falling within one standard deviation of the distribution average. Therefore, there is a 68% chance that the actual project duration will be within one standard deviation, ST, of the estimated average length of the project, t_e.

In our case, t_e=(12+16)=28 weeks and ST=5 weeks. Assuming t_e to be normally distributed, we can state that there is a probability of 0.68 that the project will be completed within 28 and 5 weeks, which is to say, between 23 and 33 weeks. Since it is known that just over 95% (0.954) of the area under a normal distribution falls within two standard deviations, we can state that the probability of the project will be completed within 28 weeks (10 is very high at 0.95).

3.11 Probability of project completion by due date

Now, although the project is estimated to be completed within 28 weeks (t_e=28), our Project Director would like to know what is the probability that the project might be completed within 25 weeks (i.e. Due Date or D=25).

For this calculation, we use the formula for calculating Z, the number of standard deviations that D is away from t_e:

$$Z = \frac{D - t_e}{S_t} = \frac{25 - 28}{5} = \frac{-3}{5} = -0.6$$

By looking at the following extract from a standard normal table, we see that the probability associated with a Z of −0.6 is 0.274. This means that the chance of the project being completed within 25 weeks instead of the expected 28 weeks is about 2 out of 7. Not very encouraging (Figure 3.14a).

On the other hand, the probability that the project will be completed within 33 weeks is calculated as follows:

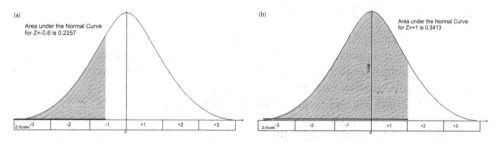

Figure 3.14 Probability of project and normal distribution standard.

$$Z = \frac{D - t_e}{S_t} = \frac{33 - 28}{5} = \frac{5}{5} = 1$$

The probability associated with Z= +1 is 0.84134. This is a strong probability and indicates that the odds are 16 to 3 that the project will be completed by the due date (Figure 3.14b).

If the probability of an event is p, the odds for its occurrence are a to b, where

$$\frac{a}{b} = \frac{p}{1-p} = \frac{0.84134}{0.15866} \approx \frac{16}{3}$$

3.12   Transportation problems (application to mining problems)

A transportation problem (TP) deals with the problem, which aims to find the best way to fulfil the needs of n demand points using the capacities of m supply points. While trying to find the best way, generally a variable cost of shipping the product from one supply point to a demand point or a similar constraint should be taken into consideration. It is a special type of LP problem where the objective is to minimize the cost of distributing a product from many sources to several destinations. Unlike other LP problems, a *balanced* TP with m supply points and n demand points is easier to solve, although it has m+n equality constraints. The reason for that is, if a set of decision variables (x_{ij}'s) satisfy all but one constraint, the values for x_{ij}'s will satisfy that remaining constraint automatically.

There are three basic methods to find the best feasible solution for a transportation problem:

1.   Northwest Corner Method (NWC)
2.   Minimum Cost Method
3.   Vogel's Method

Problem definition: Coal company has three mines that supply the coal to four power plants:

a.   The associated coal supply of each mine and demand of each plant is given in the table.
b.   The cost of sending 1 lakh tonnes coal from a mine to a plant depends on the distance travel.

A TP is specified by the supply, the demand, and the transportation costs. So the relevant data can be summarized in a transportation tableau. The transportation tableau implicitly expresses the supply and demand constraints and the transport cost between each demand and supply point.

Three mines located in area with production capacity of 50, 75, and 25 (× 1,000) tonnes per day. Each day the mines must supply coal to four thermal plants P1, P2, P3, and P4 with at least 20, 20, 50, and 60 (× 1,000) tones per day respectively. The transportation cost per unit is given in Table 3.4.

Table 3.4 An Example of Transport Costs, Supply, and Demand for a Coal Company

Mine (From)	To		Plants		Supply
	PI	P2	P3	P4	
A	3	5	7	6	50
B	2	5	8	2	75
C	3	6	7	2	25
Demand	20	20	50	60	

Table 3.5 NWC Method

Mine			Plants		Supply
	PI	P2	P3	P4	
A	3 (20)	5 (20)	7 (10)	6	50, 30, 10
B	2	5	8 (40)	2 (35)	75, 35
C	3	6	9	2 (25)	25
Demand	20	20	50 (40)	60 (25)	

Finding basic feasible solution for TP using different methods are as given below:

Northwest Corner Method: To find the solution by the NWC method, begin in the upper left (northwest) corner of the transportation table and set x_{11} as large as possible (here, the limitations for setting x_{11} to a larger number will be the demand of demand point 1 and the supply of supply point 1.)

Steps in NWC method:

1. Select the upper left hand of the table and allocate as many units as possible.
2. Adjust the demand and supply numbers in the respective rows and column.
3. If the demand in first cell is satisfied, then move horizontally to the next cell in next column.
4. If the supply for the first row is exhausted, then move down to the first cell in the second row.
5. If for any cell supply equals demand, then the next allocation can be made in cell either in the next row or in the next column.
6. Continue the process until all supply and demand values are exhausted.

Starting from North west corner, we allocate 20 tonnes to AP1
    Initial basic feasible solution of total transportation cost:

$$= AP1 + AP2 + AP3 + BP3 + BP4 + CP4$$

(where allocation is there, other cell values are zero as there is no allocation)

$$= 3 * 20 + 5 * 20 + 7 * 10 + 8 * 40 + 2 * 35 + 2 * 25$$
$$= 670 \ (\times 1,000)$$

Table 3.6 Minimum Cost Method

Mine	Plants				Supply
	P1	P2	P3	P4	
A	3	5 20	7 30	6	50 30
B	2 20	5	8	2 55	75 55
C	3	6	9 20	2 5	25 20
Demand	20	20	50 20	60 5	

Minimum Cost Method: Follow these steps (Table 3.6)

1. Identify the box having minimum unit transportation cost.

    Initial basic feasible solution:
    Total transportation cost:

$$= 2*20+5*20+7*30+9*20+2*55+2*5$$
$$= 650 \,(\times 1{,}000)$$

2. If the minimum cost is not unique, then you are at liberty to choose any cell.
3. Choose the supply value as much as possible subject to demand and supply constraints.
4. Repeat steps 1 to 3 till all the restrictions are satisfied.

Vogel's Method: Begin with computing each row and column a penalty. The penalty will be equal to the difference between the two smallest shipping costs in the row or column. Identify the row or column with the largest penalty. Find the first basic variable which has the smallest shipping cost in that row or column. Then assign the highest possible value to that variable and cross out the row or column as in the previous methods. Compute new penalties and use the same procedure. This method is based on the concept of *penalty cost* or *regret*.

–   A penalty cost is the difference between the largest and the next largest cell cost in a row (or column).
–   In this method, the first step is to develop a penalty cost for each source and destination.
–   Penalty cost is calculated by subtracting the minimum cell cost from the next higher cell cost in each row and column.

Problem definition: Coal company has three mines located in an area with the production capacity of 120, 70, and 50 T per day. Each day, the mines must supply coal to four cement plants C1, C2, C3, and C4 with at least 60, 40, 30, and 110 tonnes per day, respectively. The transportation cost (Units) is given in Table 3.7.

Table 3.7 The Transportation Cost

Mine	Cement plants				Supply
	CI	C2	C3	C4	
A	20	22	17	4	120
B	24	37	9	7	70
C	32	37	20	15	50
Demand	60	40	30	110	240

Table 3.8a The Penalty is a Cost

Mine	Cement Plants				Supply	Penalty
	C1	C2	C3	C4		
A	20	22 40	17	4	120 80	17-4=13
B	24	37	9	7	70	9-7=2
C	32	37	20	15	50	20−15 = 5
Demand	60	40	30	110	240	
Penalty	24−20 =4	37−22 =15	17−9 =8	7−4= 3		

Table 3.8b The Penalty is a Cost

Mine	Cement Plants			Supply	Penalty
	C1	C3	C4		
A	20	17	4 80	80	17-4=13
B	24	9	7	70	9-7=2
C	32	20	15	50	20−15 = 5
Demand	60	30	110 30		
Penalty	24−20 =4	17−9 =8	7−4= 3		

Distribute the coal to each plant in such a way that the total transportation cost is minimum.

Identify the boxes having minimum and next to the minimum transportation cost in each row and column, and write the difference (penalty) (Table 3.8).

Highest penalty occurs in C2 column=15. In C2 column, minimum transportation cost is 22. Therefore, allocate 40 T to this cell. Prepare the new table and remove column 2.

The highest penalty occurs in the first row for mine A. In the first row, the minimum transportation cost is 4. Therefore, allocate 80 tonnes to this cell. Prepare the new table and remove row 1.

Table 3.8c The Penalty is a Cost

Mine	Cement Plants			Supply	Penalty
	C1	C3	C4		
B	24	9 30	7	~~70~~ 40	9-7=2
C	32	20	15	50	20 −15 = 5
Demand	60	~~30~~	30		
Penalty	32-24 = 8	20 −9 = 11	15-7= 8		

Table 3.8d The Penalty is a Cost

Mine	Cement Plants		Supply	Penalty
	C1	C4		
B	24 10	7 30	~~40~~ 10	24-7=17
C	32 50	15	50	32 −15 = 17
Demand	~~60~~	~~30~~		
Penalty	32-24 = 8	15-7= 8		

Table 3.9 Final Allocation Table of Vogel's Method

Mine	Cements Plants							Supply
	C1		C2		C3		C4	
A	20		22	40	17		4 80	~~120~~
B	24	10	37		9 30		7 30	~~70~~
C	32	50	37		20		15	~~50~~
Demand	~~60~~		~~40~~		~~30~~		~~110~~	

The highest penalty occurs in C3 column. In C3 column, the minimum transportation cost is 9. Therefore, allocate 30 ton to this cell. Prepare the new table and remove column C3.

The highest penalty occurs in B and C row. In rows B and C, the minimum transportation cost is 7. Therefore, allocate 30 tonnes to this cell.

Total transportation cost:

$$= 24*10+32*50+22*40+9*30+4*80+7*30$$
$$= 3,520$$

3.13 Decision methods

We spend a significant portion of our time and psychic energy making decisions. Our decisions shape our lives: who we are, what we are, where we are, how successful we are, how happy we are, all derive in large part from our decisions In order to raise our odds of making a good decision, we have to learn to use a good decision-making process – one that gets us to the best solution with a minimal loss of time, energy, money, etc.
  Decision-making is defined as:

- Intentional and reflective choice in response to perceived needs.
- Choice of one alternative or a subset of alternatives among all possible alternatives with respect to goals.
- Solving a problem by choosing, ranking, or classifying over the available alternatives that are characterized by multiple criteria.

An effective decision-making process will fulfil the following six criteria:

i. It focuses on what's important.
ii. It is logical and consistent.
iii. It acknowledges both subjective and objective factors and blends analytical with intuitive thinking.
iv. It requires only as much information and analysis as is necessary to resolve a particular dilemma.
v. It encourages and guides the gathering of relevant information and informed opinion.
vi. It is straightforward, reliable, easy to use, and flexible.

A key to good decision-making is to provide a structural method for incorporating the information, opinions, and preferences of the various relevant people into the decision-making process. A good decision is based on logic, uses all available resources, evaluates all possible alternatives, and utilizes a quantitative method.

3.13.1 Decision-making techniques

The techniques can be used to make the best decisions possible with the information available. With these tools, you will be able to map out the likely consequences of decisions, work out the importance of individual factors, and choose the best course of action to take. Here, some simple techniques of decision-making are discussed.
  Techniques discussed are:

- Selecting the most important changes to make – **Pareto Analysis**
- Evaluating the relative importance of different options – **Paired Comparison Analysis**
- Selecting between good options – **Grid Analysis**
- Looking at a decision from all points of view – **Six Thinking Hats**
- Seeing whether a change is worth making – **Cost/Benefit Analysis**

3.13.2 Pareto analysis

Pareto analysis is a very simple technique that helps you to choose the most effective changes to make. It uses the Pareto principle – the idea that by doing 20% of work you can generate 80% of the advantage of doing the entire job*. Pareto analysis is a formal technique for finding the changes that will give the biggest benefits. It is useful where many possible courses of action are competing for your attention.

How to use techniques: To start using the tool, write out a list of the changes you could make. If you have a long list, group it into related changes. Then score the items or groups. The scoring method you use depends on the sort of problem you are trying to solve. For example, if you are trying to improve profitability, you would score options on the basis of the profit each group might generate. If you are trying to improve customer satisfaction, you might score on the basis of the number of complaints eliminated by each change.

The first change to tackle is the one that has the highest score. This one will give you the biggest benefit if you solve it. The options with the lowest scores will probably not even be worth bothering. Solving these problems may cost you more than the solutions are worth.

Example: A manager has taken over a failing service centre. He commissions research to find out why customers think that service is poor.

The manager gets the following comments back from the customers:

1. Phones are only answered after many rings.
2. Staff seem distracted and under pressure.
3. Engineers do not appear to be well organized. They need second visits to bring extra parts. This means that customers have to take more holiday to be there a second time.
4. They do not know what time they will arrive. This means that customers may have to be in all day for an engineer to visit.
5. Staff members do not always seem to know what they are doing.
6. Sometimes when staff members arrive, the customer finds that the problem could have been solved over the phone.

The manager groups these problems together. He then scores each group by the number of complaints and orders the list:

- Lack of staff training: items 5 and 6: 51 complaints
- Too few staff: items 1, 2, and 4: 21 complaints
- Poor organization and preparation: item 3: 2 complaints

By doing the Pareto analysis above, the manager can better see that the vast majority of problems (69%) can be solved by improving staff skills. Once this is done, it may be worth looking at increasing the number of staff members. Alternatively, when staff members improve their ability to solve problems over the phone, maybe the need for new staff members may decline.

It looks as if comments on poor organization and preparation may be rare and could be caused by problems beyond the manager's control. By carrying out a Pareto analysis, the manager is able to focus on training as an issue rather than spreading effort over training, taking on new staff members, and possibly installing a new computer system.

Key points: Pareto analysis is a simple technique that helps you to identify the most important problem to solve. To use it:

- List the problems you face or the options you have available.
- Group options where they are facets of the same larger problem.

- Apply an appropriate score to each group.
- Work on the group with the highest score.

Pareto analysis not only shows you the most important problem to solve, but it also gives you a score showing how severe the problem is.

3.13.3   Paired comparison analysis

Paired comparison analysis helps you to work out the importance of a number of options relative to each other. It is particularly useful where you do not have objective data to base this on.

This makes it easy to choose the most important problem to solve, or select the solution that will give you the greatest advantage. Paired comparison analysis helps you to set priorities where there are conflicting demands on your resources.

How to use technique: To use the technique, first of all list your options. Then draw up a grid with each option as both a row and a column header. Use this grid to compare each option with each other option, one-by-one. For each comparison, decide which of the two options is most important, and then assign a score to show how much more important it is. One can then consolidate these comparisons so that each option is given a percentage importance.

Follow these steps to use the technique:

1. List the options you will compare. Assign a letter to each option.
2. Set up a table with these options as row and column headings.
3. Block out cells on the table where you will be comparing an option with itself – there will never be a difference in these cells! These will normally be on the diagonal running from the top left to the bottom right.
4. Also block out cells on the table where you will be duplicating a comparison. Normally, these will be the cells below the diagonal.
5. Within the remaining cells, compare the option in the row with the one in the column. For each cell, decide which of the two options is more important. Write down the letter of the more important option in the cell, and score the difference in importance from 0 (no difference) to 3 (major difference).
6. Finally, consolidate the results by adding up the total of all the values for each of the options. You may want to convert these values into a percentage of the total score.

Example: As a simple example, an entrepreneur is looking at ways in which she can expand the business with limited resources, but also has the options which are listed below:

- Expand into overseas markets
- Expand in home markets
- Improve customer service
- Improve quality

Firstly, he draws up the paired comparison analysis table as shown in Figure 3.15.

	Overseas Market (A)	HomeMarket (B)	CustomerService (C)	Quality(D)
Overseas Market(A)	Blocked Out(Step 3)			
Home Market (B)	Blocked Out(Step 4)	Blocked Out (Step 3)		
Customer Service(C)	Blocked Out(Step 4)	Blocked Out (Step 4)	Blocked Out (Step 3)	
Quality(D)	Blocked Out (Step 4)	Blocked Out (Step 4)	Blocked Out (Step 4)	Blocked Out (Step 3)

Figure 3.15 Example of paired comparison analysis table (not filled in).

	Overseas Market (A)	Home Market (B)	Customer Service (C)	Quality (D)
Overseas Market (A)		A,2	C,1	A,1
Home Market (B)			C,1	B,1
Customer Service (C)				C,2
Quality (D)				

Figure 3.16 Example of paired comparison analysis table (filled in).

Then, he compares options, writes down the letter of the most important option, and scores their difference in importance. An example of how it could be done is shown in Figure 3.16.

Finally, he adds up the A, B, C, and D values and converts each into a percentage of the total. This gives these totals:

- A=3 (37.5%)
- B=1 (12.5%)
- C=4 (50%)
- D=0.

Here, it is most important to improve customer service (C) and then to tackle export markets (A). Quality is not a high priority – perhaps it is good already.

Key points: Paired comparison analysis is a good way of weighing up the relative importance of different courses of action. It is useful where priorities are not clear or are competing in importance.

The tool provides a framework for comparing each course of action against all others and helps to show the difference in importance between factors.

3.13.4 Grid analysis

How to use technique: Grid analysis is a useful technique to use for making a decision. It is most effective when you have a number of good alternatives and many factors to take into account. The first step is to list your options and then the factors that are important for making the decision. Lay these out in a table, with options as the row labels and factors as the column headings. Next, work out the relative importance of the factors in your decision. Show these as numbers. We will use these to weight your preferences by the importance of the factor. These values may be obvious.

The next step is to work your way across your table, scoring each option for each of the important factors in your decision. Score each option from 0 (poor) to 3 (very good). Note that you do not have to have a different score for each option – if none of them are good for a particular factor in your decision, then all options should score 0. Now multiply each of your scores by the values for your relative importance. This will give them the correct overall weight in your decision.

Finally, add up these weighted scores for your options. The option that scores the highest wins!

Example: A windsurfing enthusiast is about to replace his car. He needs one that not only carries a board and sails, but also that will be good for business travel. He has always loved open-topped sports cars. No car he can find is good for all three things. His options are:

- A four wheel drive, hard topped vehicle
- A comfortable 'family car'
- An estate car
- A sports car

Criteria that he wants to consider are:

- Cost
- Ability to carry a sail board at normal driving speed
- Ability to store sails and equipment securely
- Comfort over long distances
- Fun!
- Nice look and build quality to car

Firstly, he draws up the table as shown in Figure 3.17 and scores each option by how well it satisfies each factor.

Next, he decides the relative weights for each of the factors. He multiplies these by the scores already entered and totals them. This is shown in Figure 3.18.

This gives an interesting result: Despite its lack of fun, an estate car may be the best choice.

If the wind-surfer still feels unhappy with the decision, maybe he has underestimated the importance of one of the factors. Perhaps he should weight 'fun' by 7.

Key points: Grid analysis helps you to decide between several options, while taking many different factors into account. To use the tool, lay out your options as rows on a table. Set up the columns to show your factors. Allocate weights to show the importance of each of these factors. Score each choice for each factor using numbers from 0 (poor) to 3 (very good). Multiply each score by the weight of the factor to show its contribution to the overall selection.

Factors:	Cost	Board	Storage	Comfort	Fun	Look	Total
Weights:							
Sports Car	1	0	0	1	3	3	
4 Wheel Drive	0	3	2	2	1	1	
Family Car	2	2	1	3	0	0	
Estate Car	2	3	3	3	0	1	

Figure 3.17 Example of grid analysis showing unweighted assessment of how each type of car satisfies each factor.

Factors:	Cost	Board	Storage	Comfort	Fun	Look	Total
Weights:	4	5	1	2	3	4	
Sports Car	4	0	0	2	9	12	27
4 Wheel Drive	0	15	2	4	3	4	28
Family Car	8	10	1	6	0	0	25
Estate Car	8	15	3	6	0	4	36

Figure 3.18 Example of grid analysis showing weighted assessment of how each type of car satisfies each factor.

3.13.5 Six thinking hats

If you look at a problem with the 'Six Thinking Hats' technique, then you will solve it using all approaches. Your decisions and plans will mix ambition, skill in execution, public sensitivity, creativity, and good contingency planning.

How to use technique: You can use Six Thinking Hats in meetings or on your own. In meetings, it has the benefit of blocking the confrontations that happen when people with different thinking styles discuss the same problem. Each 'Thinking Hat' is a different style of thinking. These are explained below:

- **White Hat**: With this thinking hat, you focus on the data available. Look at the information you have and see what you can learn from it. Look for gaps in your knowledge, and either try to fill them or take account of them. This is where you analyse past trends and try to extrapolate from historical data.
- **Red Hat**: 'Wearing' the Red Hat, you look at problems using intuition, gut reaction, and emotion. Also try to think how other people will react emotionally. Try to understand the responses of people who do not fully know your reasoning.
- **Black Hat**: Using Black Hat thinking, look at all the bad points of the decision. Look at it cautiously and defensively. Try to see why it might not work. This is

important because it highlights the weak points in a plan. It allows you to eliminate them, alter them, or prepare contingency plans to counter them.

Black Hat thinking helps to make your plans 'tougher' and more resilient. It
can also help you to spot fatal flaws and risks before you embark on a course of
action. Black Hat thinking is one of the real benefits of this technique, as many
successful people get so used to thinking positively that often they cannot see
problems in advance. This leaves them under-prepared for difficulties.

- **Yellow Hat**: The Yellow Hat helps you to think positively. It is the optimistic viewpoint that helps you to see all the benefits of the decision and the value in it. Yellow
  Hat thinking helps you to keep going when everything looks gloomy and difficult.
- **Green Hat**: The Green Hat stands for creativity. This is where you can develop
  creative solutions to a problem. It is a freewheeling way of thinking in which there
  is little criticism of ideas. A whole range of creativity tools can help you here.
- **Blue Hat**: The Blue Hat stands for process control. This is the hat worn by people
  chairing meetings. When running into difficulties because ideas are running dry,
  they may direct activity into Green Hat thinking. When contingency plans are
  needed, they will ask for Black Hat thinking, etc.

A variant of this technique is to look at problems from the point of view of different
professionals (e.g. doctors, architects, sales directors) or different customers.

Example: The directors of a property company are looking at whether they should
construct a new office building. The economy is doing well, and the amount of vacant
office space is reducing sharply. As part of their decision, they decide to use the 6
Thinking Hats technique during a planning meeting.

Looking at the problem with the White Hat, they analyse the data they have. They
examine the trend in vacant office space, which shows a sharp reduction. They antici-
pate that, by the time the office block would be completed, there will be a severe shortage of office space. Current government projections show steady economic growth for
at least the construction period.

With Red Hat thinking, some of the directors think the proposed building looks
quite ugly. While it would be highly cost-effective, they worry that people would not
like to work in it.

When they think with the Black Hat, they worry that government projections may
be wrong. The economy may be about to enter a 'cyclical downturn' in which case
the office building may be empty for a long time. If the building is not attractive, then
companies will choose to work in another better-looking building at the same rent.

With the Yellow Hat, however, if the economy holds up and their projections are
correct, the company stands to make a great deal of money. If they are lucky, maybe
they could sell the building before the next downturn or rent to tenants on long-term
leases that will last through any recession.

With Green Hat thinking, they consider whether they should change the design to
make the building more pleasant. Perhaps they could build prestige offices that people
would want to rent in any economic climate. Alternatively, maybe they should invest
the money in the short term to buy up property at a low cost when a recession comes.

The Blue Hat has been used by the meeting's chair to move between the different
thinking styles. He or she may have needed to keep other members of the team from
switching styles or from criticizing other peoples' points.

Key points: Six Thinking Hats is a good technique for looking at the effects of a decision from a number of different points of view. It allows necessary emotion and scepticism to be brought into what would otherwise be purely rational decisions. It opens up the opportunity for creativity within decision-making. The technique also helps, e.g. persistently pessimistic people to be positive and creative. Plans developed using the '6 Thinking Hats' technique will be sounder and more resilient than they otherwise would be. It may also help you to avoid public relations mistakes and spot good reasons not to follow a course of action before you have committed to it.

3.13.6   Cost/benefit analysis

How to use techniques: You may have been intensely creative in generating solutions to a problem and rigorous in your selection of the best one available. This solution may still not be worth implementing, as you may invest a lot of time and money in solving a problem that is not worthy of this effort.

Cost/Benefit analysis is a relatively simple and widely used technique for deciding whether to make a change. As its name suggests, to use the technique, simply add up the value of the benefits of a course of action and subtract the costs associated with it.

Costs are either one – Off or may be ongoing. Benefits are most often received over time. We build this effect of time into our analysis by calculating a payback period. This is the time it takes for the benefits of a change to repay its costs. Many companies look for payback over a specified period of time – e.g. three years.

In its simple form, cost/benefit analysis is carried out using only financial costs and financial benefits. For example, a simple cost/benefit analysis of a road scheme would measure the cost of building the road and subtract this from the economic benefit of improving transport links. It would not measure either the cost of environmental damage or the benefit of quicker and easier travel to work.

A more sophisticated approach to cost/benefit analysis is to try to put a financial value on these intangible costs and benefits. This can be highly subjective – e.g. a historic water meadow worth 25,000 units, or is it worth 500,000 units because of its environmental importance? What is the value of stress-free travel to work in the morning?

These are all questions that people have to answer, and answers that people have to defend. We are explaining here a simple version of cost/benefit analysis. Where large sums of money are involved (for example, in financial market transactions), project evaluation can become an extremely complex and sophisticated art.

Example: A sales director is deciding whether to implement a new computer-based contact management and sales processing system. His department has only a few computers, and his salespeople are not computer literate. He is aware that computerized sales forces are able to contact more customers and give a higher quality of reliability and service to those customers. They are more able to meet commitments and can work more efficiently with fulfillment and delivery staff.

His financial cost/benefit analysis is shown below:

Costs:
(monetary units may be in local currency)
New computer equipment:

- 10 network-ready PCs with supporting software @ 1,225 each
- 1 server @ 1,750
- 3 printers @ 600 each
- Cabling & Installation @ 2,300
- Sales Support Software @ 7,500

Training costs:

- Computer introduction – 8 people @ 200 each
- Keyboard skills – 8 people @ 200 each
- Sales Support System – 12 people @ 350 each

Other costs:

- Lost time: 40 man days @ 100/day
- Lost sales through disruption: estimate: 10,000
- Lost sales through inefficiency during first months: estimate: 10,000

Total cost: 55,800

Benefits:

- Tripling of mail shot capacity: estimate: 20,000/year
- Ability to sustain telesales campaigns: estimate: 10,000/year
- Improved efficiency and reliability of follow-up: estimate: 25,000/year
- Improved customer service and retention: estimate: 15,000/year
- Improved accuracy of customer information: estimate: 5,000/year
- More ability to manage sales effort: 15,000/year

Total benefit = 90,000/year
Payback time = 55,800 / 90,000 = 0.62 of a year = 8 months (approx.)

Inevitably, the estimates of the benefit given by the new system are quite subjective. Despite this, the sales director is very likely to introduce it, given the short payback time.

Key points: Cost/Benefit Analysis is a powerful, widely used, and relatively easy tool for deciding whether to make a change. To use the tool, firstly work out how much the change will cost to make. Then, calculate the benefit you will from it. Where costs or benefits are paid or received over time, work out the time it will take for the benefits to repay the costs.

1. Cost/Benefit Analysis can be carried out using only financial costs and financial benefits. You may, however, decide to include intangible items within the analysis. As you must estimate a value for these, this inevitably brings an element of subjectivity into the process (Mind Tools, n.d.).

3.14   Subfields of systems engineering

While it is obvious that a great many specialist or 'niche' areas within the discipline of engineering may be considered to be subfields of Systems Engineering, the increasing diversity and complexity inherent in today's systems has established a greater degree of distinction between these areas. Many subfields may be considered on their own merits as opposed to existing only as a subset of a more general subject. Indeed, many of these specialisms have contributed to the development of Systems Engineering as a distinct entity, researched and refined separately.

3.14.1   Safety engineering

The techniques of safety engineering can be applied by everyday people to planning complex events to assure that the systems cannot cause harm. Most of safety engineering is just a way of making plans that cope with failures.

Usually, a failure in safety-certified systems is acceptable if less than one life per 30 years of operation (10^9 hours) is lost to mechanical failure. Most Western nuclear reactors, medical equipment, and commercial aircraft are certified to this level. This level is accepted not because loss of life is acceptable, but rather because a design near this level usually has significant mechanical redundancy and the failures will be gradual enough that repairs can be scheduled before significant loss of life can occur.

3.14.2   Reliability engineering

Reliability engineering is the discipline of ensuring a system will be reliable, i.e. failure-free, when operated in a specified manner. Reliability engineering is performed throughout the entire life cycle and relies heavily on statistics, probability theory, and reliability theory. Reliability engineering applies to the entire system, including hardware and software. It is closely associated with maintainability engineering and logistics engineering. Some reliability engineering techniques overlap safety engineering techniques, such as the failure modes and effects analysis.

3.14.3   Interface design

Interface design and specification are concerned with making the pieces of a system interoperate. For example, the plugs between two computer systems can be a fertile source of failures. Sometimes something as simple as gold-plating the plugs can lower the probability of a failure enough to save millions of dollars.

Another issue is assuring that the signals that pass from system to the next are in tolerance and that the receivers have a wider tolerance than transmitters. Another issue is that the interface should be able to accept new features. Most often, this is a transmission speed problem with a plug and jack, although it sometimes affects computer data formats. The rule of thumb is that roughly 20% of the space in an interface should be reserved for future additions.

Human–Computer Interaction (HCI) is another aspect of interface design and is a vital part of modern Systems Engineering when considering the user of a system.

3.14.4   Cognitive systems engineering

Cognitive systems engineering is a systems engineering with the human integrated as an explicit part of the system. Sometimes referred to as *Human Engineering*, this subject also deals with ergonomics in system design.

It depends on the direct application of centuries of experience and research in both *Cognitive Psychology* and *Systems Engineering*. Cognitive systems engineering focuses on how man interacts with the environment and attempts to design systems that explicitly respect how humans think. Cognitive systems engineering works at the intersection of the problems imposed by the world, the needs of agents (both human and software), and the interaction with the various systems and technologies to affect the situation.

3.14.5   Communication protocols

Interface design principles also have been used to place reserved wires, plug-space, command codes, and bits in communication protocols. Systems engineering principles are applied in the design of network protocols.

3.14.6   Security engineering

Security engineering is a high technology area concerning to data and cyber security that is viewed as a field of systems engineering. The security engineers of the mine system are vital to the mine/company and are in great demand in the industry.

3.15   Mine system engineering tutorials

Many of the world's mining operations and projects are done these days through modern software in different subject areas, e.g. mine planning and ventilation planning, some of which are listed in Chapter 6 of this book, but tutorials play a crucial role in learning the mining operation quite closely. Recent video techniques have made the tutorials very attractive and interesting to learn (see the Case Study 6.1 in this book). It is plausible to fill the learning gaps with modern tools namely computers, PowerPoint presentations, scenes captured with a freely moving camera, video frames, and software-based input–output tools. We are depicting some tutorials below for illustration only, but their numbers could be many more according to the need.

PROBLEM

Design an ore pass for a metal mine using numerical modelling. Apply software that is available to you. Given the *input parameters, geo mining condition and physico-mechanical properties* for the numerical modelling study are as follows:

- Rock type: Schist and Quartzite
- Compressive strength: 55 MPa for Schist and 70 MPa for Quartzite

- Tensile strength: 05 MPa for Schist and 6.5 MPa for Quartzite
- Density: 55 MPa for Schist and 70 MPa for Quartzite
- Elastic modulus: 10 GPa for Schist and 22 GPa for Quartzite
- Geological Strength Index (GSI): 40 for Schist and 60 for Quartzite
- Depth of ore pass: 250 m and 650 m and Ore pass diameter: 1.5 m

Solution

Before dealing with the ore pass modelling, let us understand some brief facts about the ore pass used in the metal mines mostly (Box 3.2). The ore pass is not a common term in coal mining.

Numerical modelling analysis is carried out for 2D plain strain models and solved for ore pass support design. A simulation was performed using **FLAC3D** software (ITASCA Consulting Group, USA) based on the explicit FED (Finite Element Difference) method. The simulated mining conditions have considered the rock type, its properties, material behaviour, and *in situ* stress condition for the analysis. Utilizing all this information, given in the problem, the Mohr–Coulomb criterion is used to analyse the rock mass stability in this problem.

Numerical modelling: Two-dimensional (2D) Plain Strain Models are a special case of a 3D situation. In this case, the body is confined in one direction and allowed to strain in the remaining two directions. A model of size 10 m × 10 m × 1 m is prepared with an

---

Box 3.2: What Is Ore Pass?

An ore pass is a mining shaft used to transfer ore from the mine (Figure 3.19). Ore passes are a convenient way to transport and handle materials and minimize hauling distances between levels in a mine. There are two types of ore pass systems in the mining industry: (i) a flow-through ore pass system and (ii) a full ore pass system. For both systems, ore and waste are dumped into the ore pass and end up at the draw point, or point in which the ore is transported by a loader, conveyor, or rail car to an underground stockpile.

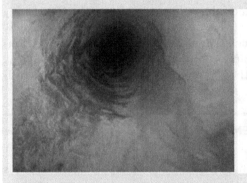

Figure 3.19 A view of the ore pass and hang up in an ore pass.

In *flow-through systems*, the ore is dumped into the ore pass and then flows down to the draw point. This system is used when ore has high levels of fines, or fine particulate, which results in ore pass hang-ups. Because the material is flowing at a continual rate, fines don't settle and hang-ups are less likely to occur. In *full ore pass systems*, a certain amount of ore is maintained in the ore pass at all times. Ore passes take advantage of gravity to move materials from one level of a mine to another. Ore passes are excavated by the drill and blast method or the mechanical method as per the encountered development condition of the mine. When building an ore pass, miners consider location, length, dimension, shape, orientation, and support. An ore pass system is considered to have failed if it does not meet its required task of non-interrupted material flow for any given time. It's important to monitor the ore pass to preserve the ore pass's structural integrity, prevent mining-related accidents, and prevent overfilling of the ore pass.

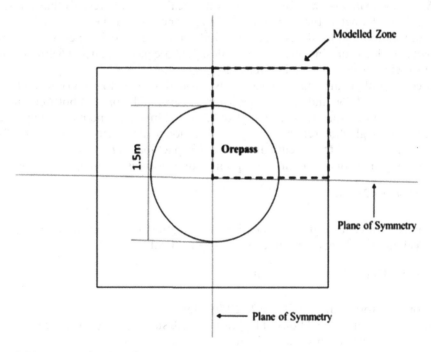

Figure 3.20 Modelled zone of an ore pass.

element size of 0.13 m in the area of interest for support design. Taking into consideration the planes of symmetry, only a quarter portion of ore pass is modelled (Figure 3.20).

After the generation of *in situ* stresses in the model, the ore pass is excavated in the elasto-plastic model and then yielding zones around the ore pass are studied. Supports required have been estimated based on the height of rock load determined from numerical models.

Based on numerical modelling results, the required support load (L) at different locations can be estimated using the following equation:

$$L = r \cdot h \tag{3.3}$$

where L is the required support density in t/m^2; r is the rock density, 2.8 t/m^3; and h is the height of the yield zone (roof).

During the installation of rock bolts in the roof, the applied load (P) can be estimated using the following equation:

$$P = n. \frac{n.c_t}{m^2} * B.a \tag{3.4}$$

where c is the bearing capacity of the rock bolts, B is the width of the drivage, n is the number of bolts in a row, and a is the spacing between two consecutive rows.

The factor of safety of the support system is defined as $F = P/L$.

Support requirement for the ore pass: Numerical modelling reveals that for an ore pass of 1.5 m diameter, the maximum yielding zone around the ore pass lies in the range of 0.12 to 0.4 m for different depths and encountered rock types, i.e. schist and quartzite. Yielded zone in and around the designed ore pass at 250 and 650 m depth for various rocks has been shown in Table 3.10.

Block contour of yielding zones in rock mass around ore pass for Quartzite and Schist rock at a depth of 250 and 650 m is shown in Figures 3.21 (top and bottom) and 3.22 (top and bottom), respectively. For a 0.1 to 0.2 m yielding zone, no major supports are required, although shotcrete can be sprayed for longevity or steel sheets of 8–10 mm thickness can be used to prevent damage from impact loading.

For a yielding zone of 0.4 m, the support requirement is calculated as follows:

$$L = 0.4 \text{ m} \times 2.7 \text{ t/m}^3$$

If 3 numbers of cement grouted rock bolts of length 1.0 m are installed in a 1.5 m × 1.5 m grid pattern along the perimeter, the applied load becomes

$$P = (3 \times 4) / (4.71 \times 1.5) = 2.12 \text{ t/m}^2$$

Therefore, Factor of safety will be $= 2.12/1.08 = 1.96$.

Hence, cement grouted bolts of length 1.0 m will suffice for the purpose that can be provided along the perimeter at an interval of 1.5 m.

Discussion on problem result (Output): Numerical modelling output shows that the excavated ore pass fits into the elasto-plastic model as far as rock mass behaviour

Table 3.10 Yielding Zone from the Numerical Modelling

S.No.	Depth (m)	Rock type	Yielded zone from model (m)
1	250	Quartzite	0.14
2	650	Quartzite	0.13
3	250	Schist	0.26
4	650	Schist	0.38

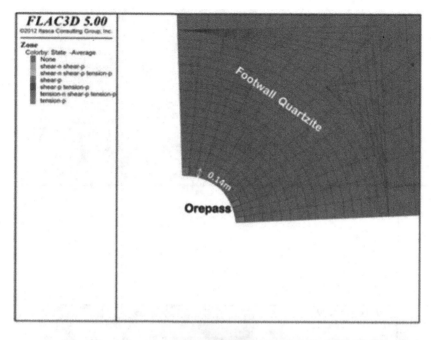

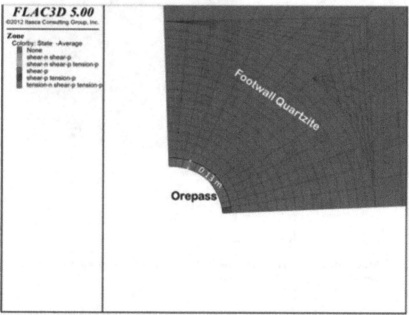

Figure 3.21 Block contour of yielding zone in rock mass around orepass at (a) 250 m depth and (b) 650 m depth for quartzite rock.

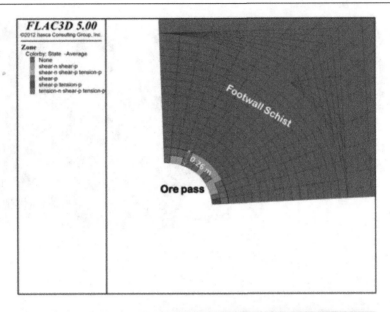

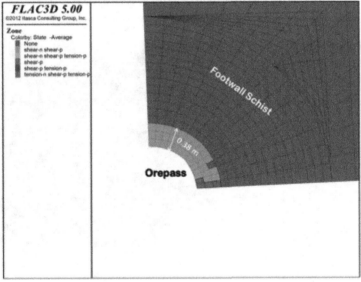

Figure 3.22 Block contour of yielding zone in rock mass around orepass at (a) 250 m depth and (b) 650 m depth for schist rock.

is concerned. The modelling output (Figures 3.21 and 3.22) indicates that the maximum yielding zone around ore-pass falls in the range of 0.12–0.4 m for different depths and two different rock types. The estimated supports required have relations with the height of rock load too.

Ore passes need to be designed to cut across ground levels and sublevels to provide easy access to underground minerals. The intervals between ore passes should preferably not exceed 150 m theory says. Ore pass inclination should be maintained between 45° and 90°, while ore pass length should be in the range of 10–300 m.

Ore-pass system is considered failed if it does not meet its required task of non-interrupted material flow at any given time. Hence, the ore passes need proper support to ensure adequate safety. *Resin-grouted short cable bolts* and *Resign-grouted rebar* are some well-known and available supports for the ore passes. In some cases, ore passes require support through *shotcrete* or *steel liners*.

References

Bahill, A. T. and Dean, F. F. (1999), Discovering System Requirements, In: *Handbook of Systems Engineering*, edited by A. P. Sage and W. B. Rouse, New York: John Wiley & Sons, pp. 175–219.

Bahill, A. T. and Gissing, B. (1998), Re-evaluating systems engineering concepts using systems thinking, *IEEE Transaction on Systems, Man and Cybernetics, Part C: Applications and Reviews*, Vol. 28, Issue 4, pp. 516–527.

Chatterjee, S., Hadi, A. S., and Prince, B. (1977), *Regression Analysis by Example*, New York: Wiley-Interscience, p. 228.

Freund, J. E. (1979), *Modern Elementary Statistics*, New Delhi: Prentice-Hall of India Private Limited.

Maier, M. W. and Rechtin, E. (2000), *The Art of Systems Architecting*, Boca Raton, FL: CRC Press.

Mind Tools. (n.d.), Decision Making. https://www.mindtools.com/pages/main/newMN_TED.htm

Wiest, J. D. and Levy, F. K. (1974), *A Management Guide to PERT/CPM*, New Delhi: Prentice-Hall of India Private Limited.

Chapter 4

Mine safety engineering

Satya-Mev-Jayate.

(Bhagwad Gita)

4.1 Introduction

Safety in mines continues to be a major concern for the mining industry despite considerable improvement in the working conditions resulting in a reduction in accidents and injury rates. Safety is of paramount importance in any activity; however, it is more so in mining which has a high potential risk of accidents as the environment and the geo-environment change continually with the progress of work. Accidents just don't happen but they are caused. Certain causes are responsible for accidents.

Mine and its safety management involve the following strategies, keeping in view various aspects of planning and control:

1. To adaptation of the safety engineering and safety management approach as a matter of policy.
2. Legislative control constitutes the government's regulatory and enforcement activities, e.g. periodic inspections to oversee compliance with safety statutes, the investigation into accidents and punishing delinquent mine officials, granting of permission to start and operate a mine working, and the mine management's responsibility to abide by the statutory safety provisions as far as practicable.

Definitions (Kalia et al., 2015):

Safety Engineering: This involves the identification, evaluation, and control of hazards in man-machine systems that contain the potential to cause injury to people or damage to property.

Safety Management: This consists of a set of safety program elements, policies, and procedures that manage the conduct of safety activity. Safety Engg. and safety management make up an integrated whole. Safety engineering is the physical and mathematical side of injury and damage prevention, whereas safety management is the administrative or software side of such prevention. Safety management provides the structure within which the techniques of safety engineering are applied.

DOI: 10.1201/9781003274346-4

Accident: An accident means a dynamic (multi-causal) event that begins with the activation of a pre-existing hazard that flows through its host system in a logical sequence of preceding events, factors, and circumstances to produce a final loss event (typically the personal injury of the system operator).

In the past, oversimplified labels such as 'unsafe acts' and 'unsafe conditions' were seen as factors. Focus on 'unsafe acts' leads to neglect of control of unsafe conditions. As a result, potentially more important root causes related to system design are overlooked.

Hazard: This is a source of potential harm, or a situation with a potential for causing harm, in terms of human injury; damage to health, property, the environment, and other things of value; or some combination of these.

Hazard identification: The process of recognizing that a hazard exists and defining its characteristics.

Risk: The chance of injury or loss is defined as a measure of the probability [likelihood] and severity of an adverse effect on health, property, the environment, or other things of value.

Risk analysis: The systematic use of information to identify hazards and to estimate the chance for and severity of injury or loss to individuals or populations, property, the environment, or other things of value.

Probability: This is the likelihood of a specific outcome, measured by the ratio of specific outcomes to the total number of possible outcomes.

Exposure: This is a measure of the number of occurrences in a given time and frequency of exposure to an event.

A consequence can be defined as the outcome of an event or situation, such as a loss, injury, or even a gain. The loss events could include death, serious injury, first aid treatments, acute or chronic disease, loss of production, equipment damage, environmental damage, and loss of reputation. The consequences (or impacts) harm health, property, the environment, or other things of value.

Vulnerability: People, property, infrastructure, industry and resources, or environments that are particularly exposed to adverse impact from a hazard event.

In risk assessment, the words *Hazards* and *Risks* are often used and it is necessary to be clear about what *hazards and risks* are:

- A *hazard* is anything that has the potential to cause harm.
- The *risk* is how likely it is that a hazard will cause actual harm.

4.2   Safety management in mines

The principal responsibility for the safety and good health of workers employed in mines rests with the management of that mine. Safety Management is a structured process composed of well-defined systems that emphasize continuous improvement in work quality, health, welfare, and productivity through the setting up of improved safety standards and their effective implementation and administration. Using the synchronized approach of management for accomplishing safety in mines lies the planning, organizing, directing, leading, executing, and controlling. All these acts are guided towards safety compliance.

Several different approaches to mine safety have been sincerely tried, but with limited success only. However, the Safety Management System approach has been tried successfully in US and Australian mines, with a notable reduction in accidents over a length of time. This is based on the concept of 'Risk management' through 'Risk assessment'. The Mines Safety Directorate in India has introduced this concept in Indian mines. This system is based on risk assessment of all hazardous operations, equipment, and machinery, considering the procedures used, maintenance, supervision, and management. The approach is having participative management involving the entire staff in the safety improvement programme with clearly defined responsibility and accountability.

The risk assessment process aims at identifying all the existing and probable hazards in the work environment and in all operations and assessing the risk levels of the hazards to prioritize what needs immediate attention for redressal. Then for managing these risks, different mechanisms responsible for these hazards are identified and their control measures, set to a timetable, are recorded pinpointing the responsibilities. The system devises monitoring and auditing at regular intervals to ensure that safe operating procedures are followed, evaluated, corrected, and standardized. A documented training procedure, for workers and executives, is carried out regularly, and commitment to health and safety is demonstrated at all levels of an organization (Heinrich, 1959).

Safety Management System could be broadly considered to include:

1. Senior Management Commitment
2. Safety Policy
3. Safety Information
4. Safety culture
5. Setting Safety Goals, clearly identifying the responsibility and accountability of various personnel
6. Hazard Identification and Risk Management
7. Establishing a Safety Reporting (MIS) System based on Accident Statistics collection, analysis & actions – preparation of method statement for each process viz. roof control and gallery development
8. Safety Audit/Assessment
9. Accident and Incident Reporting and Investigation (Tripathi, 2001)
10. Safety Orientation and Recurrent Training
11. Emergency Response Plan
12. Documentation of Safety

Accident and Incident Reporting and Investigation:

Senior Management Commitment: It plays a major role in determining the company's safety culture. Any safety programme will be ineffective without the wholehearted commitment of the management.

4.3 Safety policy

Senior management commitment will not lead to positive action unless that commitment is expressed as direction. Senior management must develop and communicate a safety policy that allocates responsibilities and holds people accountable for meeting safety performance goals. For this purpose, if necessary, job redesign can be effected by senior management for managing safety at the workplace. The responsibility and accountability towards safety have to be fixed within the organization.

Safety Policy should include, at a minimum:

- a clear declaration of commitment and objectives;
- a means for setting safety goals and regular review of safety performance;
- clear statements of responsibility applying to every department or functional area in the organization;
- clearly stated accountabilities converging at the top of the organization;
- a means for ensuring compliance with regulations;
- a means for ensuring adequate safety management knowledge and skills at all levels; and
- compatibility or integration with other management systems.

4.4   Safety information

It is a communication process wherein information is recorded, stored, processed, and retrieved for decision-making. Information processing is a very important activity. The information forms the basis for making decisions. Therefore, management must establish a system to collect and analyse safety data. This would include:

- safety goals and evaluation of progress towards those goals;
- records of accidents and incidents including internal/external investigation findings and corrective actions;
- safety concerns raised by employees including analysis and resultant action;
- results of safety reviews and audits, and when appropriate, corrective action; and
- records of all safety initiatives or interventions.

Managers and employees should also be keeping abreast of the ongoing developments in safety management, work practices, and technology development outside. This is accomplished by subscribing to safety-related publications, making relevant accident investigation reports available, and encouraging staff to participate in safety-related training, seminars, and workshops.

4.5   Safety culture

The enduring value and priority placed on workers and public safety by everyone in every group at every level of an organization. It refers to the extent to which individuals and groups will:

1. Commit to personal responsibility for safety.
2. Act to preserve, enhance and communicate safety concerns.
3. Strive to actively learn, adapt, and modify (both individual and organizational) behaviour based on lessons learned from mistakes.
4. Be rewarded or held accountable in a manner consistent with these values.

Non-prevailing safety culture fructify due to

- Lack of safety concerns
- Operational pressures
- Poor leadership
- Conflict with management
- Negative organizational climate
- Morale/job satisfaction

Compliance with regulated aspects of safety (e.g. training requirements, manuals and procedures, and equipment maintenance), as well as the coordination of activity within and between teams/units.

Going Beyond Compliance – Priority is given to safety in the allocation of company resources (e.g. equipment, personnel time) even though not required by regulations.

Organizational Commitment: The degree to which an organization's senior management prioritizes safety in decision-making and allocates adequate resources to safety (ICMM, 2009).

4.6   Hazard identification and risk management

Risk is the chance of injury or loss. This concept includes both the likelihood of a loss and the magnitude. Various approaches to risk management are prevailing. Risk probability and risk severity approaches are combined to arrive at a safety risk matrix, leading to acceptance or rejection of the practice. Every activity/operation should be split into elementary activities, the probability and severity of occurrence judged/calculated, and safety risk worked out; methodology for safe working is worked out, standard safe procedure is drawn, and use of special tools, remote operation, etc., is spelt out; this is known as the *method statement* or *safety management plan*. The whole recommended procedure is documented in the form of a manual, and work persons are trained in the procedure. The continuous improvement approach based on the PLAN-DO-CHECK-ACT strategy is more often adopted to achieve better results. Policy – Planning and Organization – Implementation – Measurement and Evaluation – Management Review model is adopted in such cases (Figure 4.1).

The various phases of the *risk management process* are elaborated below.

4.6.1   Establishing the context

A team of individuals doing, supervising, or otherwise, associated with the activity, as well as a fresh pair of eyes are selected to draw the specification of the activity, its objectives, key stakeholders, and the critical success factors. The key stakeholders may include the company its shareholders & staff, Customers for the product/service, user groups, principal, suppliers, contractors, regulatory, licensing and approval authorities, people affected, the government, and special interest groups. Using the requirement and the identified key stakeholder, critical success factors for the activity are derived. Say for instance, for safety and health aspects, the key success factor shall be to maintain the highest standards. Now, based on the critical success factors, the broad categories of risk consequences by which risk will be analysed are defined which

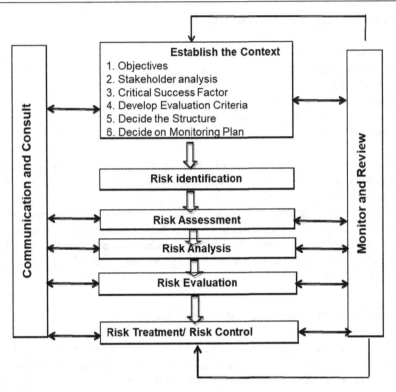

Figure 4.1 Overview of the risk management process.

include safety, environment outrages/reputation, costs, and business interruption. The list of key elements of the activity is made which is then used for structuring the risk analysis, which in turn decides on the monitoring and review method (Anon, 1999).

4.6.2   Risk identification

Risk identification is the process of identifying risks and their causes and determining what, how, and why things may go wrong. It is necessary to collect and document several items of information:

*   A brief description of the risk
*   What can happen?
*   How it can happen?
*   The class of risks (Which key element).

A list of all possible risks is generated using one of the following methods:

*   Brainstorming workshop for qualitative and quantitative risk assessment
*   Fault Tree Analysis
*   HAZOP: Hazards and Operability Studies
*   Job Hazard Analysis

- Checklists
- Specialist techniques

Risk identification is the most significant step in the risk management process. If risks are not identified, they cannot be managed.

4.6.3   Risk analysis

Risk Analysis involves estimating the likelihood that things may go wrong and the potential consequences for the objective and critical success factor of the activity. The final stage of risk analysis involves the estimation of the risk levels, i.e. risk rating. The estimation of uncertainty in the likelihood estimate, the consequence estimate, and the risk estimate are also applied. If uncertainty is high, this gives a hint that further work is required or some form of early warning technique may be required. Figure 4.2 depicts the process of risk analysis.

4.6.3.1   Consequence analysis

After identifying the range of hazards, the next step is to obtain an estimate of the severity or consequence of the events, should they occur. Consequence analysis concentrates on impacts in one or more areas. Areas of impact may include physical assets and resource base; human assets; revenues; direct and indirect cost of activity, well-being, health and safety of employees; business interruption; the environment and intangibles like reputation/image, goodwill, and quality of life. As an example, qualitative measures used for a consequence, if a hazard event occurs, are tabulated.

4.6.3.2   Likelihood analysis

The likelihood of an event is used as a qualitative description of probability and frequency. The frequency of an event is a measure of the rate of occurrence of an event expressed as the number of occurrences of an event in a given time. In risk assessment,

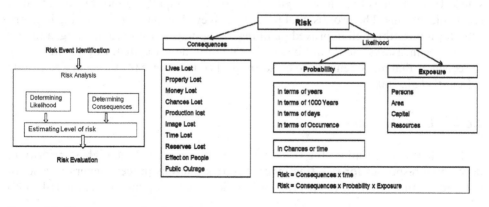

Figure 4.2 The process of risk analysis.

Table 4.1 Quality Measures used for Consequences, if a Hazard Event Occurs

Code	Descriptor	Safety	Community outrage	Environmental	Replacement or repair cost	Business interruption
1	Insignificant	No injuries	Individual concern	Minor or no pollution	< US$ 1 M	1 hour to 1 day
2	Major	First aid treatment	Involves small group concern	Some pollution	US$ 1 M to US$ 5 M	1 day to 1 week
3	Medium	Medium treatment required	Involves widespread local community concern	Serious pollution	US$ 5 M to US$ 10 M	1 week to 1 month
4	Major	Extensive injury or one fatality	Province or nationwide significance	Major environment release	US$ 10 M To US$ 20 M	1 month to 6 month
5	Catastrophic	More than one fatality	External (out of country significance)	Catastrophic environmental event	> US$ 20 M	Site Closure

the given time frame is usually taken as one year. There the frequency can be expressed as the number of occurrences per year. The probability of the event is the likelihood of that specific event, measured by the ratio of specific events to the total number of possible events. Probability is expressed as a number between 0 and 1. With 0 indicating an impossible event or outcome and 1 indicating an event or outcome is certain. Three basic approaches are used for likelihood analysis (Table 4.1).

Accidents could be understood as the results of hazard/danger acting on the 'Object'. The object may be a man, machine, or material. This action will result in damage to the man, machine, or material. Hence, to prevent an accident or injury, appropriate steps are necessary to stop the hazard/danger acting and/or stop the action of hazard/danger before it damages the object. The most valuable asset of the industry is human resources and the priority of the mining industry is to take care of their health and safety. Human life is the supreme gift of god and principally cannot be compared in economical terms. The professional progress achieved in making the mining industry safer for miners through improved planning, designing, staffing, controlling, and regulating the measures should not be overlooked or underestimated. But it is necessary to recognize that much remains to be accomplished to achieve the target of a safe mine (Michael et al., 2003).

4.6.4   Risk evaluation

Risk evaluation determines the risks that should be accorded the highest priority in developing responses for risk treatment. The risk analysis process generates a set of risk ratings that are used to set priorities. Risk evaluation aims to sort the risks into

Table 4.2 Determination of Total Risk using a Quantitative Approach

Determination of R_P		Determination of R_C		Determination of R_E		Determination of R_T	
Level of Probability	Probability risk rating (R_P)	Consequence of risk	Consequence of risk rating (R_C)	Exposure level	Exposure risk rating (R_E)	Total risk level	Total Risk
Frequent	5	Disastrous	4	Very high	5	1–20	Very low risk
Possible	4	Critical	3	High	4	21–40	Tolerable risk
Occasional	3	Marginal	2	Moderate	3	41–60	Moderate risk
Small	2	Minor	1	Low	2	61–80	High risk
Improbable	1	–	–	Very Low	1	81–100	Very high risk

groups like *extreme risk, high risk, substantial risk, moderate risk, and low risk* that determine the level of management response and effort required.

We all make a judgement about risks according to our perceptions and beliefs. Within the company, we undertake risk studies in a structured way, making use of information, judgement, and experience from a range of sources, so we can have a degree of confidence in our conclusions. Since the perceptions of risk evaluations are that of the group and may not necessarily match with that of a community, there is a possibility that one may meet unexpected resistance to the plans. Table 4.2 illustrates a risk matrix, where exposure and probability scores vary from 0 to 10 and consequence score varies from 0 to 100. In this risk rating, any score above 400 must be eliminated, and care should be taken immediately to reduce the risk. Another method of determination of Total Risk using the different ratings are Probability Risk Rating (Rp), Consequence of Risk Rating (Rc), and Exposure Risk Rating (R_E) (Table 4.2).

4.6.5 Risk treatment

Risk treatment involves establishing and implementing appropriate management responses for significant and high risks. The aim of treating risks is to bring them to a level that is acceptable/tolerable or as low as reasonably practical.

There are several measures, which can be employed to control risks, and these can be listed in order of preference as follows:

- **Elimination:** Modify the process, method, or material to eliminate the Hazard (100% effective).
- **Substitution:** Replace the process, method, or material with a less hazardous one (75% effective).
- **Separation:** Isolate the hazard from persons by guards, spaces, or time separation (50% effective).
- **Administration:** Adjust time or conditions of risk exposure (30% effective).

- **Training:** Improve skills to make tasks less hazardous (20% effective).
- **Personal protective equipment:** Appropriately designed and fitted safety apparel (5% effective).

The controlling process defines the action, including whether the action is going to eliminate the hazards, reduce the chance or severity of an event, or be of acceptable risk. Who will carry out the action, when the action will be taken, and who will check or audit the process? (Paliwal and Jain, 2001; Tripathi, 2001).

4.6.6 Monitoring and review

Monitoring and review involve continuous monitoring and review of the risk profile and risk treatment implementation progress and aim to ensure that the risk management process reacts to the dynamic nature of risks and it continues throughout the life of the activity (Figure 4.3).

$$\text{Total risk } (R_T) = R_P \times R_C \times R_E$$

where R_P is the probability of occurrence, R_C is the consequence of risk, and R_E is the exposure level to risk.

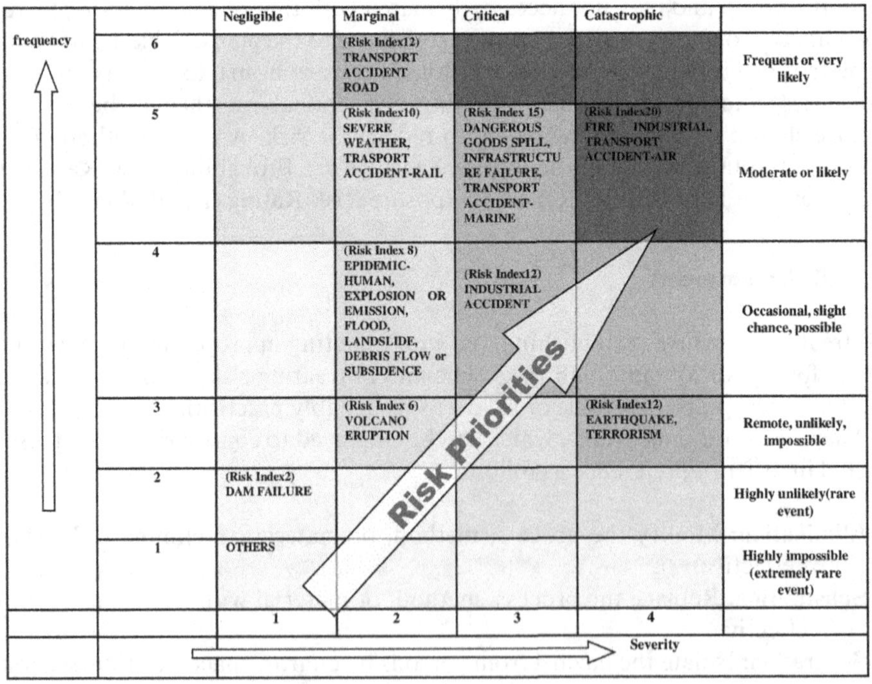

frequency		Negligible	Marginal	Critical	Catastrophic	
	6		(Risk Index12) TRANSPORT ACCIDENT ROAD			Frequent or very likely
	5		(Risk Index10) SEVERE WEATHER, TRASPORT ACCIDENT-RAIL	(Risk Index 15) DANGEROUS GOODS SPILL, INFRASTRUCTURE FAILURE, TRANSPORT ACCIDENT-MARINE	(Risk Index20) FIRE INDUSTRIAL, TRANSPORT ACCIDENT-AIR	Moderate or likely
	4		(Risk Index 8) EPIDEMIC-HUMAN, EXPLOSION OR EMISSION, FLOOD, LANDSLIDE, DEBRIS FLOW or SUBSIDENCE	(Risk Index12) INDUSTRIAL ACCIDENT		Occasional, slight chance, possible
	3		(Risk Index 6) VOLCANO ERUPTION		(Risk Index12) EARTHQUAKE, TERRORISM	Remote, unlikely, impossible
	2	(Risk Index2) DAM FAILURE				Highly unlikely(rare event)
	1	OTHERS				Highly impossible (extremely rare event)
		1	2	3	4	

Risk Priorities

Severity

Figure 4.3 Risk characterization.

4.7   Fundamentals of accident prevention

4.7.1   Basic activities

Successful accident prevention requires a minimum of four fundamental activities:

1. A study of all working areas to detect and eliminate or control physical hazards which contribute to accidents.
2. A study of all operating methods and practices.
3. Education, instruction, training, and discipline to minimize human factors which contribute to accidents.
4. A thorough investigation of accidents/injuries to determine contributing circumstances.

4.7.2   Accidents are preventable

1. Many persons, either through ignorance or misunderstanding, unfortunately, believe that accidents are the inevitable results of unchangeable circumstances, fate, or a matter of luck.
2. It must be emphasized that accidents do not happen without cause and that the identification, isolation, and control of these 'causes' are the underlying principles of all accident prevention techniques.
3. No person in a supervisory position can be effective in his job of accident prevention unless he fully believes that accidents can be prevented and constantly strives to achieve this result.

4.7.3   Causes of accidents

1. Causes of accidents are divided into three major categories:

    a. unsafe acts of people;
    b. unsafe physical or mechanical conditions; and
    c. factors beyond control or unavoidable.
        Statistics indicate that nearly 88% of all accidents are caused by unsafe acts of people, 10% by unsafe conditions or mechanical failures, and 2% by factors beyond the control or unavoidable.

The basic causes for high accidents or injury rates are *unsafe acts* and *unsafe conditions* or *both*. Salients of these are as given hereunder.
    Unsafe acts: Any unsafe act(s) mainly arise through human error or behaviour-related causes due to ignorance or lack of alertness.

*   Failure to follow instructions of proper job procedure
*   Operating without authority, failure to secure or warn
*   Cleaning, oiling, adjusting, or repairing equipment that is moving, electrically energized, or pressurized
*   Failure to secure or warn, particularly during blasting

- Operating or working with unsafe speed
- Making safety devices inoperative for the sake of personal convenience
- Using unsafe equipment, hands instead of tools or using equipment unsafely
- Unsafe loading, placing, mixing, combing, etc.
- Taking an unsafe position or posture
- Working on moving or dangerous equipment, on running conveyor, train, etc.
- Using tools or equipment known to be unsafe
- Driving and operating errors
- Improper attitude, distracting, teasing, abusing, instigating, provoking, etc., while working
- Failure to use safe attire or Personal Protective Equipment (PPE)
- Lack of knowledge, skill, coordination, or planning
- Temporary lack of safety awareness at the time of the accident
- Physical or mental defects
     Personal Characteristics like age, job experience, visual function, perception, reaction time, cardiovascular disorders, neuro-psychiatric disorders, and other physiological disorders may lead to unsafe acts.

Unsafe conditions: Unsafe conditions may occur due to insufficient mine design, unanticipated rock behaviour and geology, ill-maintained equipment, inadequate supervision, or a combination of these factors. Improvement or elimination of hazardous conditions is necessary for reducing injuries and improving safety. Most unsafe (or hazardous) conditions can be grouped into one of the following classifications:

- Inadequately guarded or unguarded machinery, equipment, work areas and unsupported workforce
- Defective condition, rough, sharp, slipper, decayed, corroded, frayed, cracked, etc.
- Defective, inferior, or unsuitable tools, machinery, equipment, or materials
- Unsafe design and construction
- Hazardous arrangement processes like piling, storage, restricted space, exits, layout, overload, misalignment
- Unsafe illumination (inadequate or unsuitable)
- Unsafe ventilation (inadequate or unsuitable)
- High temperature, humidity, noisy, and dusty environment
- Steep inclination, wet and slippery mine roadways
- Low or excessive heights of the face
- Unsafe dresses, clothes
- Unsafe method, process, planning, etc.
- Placement hazards (person not mentally or physically compatible with job requirements).

Enforcement of the safety legislation improves safety. (Kejriwal, 1994; Prasad and Rakesh, 1990)

Regulation: Legal mandatory requirements (Acts, Regulations, and Rules) concerning such matters as general working conditions in mines, the design, construction, maintenance, inspection, testing and operation of machines and equipment, the duties of employers, supervisors, miners, medical supervision, first aid, and medical examination.

Education and training: Providing the knowledge of safety in formal and non-formal education and imparting the training to the miners, particularly for the new entrants and refresher training.

Persuasion: This is done by employment of various methods of publicity, campaign, appeal, mock, meetings, safety weeks, etc.

Leadership: Good leadership of safety officials and mine managers certainly matters for improving safety. The leader is ultimately responsible for providing the leadership, systems, and processes for the prevention of fatalities. The actions of leaders are fundamental to the elimination of fatalities. Strong and consistent leadership that demonstrates every day a continuous commitment to safe and fatality-free production will drive us to zero fatalities (Heinrich, 1959; Michael et al., 2003).

4.8 Statutory compliance

Over the past two decades, there has been a growing appreciation of the many and varied ways that people contribute to accidents in hazardous industries like mining or simply in everyday life. Not long ago, most of these would have been lumped together under the catch-all label 'human error'. Nowadays, it is apparent that this term covers a wide variety of unsafe behaviours. Most people would agree with the fact that old age errs humans and human beings are frequent violators of the 'rules' whatever they might be. But violations are not all that bad – through constant pushing at accepted boundaries, they got us out of the caves! Assuming that the rules, meaning safe operating procedures, are well-founded, any deviation will bring the violator into an area of increased risk and danger. The violation itself may not be damaging, but the act of violating takes the violator into regions in which subsequent errors are much more likely to have bad outcomes. This relationship can be summarized quite simply by the following equation:

Violations + errors = Injury, death, and damage

The resultant situation can sometimes be made much worse because persistent rule violators often assume, somewhat misguidedly, that nobody else will violate the rules, at least not at the same time! Violating safe working procedures is not just a question of recklessness or carelessness by those at the sharp end. Factors leading to deliberate, non-compliance extend well beyond the psychology of the individual in direct contact with working hazards and include such organization and related issues such as:

- The nature of the workplace
- The quality of tools and equipment
- Whether or not supervisors or managers turn a 'blind eye' to getting the job done

Concerning compliance, getting to grips with the human factor; the organization's overall safety culture, or indeed its absence; the quality of the statutes, i.e. rules, regulations and procedures, matters a lot and is important too.

Table 4.3 Errors and Violations

Errors	Violations
Stem mainly from *informational* factors: Incorrect or incomplete knowledge, either in the head or in the world.	Stem mainly from *motivational* factors: Shaped by attitudes, beliefs, social norms, and organizational culture.
They are *unintended* and may be due to a memory failure (a 'lapse') or an attentional failure (a 'slip').	They usually involve *intended* or deliberate *deviations* from the rules, regulations, and safe operating procedures.
They can be explained by reference to how *individuals* handle information.	They can only be understood in a *social context*.
The likelihood of mistakes occurring can be reduced by *improving* the *relevant information*: training, roadside signs, the driver–vehicle interface, etc.	Violations can only be reduced by *changing attitudes,* beliefs, social norms, and organizational cultures that tacitly condone non-compliance (culture of evasion).
Errors can occur in any situation. They need not of themselves, incur risk.	Violations, by definition, bring their perpetrators into areas of increased risk, i.e. they end up nearer the 'edge'.

Violations are usually deliberate, but can also be unintended or even unknowing. They can also be mistaken in the sense that deliberate violations may bring about consequences other than those intended. The distinction between errors and violations is often blurred, but the main differences are shown in Table 4.3.

Classifying errors: Error types committed by us can be classified at three levels as described below:

- At the *skill-based level,* we carry out routine, highly practised tasks in a largely automatic fashion, except for occasional checks on progress. This error is what people are very good at most of the time.
- We switch to the *rule-based level* when we notice a need to modify our largely pre-programmed behaviour in line with some change(s) in the situation around us. This problem is often one that we have encountered before and for which we have some pre-packaged solutions. It is called rule-based because we apply stored rules of the kind if (this situation), then do (these actions).
- The *knowledge-based level* is something we come to very reluctantly. Only when we have repeatedly failed to find a solution using known methods, we do resort to the slow, effortful, and highly error-prone business of thinking things on the spot conditions. In an emergency, in a scary & fearful situation, or when very strong emotions persist, a knowledge-based error is committed. Given time, trial & error, learning can often produce good solutions.

Classifying violations: Case and field studies suggest that violations can be grouped into four categories namely routine violations, optimizing violations, situational violations, and exceptional violations. They are described below:

Routine violations: Almost invisible until there is an accident (or sometimes as the result of an audit), routine violations are promoted by a relatively indifferent environment, i.e. one that rarely punishes violations or rewards compliance – 'we do it like this all the time and nobody even notices'.

Optimizing violations: Corner-cutting, i.e. following the path of least resistance, sometimes also thrill-seeking – 'I know a better way of doing this'.

Situational violations: Standard problems that are not covered in the procedures – 'we can't do this any other way'. An excellent example concerns *railway shunters.* The rule book prohibits railway shunters from remaining between wagons when wagons are being connected. Only when the wagons are stopped, can the shunter get down between them to make the necessary coupling. On some occasions, however, the shackle for connecting the wagons is too short to be coupled when the buffers are fully extended. The job can only, therefore, be done when the buffers are momentarily compressed as the wagons first come in contact with each other. Thus, the only way to join these particular wagons is for the shunters to remain in between two wagons during the connection. This violation result can be fatal.

Exceptional violations: Such violations are a result of unforeseen and undefined situations – 'now this is what we got trained for'. A simple example illustrates this violation, e.g. a pair of engineers were inspecting a pipeline on an oil rig. One of them jumps into an inspection pit and is overcome by H_2S (hydrogen sulphide) fumes. His companion fully trained to handle such situations raises the alarm, but when he jumps down to help his partner, he too is overpowered. Exceptional violations often involve the transgression of general survival rules rather than specific safety rules. Survivors of such exceptional violations are often treated as heroes and get rewarded for exceptional violations. Exceptional violations can also be seen as an exercise of initiative even if sometimes you get away with it (Hansen, 2009).

The relationship between these two with the performance levels is summarized in Table 4.4.

The safety manager or safety officer knows the error and violations yet does not enforce the hazards. As a standard safety practice

- Everyone is fallible and capable of bending rules.
- All system has technical and procedural shortcoming that endangers safety.
- Whatever you do, there's always something beyond your control that can hurt you or endangers your safety.

Seven Avenues: Through these avenues (seven), we can initiate countermeasures for safety. None of these areas overlap and they are:
i. Safety management errors: Training, Education, Motivation, and Task design.
ii. Safety programme defect: Revise information, Collection, Analysis, and Implementation.
iii. Management/Command Error: Corrected through unity of command, training, providing information, setting proper communication channels, and System defects.

Table 4.4 Performance, Error, and Violation Types

Performance levels	Error types	Violation types
Skill-based	Slips and lapses	Routine violations
		Optimizing violations
Rule-based	Rule-based mistakes	Situational violations
Knowledge-based	Knowledge-based mistakes	Exceptional violations

iv. Design Revision: Via – Sop, Regulations, Policy Letters, and Statements.
 v. Operating Error: Engineering, Training, and Motivation.
 vi. Mishap: Protective Equipment, Barriers, and Separation.
vii. Result: Containment, Firefighting, Rescue, Evacuation & First Aid.

An accident or an injury is the result of various causes acting in a particular way and sequence. To make mine a safe working place, accident prevention and control measures are must. This can be accomplished by controlling the cause of an accident or the root from where it is triggered.

4.9   Control of accident causes

There are three main methods utilized in the control of accident causes. They are (a) Engineering, (b) Education and training, and (c) Enforcement. These three methods, sometimes referred to as the three E's of safety, are outlined below.

Engineering: Environmental causes of accidents or unsafe conditions can be eliminated through the application of engineering principles. When an operation is mechanically and physically safe, it is unnecessary to be as concerned about the uncertain behaviour (unsafe acts) of people. Machines are less apt to fail than men. It may be necessary to make mechanical revisions or modifications to eliminate existing unsafe conditions, and in some cases, to prevent unsafe acts. The design of machine guards, automobile brakes, traffic signals, pressure relief valves, and handrails are varied examples of safety engineering at work.

Education and training: Just as safety engineering is the most effective way of preventing environmental accident causes (unsafe conditions), safety education is the most effective tool in the prevention of human causes (unsafe acts). Through adequate instruction, personnel gain useful knowledge and develop safe attitudes. Safety consciousness developed in personnel through education will be supplemented and broadened by specific additional instruction in safe working habits, practices, and skills. Training is a particularly important accident prevention control; it gives each employee a personal safety tool by developing in him the habits of safe practice and operation.

Enforcement: Usually, accidents can be prevented through adequate safety engineering and education. However, some people are hazards to themselves and others because they fail to comply with accepted safety standards. It is these persons for whom the strict enforcement of safety practices is necessary, backed by prompt corrective action. No organized accident prevention effort can be successful without effective enforcement because accidents are frequently the direct result of violations of safety principles. This is particularly true of vehicle accidents, many of which are caused by unsafe acts which constitute traffic law violations. Heads of departments and supervisors are responsible for enforcing safety standards and regulations. Failure to do so would be condoning conduct that leads to preventable accidents.

To be completely effective, accident prevention controls cannot be applied 'hit or miss'. All engineering, education, training, supervision, and enforcement measures will be directed towards the solution of specific problems based on the collection of facts relating to unsafe acts or unsafe conditions.

4.9.1 Control of work habits

1. Regardless of the degree of safety built into a job, unsafe actions on the part of human beings will always be a cause of injuries. Teaching employees good work habits means showing them how to do their tasks with less risk to themselves, less spoilage (of materials), and less damage to equipment. Much of this instruction can be reduced to a few simple principles or job rules. By concentrating on these, by showing the 'why' as well as 'how', and by constantly supervising to correct promptly, safe work habits can obtain acceptance by employees.
2. Whenever possible, actual demonstrations of right and wrong ways of doing tasks should be conducted, always accompanied by the basis for preferring one work habit to another. Fully important as the initial instruction is the watchful eye on subsequent performance. When the right way has been presented and agreed to by the individual worker, failure to comply must be noted.
3. It may be desirable to insist that a certain step be repeated or a job is redone, simply to emphasize the seriousness with which safety rules should be met with appropriate disciplinary action, including discharge if necessary. No matter how skilled an employee may be in performing their duties, if he or she does not perform them safely, they are not considered to be a worthy employee.

4.9.2 Safety orientation of new employees

1. When a new employee comes to work, he immediately begins to learn things and form attitudes about the job, their boss, and fellow employees. If the department head, supervisor, and fellow employees appear to be unconcerned about accident prevention, she/he will most probably believe that safety is unimportant.
2. To form a good safety attitude, the new employee must be impressed by everyone's concern for the prevention of accidents at the time she/he starts to work. Workers must be told that unsafe workers will not be tolerated and that they will be required to obey safety rules and instructions, wear protective equipment whenever required, and attend safety meetings to continue as an employee of the organization.
3. It will never be taken for granted that previous experience and apparent qualifications mean that 'somewhere along the way' the new employee has learned to experience does not automatically exempt a newly hired vehicle operator from being thoroughly instructed in safe driving practices. He/she must be made aware of what is expected of him/her in their capacity of operating a company vehicle and/or equipment, and they must be checked to assure that they can operate the vehicle and/or equipment properly.
4. The supervisor will review safety rules and procedures with the new employee, pointing out the possible hazards involved in doing the job. If possible, the new employee should be assigned to work with a safety-minded employee during the first few weeks. The new employee should be checked at frequent intervals, asked about any problems that may have arisen, and reminded of safe practices. Any tendency to overlook safety procedures should bring a prompt and vigorous warning or other appropriate action.

From the foregoing description, it is quite clear that a large number of accidents and injuries are preventable if proper attention is given to various levels of management of the mining company to implement the safety management concept and develop the safety habit and culture while taking compliance into consideration (Heinrich, 1959; Kalia et al., 2015).

4.10   Cost of accidents/mine accidents

A: Costs to the employer
1. Interruption and loss in production, loss of revenue, loss of contract and market.
2. Sickness or other payments made including ex-gratia immediately or subsequent absence of miner due to injury.
3. Compensation paid under legal liability to the man or dependents.
4. The expense of investigating the accident, and in some cases, preparing and defending a case.
5. The cost of obtaining, training, and deploying a replacement for an injured man and loss of efficiency in production resulting from new manpower.
6. Loss or damage to plant and equipment caused by accident.
7. The extra cost of remedial measures to insure against further accidents.
8. The cost incurred or losses sustained due to lower morale or friction with work people. Workers may develop fear psychosis and abstain from work. The peace of the mine may get disturbed.

B: Costs to the state
1. Loss of royalty, cess, and tax due to the loss in production and reduced profitability of the employing undertaking.
2. The cost of hospital or other medical treatment including rehabilitation and retraining.
3. Tax refund if paid earlier in advance.
4. The number of supplementary benefits payable to the man and his dependents.
5. The cost of public services of investigating the accident, trying any legal issues, and enforcing the judgement.
6. The amount of any benefit paid under the industrial injury scheme and sickness benefit.

C: Costs to society at large
1. The net loss, temporary or permanent, of goods or services provided by the injured person or by the others whose services are required to look after him.
2. Costs fall on the injured man himself, friends, dependents, and insurance companies.
3. The cost of investigations and cases for claims of damages if they are not provided by the state or employer.

4.10.1   Classification of accidents costs

Direct Costs for the Employee:

−   Lost wages and overtime
−   Doctor and hospital bills

Indirect Costs for the Employee:

– Physical pain and suffering
– Mental anguish
– Lost time with family and friends
– Loss of productivity on and off the job
– Relationship strain

Direct Costs for the Employer:

– Medical bills and workers' compensation claims
– Legal costs
– Insurance costs
– Property damage costs
– Wages being paid for a sideline worker

Indirect Costs for the Employer:

– Loss of a valued employee
– Loss of productivity
– Replacing the lost worker (e.g. hiring and training costs)
– Damage to equipment or tools
– Time it takes to handle the injury claim
– Decrease in employee morale over the loss of an employee

4.11   Accident investigations and analysis

Mining is a hazardous occupation and accident control constitutes a very vital part of mining operations. An accident prevention programme is meaningless unless the root causes of the accidents are known. An intelligent and unbiased investigation into an accident answers a very important question 'why the accident occurred'? Once an answer to this basic question is known, then it becomes much easier to arrive at accident prevention strategies for improving safety standards in mines. Therefore, data generated from accident investigations provide important input in the process of formulating programmes for improving safety standards.

For every accident, there are many contributory factors. These factors combine in a random fashion causing accidents. Further, it must be clearly understood that the accidents are not problems but are only the symptoms of certain problems existing at the mine/mines. To eradicate accidents, these problems need to be identified and then corrective action should be taken. An in-depth investigation into every accident helps in the identification of such problems. Therefore, the basic objective of an accident investigation should be preventing the recurrence of similar accidents in the future and not finding out the lapses and pinpointing responsibility for taking action against guilty persons. The emphasis of the investigation should be on the technical aspects and not the administrative aspects. If this is kept in view, then the investigation will yield valuable information that will help improve safety standards.

4.11.1   Problems associated with accident investigations

Accidents have been compared with criminals who leave little evidence of their crimes. An investigator has to reconstruct the scenario existing before the occurrence. The means available to him are witnesses in some cases, an inspection of the site of the accident, documents and reports (statutory and non-statutory), plans, etc. The site of the accident is normally disturbed for carrying out rescue and recovery operations and to prevent further danger. The eyewitnesses are sometimes tutored to save officials from getting the blame and even records may be manipulated to mislead the investigating officer. Therefore, an investigation into an accident is a tricky process with many pitfalls. If the investigation is commenced soon after the occurrence, there are better chances of getting an undiluted view. Therefore, the investigation must start as soon as possible. At least the investigator can visit the site of the accident immediately and examine the witnesses later.

4.11.2   Who should conduct an investigation?

As elaborated above, accident investigation is a tricky affair that needs a lot of skill and experience. Hence, the investigation should be entrusted to an intelligent person with an analytical approach and work experience in conditions similar to those of the mine where the accident occurred.

4.11.3   Procedure: Following is the standard procedure for a normal accident investigation

1. Collection of basic information: The basic information required includes:

    *   Date and time of the accident and the hours of work put in by the victim
    *   Name, age, sex, and nature of the job of the victim
    *   Details of training undergone by the victim
    *   Place of accident
    *   Details of mine workings and operations related to the accident
    *   System of supervision and names of supervisors
        This data can be collected from the mine office.

2. Inspection of the site of the accident: The site of the accident should be inspected as early as possible after the accident so that intentional or unintentional tampering with the evidence is the least. The investigators should make a very close inspection and note down then and there the minutest details. It is helpful to prepare a hand sketch of the site and note down the location of various relevant objects on it. The details to be looked for will depend upon the type of accident. For example, if it is an accident due to a roof fall, the position and length of props or roof bolts may be noted together with the height of the place of the accident. The condition of props or bolts will give an idea of whether the same has been kept there after the accident or was there earlier. The position of slips or planes of weakness should be noted. In case of an accident due to blasting, the details to be noted will be different.

It is always useful to have some eyewitnesses accompany the investigator so that they can pinpoint the exact location of operations and different persons at the time of the accident at the site.

A surveyor should invariably accompany the investigating officer so that he can be given necessary instructions regarding plans and sections to be prepared and details to be recorded thereon. Plans and sections not only make the job of writing a report easier but are also vital for understanding the details of the accident easier by one who has not visited the site. An accident report must have a site plan and section where required. If possible, a few photographs of the site of the accident may also be taken. In this connection, the provisions of the Regulations are as follows:

'Place of accident not to be disturbed – (1) Whenever there occurs in or about a mine an accident causing loss of life or serious bodily injury to any person, the place of the accident shall not be disturbed or altered before the arrival or without the consent of Chief Inspector or the Inspector to whom notice of the accident is required to be given under subsection (1) of Section 23 of The Mines Act, India, unless such disturbance or alteration is necessary to prevent any further accident, to remove bodies of the deceased, or to rescue any person from danger, or unless discontinuance of work at the place accident would seriously impede the work of the mine: Provided that, where the Chief Inspector or the said Inspector fails to inspect the place of accident within 72 hours of the time of the accident, work may be resumed at the place of accident. (2) Before the place of an accident involving a fatal or serious accident is disturbed or altered due to any reason whatsoever, a sketch of the site, illustrating the accident and all relevant details, shall be prepared (in duplicate) and such sketch shall be duly signed by the manager or assistant manager, safety officer, surveyor, and the Workmen's Inspector, or, where there is no Workmen's Inspector, by any workperson present at the place of accident. Such sketch shall also be supported by the photographs of the place of accident'.

3. Examination of records, reports, documents, and plans: A close examination of records, reports, other documents, etc., provides valuable information about the conditions prevailing before the accident. Inspection reports may reveal whether any unsafe condition existed. Ventilation records will show the state of ventilation and the occurrence of inflammable gas if any. The records will also reveal the status and standard of supervision. There are a multitude of records and plans to be maintained. It is not necessary to examine all the records. Only those records which are relevant to the accident need to be scrutinized. An investigator should look for missing pages, overwriting, corrections, and alterations in the entries; though it is a time-consuming process, the time spent is necessary to get at the truth.

4. Examination of material, equipment, and testing: Where failure of any material, appliances, or equipment appears to be the cause of an accident, these should be examined by experts, and if required, be tested at a reliable test house or laboratory.

5. Examination of witnesses: It is a crucial part of the investigation. As far as possible, only the eyewitnesses or those who can shed light on the conditions existing before the occurrence should be examined. Witnesses giving hearsay information may be avoided. The following procedure may be followed for the examination of witnesses:

a. The witnesses should be examined in a definite sequence. The injured should be examined first, then the other workers who were present in the vicinity followed by supervisory officials, the under-manager, and finally the manager. If this ascending order of hierarchy is not followed, the depositions of the witnesses are likely to be influenced by the views expressed by their senior officials.

b. The witnesses should be examined individually and separately so that each one can depose without fear of victimization and their evidence is unbiased.

c. If conflicting statements are made by some witnesses, they may be confronted by each other.

    While examining witnesses, the investigator has to be very fair and unbiased. He should not threaten the witnesses or give any allurements.

6. Analysis of evidence: After completing the above operations, the evidence should be carefully analysed and compared with the observations and test reports if any. The conclusions may be drawn about the unsafe conditions or acts that led to the accident. It should also be analysed as to why the unsafe conditions or acts were permitted including shortage of men and materials, safety equipment or appliances, laxity in supervision, and the reasons thereof.

7. Enquiry report: Presenting the findings of the Enquiry is equally important. The report has to be such that a person who is not conversant with the particular mine where the accident took place can visualize why and how the accident took place.

The report also provides very important inputs to statisticians and policymakers. Therefore, it must show detailed data on the time and place of the accident, level of supervision, age, and occupation of the victim (Kejriwal, 1994; Prasad and Rakesh, 1990).

4.12   Emergency, rescue, and management

A major challenge for the mining sector can be seen in the increasing depth of operation. The most important issues are related to the safety and health of the mine workers. The increasing depth of a mine not only causes day-to-day burdens of safety but also affects the emergency. In opencast, slope failure of the benches and dump is commonly observed, and many occasions lead to the emergency of falling debris and entrapping workers and equipment. In an underground deep mine, many emergencies occur due to fire, roof fall, inundation, the eruption of gases, etc., and then rescue operations are essential to reduce the severity and entrapped workers are rescued. Mine should have more escape ways for the workers to surface. The same applies to the increasing distance between mine rescue teams and the underground workplaces.

4.12.1   Identification of emergency

Mine workers must always be on alert for unusual occurrences or emergencies. Early identification of a problem and the response to it can mean the difference between life and death for everyone in the area. Every worker must be able to recognize the early signs of an imminent emergency, such as a fire, inrush of water, severe fall of ground, or an unusual gaseous condition.

The following are some of the signs or indications of possible emergencies:

- sudden changes in ventilation,
- ir blasts caused by the fall of ground or inrush of water,
- he odours of smoke or other contaminants,
- nusual noises or explosions,
- nterruption of normal services such as power failures,
- ire alarm/emergency warning systems such as stench gas,
- isual or audio warning, and
- unusual haste of workers.

Any of these signs could mean that something irregular or dangerous has happened and that quick action may be necessary to prevent loss of life and damage. If the situation is such that the worker believes that evacuation is warranted, or there is a sign of fire all work should be stopped. Workers should, without delay, implement the emergency procedures as per their training and informed the competent authority immediately. During a mine emergency, workers with basic knowledge and proper emergency training, who act in a calm, rational manner, have an excellent chance of surviving.

In case of being trapped due to an untoward incident or the occurrence of an emergency, workers should follow the laid down protocol and use the safety devices for their survival. 'Self-Rescuers' are one-time use devices used for escape purposes only. The types commonly used are MSA W-65 and Drager FSR-810.

Rugged construction allows both to be carried by mine workers or mounted on mobile equipment ready for instant use. As part of underground orientation for workers, the mine management must familiarize each worker with the emergency procedures and the rescue equipment available. Typical usage of these apparatuses is given in chart form (Figure 4.4) and descriptive form (Figure 4.5).

In addition, rescue teams should be ready with trained personnel, team leader, and other rescue team members in case of emergency and made active to reduce the severity of an accident (SMR, 2017).

4.13   Zero-accident plan (ZAP) and court of enquiry for mining accidents

Mining activity has heightened accident potential and therefore Zero-Accident Plan (ZAP) is a good practice. It signs well for the safety commitments of the organizations. Though it seems theoretical, achieving zero accidents and planning for them is achievable with concerted efforts. The ZAP assigns authority and accountability, ensures compliance, and establishes two-way communications. ZAP provides for periodic safety inspections and corrects hazards when observed or discovered. It specifies general and job-specific safety training for the miners and people at its workplace.

While reviewing the status of safety in Indian mines and the Indian mining industry from an accident angle, the trend and statistics of accidents in the Indian coal mines in the post-independence period have been depicted in Tables 4.5a and 4.5b. This gives a glance towards the fluctuating industry safety records.

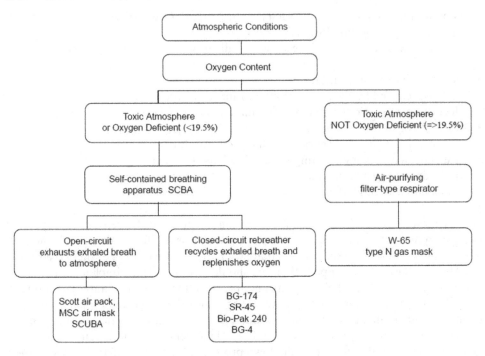

Figure 4.4 Usage of respirator and breathing apparatus.

The zero-accident culture emphasizes that accidents can be prevented from happening if the safety culture was adopted into practice and implemented on the ground. Several proactive steps need to be taken.

In the case of major mine accidents (involving fatalities and losses of large magnitude), the *Court of Enquiry* is set up, which is like an SIT (special investigation team) to enquire about the cause of the mining accidents. When any major disturbing accident of large size takes place, utilizing the power vested with the concerned government, a court of enquiry can be appointed, e.g. in India, under Section 24 of The Mines Act, 1952, such enquiries are legal entities to look into the cause and remedies of accidents. Even punishment for the defaulters can be framed legally by such enquiries. One can plan the best strategy for safety control in mines through such investigations. For minor accidents and injuries, standard safety management principles and practices shall remain applicable.

4.14   Mine safety tutorials

A deep effective learning approach can be fostered in a learner, either a student or a professional, by a careful selection of tutorials that can stimulate a better and correct way of managing. A tutorial is one of the best teaching–learning tools for enhancing intellectual, communication, and social skills. They are like *action-oriented research projects*, conducted for a specific topic of the subject. Its outcome brings changes in engineering practices, giving readers a much better opportunity for an in-depth understanding of a specific topic of interest. From the standpoint of a mining engineer, it helps develop technical & scientific skills (empirical/observational/diagnostic), general analysis, task

Filter-type self-rescuer

This (MSA) type of rescuer consists of a small canister with a mouthpiece directly attached to it. The wearer breathes through the mouth, while the nose is closed by a clip. Filter-type carbon monoxide self-rescuers do not protect against noxious gases or a deficiency of oxygen. The presence of carbon monoxide in the air is indicated by heat generated in the self-rescuer when it is worn. Both types of respirators will provide adequate protection for 60 minutes in the air with a 1% concentration of carbon monoxide (CO_2). At one per cent or higher CO_2 concentrations, heat generated by the chemical reaction with the hopcalite in the self-rescuer will make breathing practically unbearable. All units have a

Self-contained breathing apparatus

The quantity of oxygen consumed by the body varies with the amount of energy expended. A person at rest uses approximately one m^3 of air per hour. During strenuous exercise, the consumption of air may increase to more than 8 m^3/hr, but the body uses no more oxygen than it requires. One of the most important functions of any closed-circuit, self-contained breathing apparatus is the elimination of carbon dioxide present in the atmosphere and providing pure air to the wearer. In a self-contained, open-circuit, pressure-demand type of apparatus using compressed breathing air, the exhaled air passes through a valve to the outside atmosphere. Drager BG-174's self-contained, closed-circuit breathing apparatus enables the mine rescue worker to enter unbreathable and toxic atmospheres. The apparatus permits the wearer to breathe independently of the atmosphere and enables him to effect rescues and recoveries under extremely arduous conditions. The apparatus is light in weight (12.3 kg), but its construction is rugged and highly resistant to mechanical shock. Exhaled air is freed of its carbon dioxide in a regenerative canister and passed into a breathing bag. The air, purified in this way, is withdrawn from the breathing bag during inhalation. The oxygen consumed during respiration is replaced by a cylinder of compressed oxygen through a constant flow-metering opening at the rate of 1.5 litres per minute (lpm). If this amount of oxygen is not sufficient due to strenuous exertion, additional oxygen is provided by an automatic demand valve controlled by the user's lungs. When the apparatus is first turned on, the circuit is automatically flushed with approximately six litres of oxygen. Other than occasionally checking the oxygen supply by observing the pressure gauge, the apparatus requires no further attention during use.

Figure 4.5 Rescuer and breathing apparatus.

Table 4.5a Earlier Trend of Mine Accidents in India

Type of mines	Year/Decade	Noticed/Reported accident statistics	Year/Decade	Noticed/Reported statistics
Coal and Lignite Mines (Fuel Mineral)	1940/1950	400 lives are lost every year; 900 are seriously injured	2000/2010	100 lives are lost every year; 750 are seriously injured
Metal Mines (Non-coal Mines)	1940/1950	200 lives are lost every year; 700 are seriously injured	2000/2010	80 lives are lost every year; 550 are seriously injured

Note: The data given above is meant to show trends only and the figures indicated are approximate only.

Table 4.5b Major Accidents in the Indian Coal Mines in the Post-Independence Period

S. No.	Accident date	Name of the mine	Fatalities	Cause
1	12/07/1952	Dhemomain Colliery	12	Roof fall
2	05/08/1953	Majri Colliery	11	Inundation
3	14/03/1954	Damra Colliery	10	Explosion of firedamp.
4	10/12/1954	Newton Chikli Colliery	63	Inundation
5	05/02/1955	Amlabad Colliery	52	Explosion of firedamp.
6	26/09/1956	Burra Dhemo Colliery	28	Inundation
7	19/02/1958	Chinakuri Colliery	175	Explosion of firedamp.
8	20/02/1958	Central Bhowra Colliery	23	Inundation
9	05/01/1960	Damua Colliery	16	Inundation
10	28/05/1965	Dhori Colliery	268	Coal dust explosion
11	11/04/1968	West Chirmiri	14	Premature collapse of workings
12	18/03/1973	Jitpur Colliery	48	Explosion of firedamp.
13	08/08/1975	Kessurgarh	11	Roof fall
14	18/11/1975	v	10	Inundation
15	27/12/1975	Chasnala Colliery	375	Inundation
16	16/09/1976	Central Saunda Colliery	10	Inundation
17	04/10/1976	Sudamdih Coal Mine	43	Explosion of firedamp.
18	22/01/1979	Baragolai	16	Ignition of firedamp
19	24/08/1981	Jagannath Colliery	10	Water gas explosion
20	16/07/1982	Topa Colliery	16	Roof fall
21	14/09/1983	Hurriladih Colliery	19	Inundation
22	13/11/1989	Mahabir Colliery	6	Inundation
23	25/01/1994	New Kenda Colliery	55	Fire/suffocation by gases
24	26/09/1995	Gaslitand Colliery	64	Inundation
25	06/07/1999	Prascole Colliery	6	Fall of roof/collapse of workings
26	24/06/2000	Kawadi Colliery	10	Failure of OC bench
27	02/02/2001	Bagdigi Colliery	29	Inundation
28	05/03/2001	Durgapur Rayatwari	6	Collapse of partings/ workings
29	16/06/2003	Godavari Khani-7LEP	17	Inundation
30	16/10/2003	GDK-8A	10	Roof fall
31	15/6/2005	Central Saunda	14	Inundation

Source: Soni and Suman (2012).

realization, discussion, and synthesis of a particular area or subject. It is observed that conducting tutorial in safety strengthens the theoretical foundation, develops the research interest, and stimulates the thinking process of an individual. Thus, the role of tutorials in improving skills is a sacred way of learning that is well recognized; hence, in this section of chapter, two tutorials, one on risk assessment and another on the cost of accidents, have been described as they are essentials for mine safety and its management.

PROBLEM 4.1

(Tutorial on risk assessment)

Find/assess the risk of the roof fall hazard risk using the Empirical approach that is responsible for the maximum accident in mines.

Solution

The *risk assessment* shall be done as per the procedure, laid down either by the mining company or by the standard procedure. Before any risk assessment can proceed, a team has to be selected, background information has to be gathered and processed, and the team has to be prepared for the task ahead. All of these, as well as the subsequent hazard identification and risk assessments, are carried out within a participative framework by involving all stakeholders and their representatives (Box 4.1). It is important that those conducting risk assessments should be competent and should receive formal training to execute and disseminate.

Box 4.1: Risk Assessment

Many MDOs (mining operators) carry out risk assessments on a day-to-day basis during their work. In mining, safety risk assessment is the systematic identification of potential hazards in the workplace as a first step to controlling the possible risk involved. Unsafe conditions in mines lead to several accidents and cause loss and injury to human lives, damage to property, interruption in production, etc. But the hazards cannot be completely obliterated, and thus, there is a need to define and reckon with an accident risk level possible to be presented in either a quantitative or qualitative way.

MDOs note changes in working practices, recognize unsafe working conditions and practices as they develop, and take the necessary corrective actions. This process should be more systematic and recorded regularly so that the results are reliable and the analysis is complete. In particular, employers will have to undertake a systematic general examination of the work activities and then record the significant findings of the risk assessments conducted. The risk assessment process at a mine should be continuous and should not be regarded as a one-off exercise.

The risk assessment is a methodology to determine the nature and extent of risk by analysing potential hazards and evaluating existing conditions of vulnerability that could pose a potential threat or harm to people, property, livelihoods, and the environment on which they depend. The standards define Risk as 'the chance of something to happen that will have an impact upon objectives'.

It is measured in terms of probability, exposure, and consequences and the total risk is calculated using the formula: $RT = RP \times RC \times RE$, where RT is the total risk, RP is the probability of occurrence of an event, RE is the exposure level to risk, and RC is the consequence of risk.

In underground mines, either for coal or metallic ore, roof fall is the major contributor towards fatal and serious accidents. Therefore, risk assessment methodology is discussed here which can be used by practising mining personnel. A risk assessment technique is frequently used worldwide to manage and reduce the consequences of roof falls. The mining conditions change with time after the initial exposure of the roof and sides. The risk assessment of roof fall, done before or immediately after the drivage of the gallery, must also be updated from time to time with changing mining conditions. Close monitoring of the strata using field instruments enables to update the Risk Rankings and thus act on time to manage the risk.

The gathering and analysis of information from the mine sources and external sources is an essential task before the risk assessment can start. This would normally be conducted by the safety professional of the mine. These persons should access the mine databases to assess the types and major underlying causes of past accidents and incidents. They should also review accident reports and investigations together with other records such as those maintained by engineering staff, log books, and audit reports. Externally, gather the information from government and industry organizations or publications and databases. Increasingly, the internet is a valuable means of gathering international data. All this data needs to be assimilated and converted into a useful format to prepare the team who undertakes risk assessment.

Hazard identification, probability of occurrence, exposure, and consequences are considered by the team while visiting the workplace and seeing things more clearly through a systematic approach to ensure a comprehensive collection of information. There is a variety of tools available, from simple checklists to the most sophisticated quantitative techniques, to assist the team in identifying the hazards. Once the hazards are identified, a mechanism is evolved, and based on predicted probability, exposure and consequence ratings are allotted to each hazard based on the mechanism and total risk is calculated.

The quantitative risk assessment approach which gives attribute values to the probability, exposure, and consequence groupings is the recommended approach for risk assessment (Tables 4.6 and 4.7). Several combinations of the rating scale are possible. The following ratings have been considered for the risk assessment in this assigned problem.

Using Table 4.6, the major hazard of roof fall risk rating allocation and risk calculation has been done and calculated results, as analysed for the encountered condition in the mine (Say mine 'A') being evaluated, are summarized in Tables 4.8a and 4.8b.

Table 4.6 Rating Allocation Table for Probability, Exposure, and Consequences

Probability	Score	Exposure/Frequency	Score	Consequence	Score
Expected	10	Continuous	10	Multiple Fatality/Major loss	100
Quite Possible	7	Frequent (daily)	5	Serious Injury/ Permanent Disability/Significant loss	40
Unusual but possible	3	Seldom (weekly)	3	Lost time Injury/ Moderate loss	15
Remotely Possible	2	Unusual (monthly)	2.5	Medical Treatment /Marginal loss	10
Conceive but unlikely	1	Occasional (yearly)	1.2	First Aid/minor loss	5
Practically impossible	0.5	Once in 5 years	1.5		
Virtually impossible	0.1	Once in 10 years	0.5		
		Once in 100 years	0.02		

Table 4.7 Risk Characterization: Assigned Empirical Values

Risk level	Calculated risk	Remarks/Actions
Disastrous risk	10,000	Red zone: Closed the work unless corrected
Critical Risk	1,000	Red Zone: Immediate action. Early detection procedures evolved
Significant risk	500	Yellow Zone: Stop working & correct the situation
Moderate risk	200	Yellow Zone: Alarming & early action
Minor risk	100	Green zone: Corrective action should be initiated
Tolerable Risk	10 or less	Green Zone: Warn safety personnel

Table 4.8a Calculated Risk for Roof Fall Hazard Based on the Strata Condition in an Underground Mine

Cause and mechanism			Calculated total risk		
Identified hazard type	Mechanism	Prob.	Expo.	Cons	Risk
	Failure to identify bad roof	10	10	5	500
	Improper dressing	7	10	5	350
	Improper supervision	7	10	5	350
	Poor workmanship	7	10	5	350
Strata failure	Non-superimposition of some pillars in contiguous working	7	10	5	350
	Inadequate support design	3	10	5	150
	Poor quality of support material	3	10	5	150

Since roof fall hazards occur due to strata failure at different places in the mine, total risk evaluation must consider both development and depillaring areas.

Based on Tables 4.8a and 4.8b and assigned empirical values of risk characterization (Table 4.6), each activity has been assessed for the risk involved. A comparison of

Table 4.8b Calculated Risk for Roof Fall Hazard at Different Places of an Underground Mine

Places and mechanism			Calculated total risk		
Identified hazard type	*Mechanisms*	*Prob*	*Expo*	*Cons*	*Risk*
Roof falls in development	Poor supervision of mining operations	7	5	15	525.0
Roof falls in development	Poor workmanship	7	5	15	525.0
Roof falls in development	Weak roof strata	7	5	15	525.0
Roof falls in development	Inferior quality wooden supports	7	5	15	525.0
Side falls	Weak sides	7	5	15	525.0
Roof falls in development	Slip planes	7	3	15	315.0
Roof falls in development	Direction of drivages parallel to cleat	7	2.5	15	262.5
Side falls	Direction of drivages parallel to cleat	7	2.5	15	262.5
Roof falls in development	Delays in stowing in the bottom lift	2	2.5	40	200.0
Roof falls in development	Inadequate stowing of the bottom lift	2	2.5	40	200.0
Roof fall in filling sections	Inadequate stowing	2	2.5	40	200.0
Roof fall in filling sections	Delays in stowing	2	2.5	40	200.0
Roof fall in extraction (caving) sections	Withdrawal of support not to plan	3	3	15	135.0
Roof falls in development	Delayed erection of supports	3	3	15	135.0
Roof fall in extraction (caving) sections	Slow extraction rate	3	3	15	135.0
Roof falls in development	Delays in the supply of roof support material to face	3	3	15	135.0
Roof falls in development	Poor work attitude	3	3	15	135.0
Side falls	Inadequate side support	3	3	15	135.0
Roof falls in development	Presence of faults/dykes	3	2.5	15	112.5
Side falls	Presence of overhanging sides	3	2.5	15	112.5
Roof falls in development	Synclinal structure	2	3	15	90.0

calculated and assigned values in these tables quantifies the risk empirically thereby enabling us to judge where we are. Based on the risk characterization, mitigative measures shall be adopted that reduce the risk to an acceptable level.

PROBLEM 4.2

(Tutorial on accident and its cost)

Determine the cost of an accident that occurred in an underground coal mine due to a roof fall injuring two workers. These workers suffered leg injuries of serious nature. Assume a rate of loss that occurred in the local currency applicable where the mine is situated.

Solution

Accidents are more expensive than most people realized because of the associated direct and indirect costs also referred to as the hidden costs of accidents. Unless mining companies systematically and accurately evaluate the true costs of accidents that occur, they most likely do not know how costly these accidents are. Because of the significant cost for every mine, it is important to emphasize safety on and off the job. The employee who is affected or injured will be the one who pays the most. The costs associated with an accident are always more than just money and can be classified into two broad categories namely *Direct cost* and *Indirect cost*. Beyond this, the cost to the person injured, the cost to the organization and management, and the cost to the state and society are in addition.

The direct costs of accidents are those costs incurred due to the treatment of an injury that is normally reimbursed to the workers viz. medical costs, ex-gratia payment to workers and family, compensation, liability, and property losses. Direct costs are easy to establish and quantify. These costs are payable (financially) by the employer or the insurance company (on behalf of the employer).	**Indirect costs** are generally those costs attributed to the loss of productivity of the injured worker and the crew, transportation costs to the nearest medical treatment facilities, and time spent to complete various processes related to the injury. The non-insured losses from damage to buildings, plant and machinery, equipment, tools, products and materials; interruption of business operations; overtime to make up for delays caused; inefficiency of backup employees; cost of training to new crew/employees; increased insurance premiums; and damage to a company's reputation are all included in this cost. Usually, indirect costs are non-recoverable.

Having understood the theoretical aspects of accidents and their cost, a template for accident cost determination is first prepared as given below. This *accident cost calculator* reflects the collected data and its analysis. Each activity requires some time, and this time spent on each activity is calculated in hrs/minutes/seconds and assessed in financial terms. Based on these two data, the accident cost is calculated. For a mine located in India, we will assume the rate of INR 1,000 per hour for the losses. For the given problem, the accident cost calculator is given below:

Answer: The leg injury cost per person in India is Rs. 1,67,500/- (INR) as of 2022.

Accident Cost Calculator

Date of incident: dd/mm/yy   Report by: ***
Time of incident: Shift Time: Date: ***
No. of the person involved: 2   Type of accident: Fall of roof

Severity of accident: Leg Injury
Dealing with the accident (immediate action)

Actions	Time spent (hh: mm)	Cost (INR)
First aid treatment	1.30	1,500
Taking injured person home or to hospital	1:00	1,000
Making area safe	3:00	3,000
Immediate downtime of work	5:30	5,500
Other miscellaneous costs	2:00	2,000
Total	12.30	12,500

Investigation of accident

Actions	Time spent (hh: mm)	Cost (INR)
Time for inspection and investigation	3:00	3,000
Meetings to discuss incidents/accidents	4:00	4,000
Time spent with officials/inspector	3:00	3,000
Attorney/Advocate fees to assist the company in the investigation	10:00	10,000
Other miscellaneous costs	10:00	10,000
Total	30:00	30,000

Restoration of work

Actions	Time spent (hh: mm)	Cost (INR)
Assessing/rescheduling work activities	5:00	5,000
Recovering work/production	6:00	6,000
Cleaning up site and disposal of debris, equipment, etc.	4:00	4,000
Bringing work back to the normal	8:00	8,000
Repairing damage/faults	4:00	4,000
Replacement of tools, equipment, plant, services, etc.	30:00	30,000
Other miscellaneous costs	15:00	15,000
Total	72:00	72,000

Productivity costs

Examples	Time spent (hh: mm)	Cost (INR)
Costs of the ill or injured person while off work	10:00	10,000
Cost of replacement workers (Salary)	10:00	10,000
Lost work time (people waiting to resume work, delays, reduced productivity, effects on other people's productivity, etc.)	15:00	15,000
Overtime costs	–	–
Recruitment costs for new employees	–	–
Contract penalties	–	–
Cancelled and/or lost orders	–	–
Other miscellaneous costs	5:0	5,000
Total	40:00	40,000

Preventative measures

Examples	Time spent (hh: mm)	Cost (INR)
Implementing safety practices	5:00	5,000
Cost for safety devices and safety measures	2.00	2,000

(Continued)

Accident Cost Calculator

M	–	–
Total	7.00	7,000

Sanctions and penalties

Examples	Time spent (hh: mm)	Cost (INR)
Compensation claim payments	–	–
Legal expenses	–	–
Fines and costs imposed after legal proceedings	6:00	6,000
Increase in insurance premiums	–	–
Other miscellaneous costs	–	–
Total	6:00	6,000
The total cost of the accident	**Total time spent = 167.30 hrs.**	**Total cost = Rs. 1,67,500 INR**

References

Anon. (1999), Guidance for Carrying Out Risk Assessment at Surface Mining Operations, Doc. No 5995/2/98-EN, dated 25.3.99 Committee on Surface Workings, USA.

Hansen, E. (2009), Human Error Categories TECH 434, Human Factors in Industrial Accident Prevention, MSHA.

Heinrich, H. W. (1959), *Industrial Accident Prevention – A Scientific Approach*, New York: McGraw-Hill Book Co.

ICMM. (2009), *Leadership Matters: The Elimination of Fatalities*, London, UK: The International Council on Mining and Metals (ICMM), pp. 3–4.

Kalia, H. L., Singh, A., Ravishankar, S., and Kamat, S. V. (2015), *Essentials of Safety Management*, New Delhi: Himalaya Publishing House, p. 238.

Kejriwal, B. K. (1994), *Safety in Mines*, Dhanbad: Gyan-Khan Prakashan.

Michael, J., Brnich, Launa, and Mallett, G. (2003), Focus on Prevention: Conducting a Hazard Risk Assessment, Pittsburgh, PA: U.S. Department of Health and Human Services, NOISH, July.

Paliwal, R. and Jain, M. L. (2001), Assessment & Management of Risk – An Overview, *The Indian Mining & Engineering Journal*, Vol. 40, Issue 10, pp. 79–86.

Prasad, S. D. and Rakesh, (1990), *Legislation in Mines: A Critical Appraisal*, Vol. I & II, Varanasi: Tara Book Agency.

SMR. (2017), The Saskatchewan Mine Rescue (SMR) Manual, July, p. 108. http://saskmining. ca/ckfinder/userfiles/files/Mine%20Rescue%20Manual%202017%20Update%20%2011%20 January%202018.pdf; Accessed on 15/09/22.

Soni, A. K. and Suman, Kiran. (2012), The Road to Zero Harm: Safety Management and Safety Engineering in Context to Underground Mines, In *Workshop on Safety Management in Mines*, Nagpur, pp. 1–10.

Tripathi, D. P. (2001), Qualitative and Quantitative Approaches to Risk Assessment, *Mining Engineers Journal*, January. http://www.tc.gc.ca/civilaviation/SystemSafety/Pubs/tp13739/ Features/accident.htm

Chapter 5

Emerging areas of technovations in mining

> Choosing the right over the pleasant is a sign of power
>
> (Bhagwad Gita)

Technology and *Innovations* called together, 'technovations', emphasize those new emerging engineering areas concerning the mines and processing plants from where coal or ore extraction takes place. The mining, minerals, and metals industry is poised for significant growth in the years ahead. Hence, cutting-edge technologies guided by discovery and innovations and smart devices will be replacing conventional practices slowly and steadily. In this chapter, our focus will be on the new and evolving areas of concern for the mining of minerals as a whole. To be innovative and forward-looking, we have to track all such fields of science and engineering and keep abreast with them. For a mining engineer, some key S&T areas are information and wireless. The digital innovations that, over the next decade, have the greatest potential to create value for the industry, its customers, and society are:

a. Automation, robotics, and operational hardware
    (*e.g. remote operation of the machinery and equipment*)
b. Integrated enterprise, platforms, and ecosystems
    (*e.g. GPS operation of mine fleets*)
c. Digitally enabled software tools
d. Next-generation analytics and decision support (*e.g. big data tools*)

Most of these areas of communication and technology, which will be described later in the chapter, are either fully grown or being developed. From the viewpoint of their implementation, the mining industry is fully geared up to embrace them. All such areas are hence emerging, new and involve subject area expertise for candid industrial success. For their field applicability in the mines, i.e. the real ground truth, the moderation of technology is a must. Sometimes, tailor-made solutions have to be evolved, and a combination of modern and conventional practices works better, e.g. a *Hydrogen-powered mine haul truck* is an example of technovation replacing diesel trucks with the hydrogen model that reduces carbon dioxide emissions out of the atmosphere significantly. A hydrogen-powered mine dumper (496/290 tonners) designed to operate in everyday

DOI: 10.1201/9781003274346-5

mining conditions is capable of carrying a 290-tonne payload, which has the 2 MW hybrid hydrogen-battery generating more power than its diesel predecessor. It provides a fully integrated green hydrogen system (production, fuelling, and haulage) for bulk mineral/ore transportation which is deployed in South Africa The described mining truck has entered into the actual mining operations at Mogalakwena PGMs mine of South Africa where hydrogen is produced at the mine site itself [31; Nagaraj, 2022]. In this way, the ensuing chapter guides us to practical solutions.

5.1 Digital mine

In the mining business, the buzzword 'digital mine' comes across quite often. But we do not know exactly what it means. What is a Digital Mine? Whether the mine handled is a digital one? Or has the mine a scope for digitization or conversion from conventional to digital mine? What kind of budget or team is required to become digital? Can we start investment for conversion to achieve better production and productivity results? Whether investment will be fructified in future or not. Answers to all these questions are wide, practical, and so varied that covering them in limited pages is not justifiable. What all technologies a 'Digital Mine' have are briefly and theoretically answered in this sub-section.

A 'digital mine' is a mine that will bring together real-time data on digital platforms across the mining value chain in multiple time horizons to improve planning, control, and decision-making. This results in optimized production volume, reduced cost and capital expenditure, and improved safety at the workplace.

a. A digital mine has digitized geological data to discover and establish mineral/ ore reserves.
b. A digital mine has Internet of Things (IoT)-based sensors for real-time data capture during the exploitation phase of the mine.
c. A digital mine has digital equipment and is wearable for field maintenance and operator safety during mining and ore beneficiation operations.
d. A digital mine has a digitized engineering solution to move ROM and waste, i.e. bulk handling.
e. A digital mine has drones for inspection (survey), stock, and safety monitoring.
f. A digital mine has integrated remote operations for workplace and workmen's safety.
g. A digital mine has integrated operational planning, control, and decision support. (diverse, connected, and mobile workforce for mining).
h. A digital mine has cyber security in place to mitigate the risks of greater connectivity as more equipment and devices are connected to the internet (meshing technology).

All core processes (1–5) and support processes (10–14) of mining are the nerve centre of a digital mine as depicted in Figure 5.1. Identifying and prioritizing the opportunities for digital technologies (as depicted in different quadrangles) gives an added value to the mining business to assess in terms of its value, i.e. potential benefits for a digital mine (Figure 5.1). In this decade, digitalization is poised to affect the mining

Opportunity prioritisation

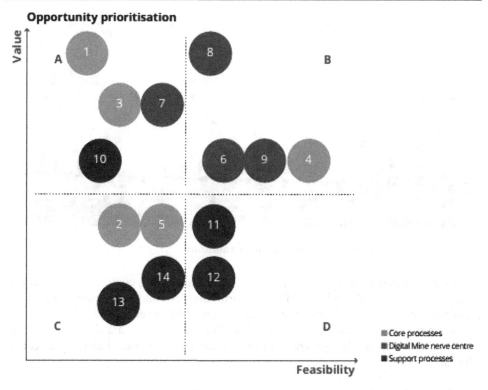

Figure 5.1 Identifying and prioritize the opportunities for digital technologies.
Source: Paul and Steve (2017).

industry, and if the digital solutions are deployed to perform or improve activities that have traditionally been carried out manually or with human-controlled machinery, it will provide real-time decision support and future projections quite efficiently and correctly. India, a major player in the mining industry, is slowly catching up with 25% of the mining companies looking for remote operations and monitoring centres. Globally, around 69% of mining companies are searching for smart solutions (Figure 5.2). There is a wide gap between digital solutions and adopting the smart philosophy in many mineral-producing economies.

To meet the global benchmark, still several countries have a long road ahead. Examples of mining companies operating on Indian soil, namely Tata Steel and Vedanta (HZL) in the private sector and iron ore miner and National Mineral Development Corporation Ltd. (NMDC) in the government sector can be quoted that have adopted digitization in their mines.

5.1.1 *Snags (bottlenecks) of digital mining*

Indeed, 'Digital mining' is a form of empowerment for strengthening the mining industry in this 21st century. Two important stages namely *formulation* and *implementation* have to go hand in hand continually for digital mining/digitization to be successful

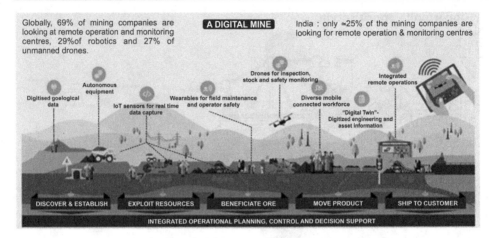

Figure 5.2 Digital technologies provide smart solutions.

and take firm shape on the ground. Thus, interested mining companies have to leverage their digital tools and capabilities to promote sustainable resource use and improve mine performance despite the existing snags whatsoever they may be.

Some bottlenecks, which we are referring to here as 'snags', are those operational hurdles which impede the digitization process and are encountered as ground reality. They are:

- High initial upfront costs
- Lack of skilled manpower
- The scale of mining operations
- Online connectivity of equipment and processes through IoT and Industrial Internet of Things (IIoT)
- Data privacy
- Cybersecurity

To generate real-time data, electronic gazettes, sensors, remote sensing methods, GIS and GPS techniques are the major tools deployed in mining operations and productivity improvements. The entire mining value chain can be revolutionized through AI (artificial intelligence)- and ML (machine learning)-based solutions.

Robotics – Devices & Techniques, which is a digital mining solution, is capable to eliminate several bottlenecks, which are key issues of concern, e.g. Repetitive unit operations of mines, namely loading, precise inclined drilling, blast-holes charging, handling of ANFO or other chemical explosives, can be performed more accurately thus providing better safety in mining and other related hazardous operation. Thus, the successful implementation of digital technologies through multi-stage snag removal (trouble shooting) will be able to increase mineral production and mine performance.

5.2   Process-oriented opencast mine

Existing conventional mines can be converted into a *Computerized Process-Oriented Opencast Mine* of the 21st century (Soni, 2017a). The mining process and its optimization; automated mine planning; digitized mine surveying; equipment operation and material control; plant utilization and raw material quality control are the major points of improvement that are embedded in the process-oriented opencast mine (Soni, 2017b).

The application of modern and *advanced process systems* improves the process sequences and the mining efficiency for the dynamic mining operation. The technical capabilities are available both at the national and international level depending on the specific engineering need of a mine or mineral extraction plant. For developing countries and planning, a future vision for the mining industry, i.e. 2030 or beyond, is always helpful. This could be in terms of technological prowess like 'digital mine' or 'process controlled mine'.

In mining, our major focus is laid on the production of minerals/coal. To raise it, conventional operations of drilling, blasting, loading, and transportation in the mining method need to be computerized and mechanized. The Satellite supported excavators (GPS-based dumpers) combined with more automated equipment, such as computerized automatic drill machines, automated blasting devices, and spreaders with automatic spread operation control, are some established examples. In all deployed heavy earth-moving machines and various mining operations, *closed-loop circuits* for repetitive process and *process scheduling* are easily possible and can be done with available know-how. The process scheduling chain consists of sub-components involving seam/deposit → shovel/excavator/surface miner → conveyor → stockpile/overburden dump, etc. Similarly, for mineral processing plant operation & quality control in CHP (coal handling plant)/Mine mill, the computerization of many processes is feasible with digital techniques.

To economize the overall mining process, reduce expenditure on excavation and induct optimization. In various unit operations, wherever computerization and automation are applicable, adopt them to economize the per ton excavation cost. Optimization of operational processes in the opencast mines has the following inevitable advantages:

- Minimization of drilling time and cutting of loading-losses.
- Reduction in travelling time or shunting head movements at the ore/material distribution points.
- Less wear and tear of equipment in operation.
- Stepping up the utilization time of the high-capacity equipment and heavy earth-moving machinery.
- Quality improvement in the exploitation of minerals close to the surface (*More exact positioning of the excavating machine with material boundaries including the digging of the selective content of the deposit hence improved quality and better control*).
- Reduced waste generation at source.

5.3 Artificial intelligence (AI) and machine learning (ML) in mining

AI or artificial intelligence is a modern tool, most talked about these days in mining. AI is 'the science and engineering of making intelligent machines, especially intelligent computer programs'. It is a way of making a computer, a computer-controlled robot (or a programmed machine) that can think intelligently. Today, AI is a frontier technology in the engineering field and John McCarthy is the Father of Artificial Intelligence. AI came into its first existence in the year 1957. Many researchers have worked on AI techniques and developed various methods and tools for better analysis of data resources. Some of the most common methods and tools that are used in AI are:

- ANN (Artificial neural network),
- Support vector machines,
- Bayesian classifiers,
- Regression techniques, and
- Clustering methods.

Machine Learning (ML):
   According to Tom Mitchell of Carnegie University, Pittsburg, Penn state, USA, Machine learning (ML) is the study of computer algorithms that allow computer programs to automatically improve through experience, e.g. Google search shows the topic of your interest based on the keywords you have used in your past searches in the browser.

5.3.1   Areas where AI and ML can be used in mining

In mining, AI and ML can be applied to many areas wherever a large magnitude of actual data is available. This data can be analysed through *data analytics* to know what is happening (data), what happened (descriptive), why did it happen, and what might happen (diagnostic). Broadly, AI and ML can be applied in the following magnified mining engineering areas:

1. **Mineral exploration**: Advancements in aerial imaging, along with drones, improved surveying, geological prospecting, and ascertaining drill target areas. These emerging technologies enable the development and innovative exploration solutions. The risk to access dangerous and unstable terrain that involves a human workforce is minimized.
2. **In the mining process:** The application of AI in mine planning optimization, mine scheduling, HEMM predictive maintenance, and enhancing the efficiency of material handling systems in coal and mineral processing and process metallurgy has huge scope.

   AI utility in mining space gives an idea to think for the correct size and dimension of the mining operation. Mining methodology selection including equipment selection as per the type of deposit/cost parameters/geotechnical features is an area where AI is advantageous. Most optimum pit uses optimization and dynamic conditions with many variable parameters using software and simulation. HEMM predictive maintenance monitoring is yet another major area where AI is extremely useful before breakdown occurs. HEMM utilization percentage can be increased by decreasing annual idle hours. Predictive maintenance focuses on failure events, therefore, it makes sense to start by collecting historical data about the machines' performance and maintenance records to form predictions about future failures. Usage history data is an important indicator of equipment condition.

3. **To improve mine operations**: Improving the overall efficiency and economics of mine using AI is a solution that creates a win-win situation for both MDOs and management. With the help of AI, machine learning, and autonomous technologies, the exposure of workers to hazardous/ dangerous underground and surface operations can be minimized. Machines can autonomously monitor the atmosphere, send signals and warnings, locate problematic areas, and work continuously even in dangerous situations.

    a. **Stoping:** Ore extraction sequencing and methodology, various mining operation in stopes, optimization of ore extraction in stopes.

    b. **Ground control:** Support and its design, paste filling, barricade design, dam design for ground support and control, and slope designing in open-pit metal mines.

    c. **Mining cycle:** In this segment, possibilities exist for AI and ML applications in: (a) Estimating the mining cycle from real-time data in stopes; (b) analysing the roof & floor conditions in stopes, crosscuts, and main roadways; (c) predicting the behaviour of raises and winzes from mining cycle data; and (d) designing the underground excavation using ANN.

    d. **Self-driving haulers:** Nearly all actual equipment used in the field make the mines, a heavy industry that have big-size hauling, loading, and transporting machines. Most of these operate in well-defined controlled areas, and AI and ML can be applied for their improved performance. Open mines are the ideal place for commercial self-driving vehicles like dumpers. Both heavy-duty mine dumpers (> 35-tonne capacity) and large-capacity trucks, which tend to travel relatively slow, are part of self-driving haulers in surface mines of coal and metals.

    e. **Autonomous drills, loaders, and trains:** Drilling system and drilling rigs to drill into the ground, automated loading using loaders into the self-driven haulers and trains (rail) lets one remote operator control. AI and ML apply to these equipment and areas.

    f. **Behaviour of goaf using caving method:** Using the data analytics to predict the roof fall in underground mines in advance. Periodic vetting of different types of data such as RMR ratings, size of pillars, movement of strata vertically / horizontally, and magnitude /interval of different roof falls can be used to make a system of correct prediction.

    g. **Prediction of flyrocks in OC mines:** By studying the behaviour of different types of rocks and varying rock qualities in blasting, the pattern of blasting, explosive types, delays and type of detonators, etc., we can develop a system of prediction to minimize the flyrocks in opencast mines.

    h. **Image recognition:** It is used in identifying the mineral grade. The error percentage reduces to zero or very negligible compared to the manual recognition done by humans.

    i. **Location finder of misfired shots in underground mines:** Data can be generated by a blasting foreman who physically locates the misfire after every shot fired. Various types of data, such as the blasting pattern, location of the misfire, and explosives used coal quality area (selective mining area), are useful for data analytics in AI to make a predictive model.

    j. **Haul road simulation (Route modelling):** As the depth of the mine increases, the length of the haul road, the cost of diesel consumption, and the wear and tear of transport machinery increase significantly. Therefore, with the help of AI using various types of data such as the gradient of the road, type of machines, and length of lead, we can develop a better efficient model to select the optimum haul length or the best and optimized path (road route) for the mineral transportation. An optimum route is for better mine efficiency.

k. **Mine ventilation system:** It is one of the key areas for creating safe working conditions in underground mining operations. Different aspects of ventilation planning and design (natural and mechanical ventilation) may be developed for theoretical and practical use in laboratory and real-field applications. AI-enhanced automatic adjustment of ventilation systems using integrated process control assists with significant energy cost reductions. An effective and efficient mine ventilation system brings good environment health and safety underground.

l. **Accumulation of methane gas and sensor-based monitoring using AI for underground coal mines:** The use of direct methane concentration values measured by a system of sensors located in the underground mines helps know work environment conditions. Further, AI systems can be developed to inspect the worksite ahead of workers' entry using robots and by collecting data from pre-installed monitoring stations. These stations can trigger alarms, give warning signals, and block the affected area to decrease the further expansion of hazards.

4. **Mining environment (ambient):** Heavy metal pollution in land and water, extraction of heavy metals from wastewater/mine water and air/water measurements and their simulation can be dealt with sophistically using AI-based instruments and equipment.

5. **Slope failure (real-time data):** Identification of potential failure planes on rock surfaces, using hardware and software, to analyse rock surfaces and using latest generation sensors provides a basis to the data-user a stability solution against failure within very less time. Capturing real-time slope failure data using AI and ML is the new and easy way out.

6. **Mine safety:** Many aspects related to underground mines safety with a safety engineering approach and zero accident potential can be done using AI and ML. Occupational safety analysis of one mine or a cluster of small mines can be dealt with.

7. **Rescue and relief:** For trapped workers in a mine, technology interventions using AI and ML provide great support and help for rescue and relief operations. The use of robotics for rescuing trapped miners in underground and hazardous places where rescue teams cannot reach is a viable and safe solution alternative to the manual rescue.

8. **Simulator-based training platforms:** Training for mine professionals in a simulator-based environment is easy, better, and educative as well. Wherever wearable sensors and continuous monitoring generate data, that is automatically analysed to spot problematic behavioural trends and recommend focused remedial measures. This enables mine managers to proactively respond to latent or emergent gaps in operator capabilities, optimizing training intervention to best match individual worker needs, and ultimately raising the performance standards across the entire mining operation. This is helpful for the operator as well as workforce efficiency.

9. **Mineral processing:** In this metallurgical-dominated area of 'ore processing', extracting metal from ore has many possibilities for AI and ML applications. To enhance and improve further, different equipment and processes can be targeted. Whole or partial mineral processing operations can be intelligently handled to get the required output within a minimum timeframe.

10. **Miscellaneous application areas:** For office/company functioning (ERP – Employee resource portal), data management, material purchases, data sizing, asset maintenance and plant management in production, maintenance, and planning of data throughout the company.

Though, AI and ML concept concerning mining is widely described in the literature however for limited further readings two references are suggested – Singh, 2020 and IS, 2017. The first one deals with the mining application part (Singh, 2020) for AI and ML and the second one is about the policy matters such as principle, policy, privacy, safety, legal accountability, open governance, and social/economic impacts (IS, 2017).

5.4 Value engineering in mining

Value engineering is a systematic, organized approach to providing necessary functions in a project at the lowest cost. Value engineering promotes the substitution of materials and methods with less expensive alternatives without sacrificing functionality. The main purpose of knowing value engineering concerning mining is to reduce the 'operational costs' without sacrificing safety, sustainability, and employee rights. Value engineering in mining has emerged from large-scale civil infrastructure projects where the cost involved is hefty and its visibility is more. Sometimes, value engineering is also referred to or called 'Value Analysis'.

A mine has *operational costs* which are high and keep on increasing continuously with time and production. It is, therefore, necessary to create a new way of doing things where we manage the mine together with our employees and optimize operational costs, improve efficiencies, and extend the mine's life to ensure sustainability for many years to come. This new way of thinking will imply the cost and economics of various mining operations. To achieve this, we need to take managerial steps that regulate cost economics. The involvement of mine management and employees right from the top to the grass-root level is thereby necessary to ensure that maximum efficiency with minimum wastage of money is targeted. The involved team, including the management team, must be exposed to the *principles of value management* and the benefits associated with it. A strategy to reduce operational costs by 10%–15% in a given timeframe shall be established. Through value analysis, mine sustainability is not compromised because the cost reduction is not very large. Slowly, higher cost-cut targets can be fixed in a shorter period and the opportunity for engineering knowledge be harnessed.

How to do it?

The success of this turnaround intervention in an operating mine can be understood with the key points mentioned below:

- **Step 1:** Fix the objective first, i.e. How much operational cost per ton of ROM production you are interested to reduce overall and in percentage terms.

- **Step 2**: Identify what results you want to achieve. Given that take appropriate action to measure (*unit consumption*), optimize (demand), correct (*cost allocation*), and revisit (involved parties). This will reduce, schedule, improve, rationalize, and educate the user. As a management personnel, one should also know what results to prevent in terms of production loss, retrenchment, prevent disasters, mine closure, and inadequate supply of power and water.

  What are the available resources in terms of maximum demand, water availability (mine water for reuse), skilled workforce, equipment, people/miner for deployment, and planned maintenance that must be known and lined up? The constraints in terms of capital, resources, time, actual mine plan, government legislation, trade unions, lack of education and motivation, ignorance of power and water cost, and any other factors should also be taken into consideration.

- **Step 3**: Work out strategy. The major strategies to be identified are: (a) Ensure correct resource management; (b) Manage contractors and service agreements properly; (c) Reduce equipment abuses; (d) Optimize power use; (e) Ensure full water utilization; (f) Eliminate waste – Stores and salvage; (g) Align labour costs (Bonuses and overtime); (h) Eliminate theft and fraud; (i) Introduce value-added services in various mining operations during various phases; (j) Ensure quality health services; and (k) Resolve grading inconsistency and ensure optimum quality of ore.

  Each of the above-mentioned project-related matters, i.e. *steps and strategy*, include the basic value management principles involving the application of common sense, sincere efforts, and scientific know-how, however, record-keeping must be done for this. Real and perceived issues as well as concerns, e.g. leakages in water pipes; leaks in compressed air pipes; excessive use of equipment such as fans (too many), pumps (too many), pipes (pressure), and winders (unnecessary trips); wrong schedule of pumping hours (peak periods); unnecessary stops of mills and unscheduled startups (peak period); open valves (lack of control); lights burning at day time; unnecessary use of air conditioners/heaters, must be addressed in order to significantly reduce the power consumption and water cost. In the mine, either owned by a government organization or a private sector company, basic value management principles are equally and easily applicable.

- **Step 4**: Determine results in terms of reduced power consumption and water cost at a mine. Calculate reduced ROM cost.

Thus, by taking these major steps, viz. optimizing demand (maximum), reducing freshwater consumption, economizing power, and machine utilization, the objective of value analysis can be achieved. With mature management thinking and implementation of strategy, a substantial reduction in operating costs per month/cost per ton of ore/mineral produced from the mine is possible. No employee or person is retrenched, and the mine's life is extended by allowing cost-effective mining in less profitable sections. Implementation of improvements by the mine's management will bring laurel and pride moments.

5.5 Reverse engineering for practical knowledge

The engineering we follow in general for a solution to an engineering problem is called 'Forward engineering' which involves logical steps in the forward direction starting from the specification of product/process>analysis>design>implementation. Contrary to this, there lies Reverse Engineering (RE). RE is a process that examines an existing product or process to determine detailed information and specifications to learn how it was made and how it works. For example: When a new machine comes to market, competing manufacturers may buy one machine and disassemble it to learn how it was built and how it works. Similarly, a chemical company may use or adopt a reverse action to know the constituents and details of the manufacturing process thereby defeating hurdles of the engineering patent from a competitor. In another example of mechanical assemblies of HEMMs, RE typically involves disassembly and then analysis, measurement, and documentation of those parts that are not common for other machines.

Reverse engineering is also known as 'backwards engineering' or 'back engineering'. RE came into prominence as a sub-discipline of software engineering because software designers/programmers were striving hard to get solutions for the complex source code easily and thereby providing support to the comprehensive analysis of software systems. Since RE can be used to interfere, reconstruct, and redesign the source code easily, there are great chances that infringement of the intellectual property laws is simply overcome. In mining, many software licence agreements are in wide use, and for their legal uses, the RE is considered helpful. RE needs tool support, which provides functionality to extract low-level facts from the systems, analyse and generate knowledge about the systems, and visualize the criticality of the product, i.e. ore quality and product design. Thus, the RE knowledge is useful for the engineers provided they can comprehend and utilize it for the positive and good aspects effectively. In brief, RE builds programme data and generates useful information.

Everyone knows that mining is a dynamic field-oriented subject that has several hidden things either related to the ground or man-made complex heavy machines that can be unfolded through RE. To cope with the complexities of mining, recover lost information, facilitate reuse, and detect uncertainties, the RE is useful for the mines either surface or underground. When a field engineer monitors the mine workings and designs the support system (as per the field condition) and not before the exposure of the rock mass in situ, it can be said as an apt case of RE to design support in a befitting manner. Mining engineers indirectly, through their field exposures, unfold a lot of ground control solutions (support systems) through RE only.

RE for *'mining machines'* and 'mining codes' (software) has tremendous utility in mining engineering for practical purposes.

5.6 Wireless technologies for the 21st century (meshing)

The mining industry continues to invest in and upgrade various technologies ranging from GPS-based fleet operations, advance remote operations for safety including monitoring, etc. The need for stable and scalable wireless communications, i.e. communications without expensive cabling, can be done possibly using 'Wireless Technologies' which forms an important communication means for the 21st Century.

Mining operations are subject to constantly changing conditions; if the network in place fails to adopt immediately and automatically, it fails to serve its basic purpose. If the network is not robust enough to withstand the dynamic working environment of the mine, it cannot utilize real-time data transfer (application) for ensuring constant productivity and safety. Smooth operation, machine downtime, cost, losses, over expenses, and hindrances of all types have been recorded and communicated in a re-al-time frame through no cable wireless communication. Its need in a mine is critical to keep various real-time equipment and machines functional and remotely operative. Therefore, the requirement of wireless technology in mining is a *network* that can support dynamic mobility. This is possible by *Meshing* also termed Kinetic mesh/Wireless mesh networks. What it is and how it relates to mining has been described here.

Meshing: This network system (*IIoT*) works autonomously to provide optimal connectivity across all nodes of the mine, both fixed and mobile points, in real-time (Figure 5.3). Full mobile broadband connectivity that is simple, instantaneous, continuous, and fail-proof in a mine for industrial uses is the beauty of the meshing technology required for the dynamic mine environment which is a production organization.

A meshing network system is a self-optimizing mesh network. 'Nodes' act/work via multiple frequency peers to peer connections and can be either *fixed* or *movable* (mobile) with full routing capabilities, e.g. connections can be made between moving assets, such as truck /dumper or people, enabling all-to-all communications between mobile equipment. In the meshing network, each node can receive and transmit data simultaneously, meaning they can be receiving information from any other connected device continuously without interruption (resilient) as a local wi-fi transmitter. This critical functionality allows the adoption of real-time location solutions for tracking machinery and personnel. The ubiquitous connectivity and flexibility of meshing technology applications are made possible through a networking protocol, which dynamically and automatically selects the fastest route for traffic and re-routes communications to the next best path if the first becomes blocked or unavailable, thereby no compromise on network performance, operational efficiency, and safety (Sagar Chandra, 2021). The advantages of meshing are:

1. Multi-frequency, rapid, and reconfigurable
2. Adaptive in real-time (Connectivity when it is needed most)
3. Resilient (uninterrupted connectivity)
4. Total mobility

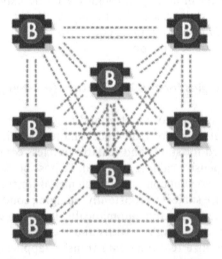

Figure 5.3 Meshing in a mine (B indicates the various nodes where machines and equipment are located).

5. Extreme ruggedness for the dynamic environment
6. Highest security
7. Maximum bandwidth utilization
8. Rapid scalability

Several IT companies and service provider facilitates kinetic mesh networks for mines and industry worldwide, e.g. 'Rajant technology' is a leading IT company in this sector. Thus, it has been made clear that wireless communication improves operational effectiveness, enhances the quality of service, drives productivity, and generates hustle free work environment.

5.6.1 Underground mines and technology transformation

The harsh climatic condition and operational nature of underground mines are some of the most challenging places to deploy technology all over the World. Underground mining workplaces and the mining environment are ever-changing as extraction takes place. It is very difficult, if not impossible, to reach every corner of an underground mine for constant, uninterrupted communication between miners and its support team, especially in emergencies. Despite breakthroughs over the years, laying the technological foundation in the entire underground mine or a part of the underground mine requires a great deal of care and management. *Fibre networks* and *Wi-Fi connectivity* (an IIoT application) are a solution to underground mine communication that replaces the wired connection. Underground mines like any opencast mines have production pressure, and the absolute need for workplace safety gets benefited from wireless connectivity. Day-by-day demand for a good work environment and increased employee facility can be fulfilled easily by this route.

When technological transformation (digitization and wireless connectivity) for an underground mine is considered, a sound approach is to:

1. Firstly, take the investment decision despite gain and loss (a business investment test) and treat the exercise as a technical challenge.
2. Search for the following:

   a. How will the transformation improve the output? Will the transformation assist to increase the production/output, reducing waste generation from the source and improving overall efficiency and productivity?
   b. Which department will benefit from the transformation, e.g. operation, processing, technical services, logistics, human resource, management, or procurement?
   c. What are the impacts of transformation?
   d. Which is the starting point of the transformation and what timeframe is needed?

3. Devise and locate the tailor-made solutions as getting a ready-made solution for the underground mine setting will be a herculean task.
4. Apply efforts for immediate and future needs to develop the technological platform for that particular mine.
5. Implement the plan and ideas into real practice and monitor the efficacy.

Without major changes in the existing platforms and processes, the transformation team can use the IIoT to assess the ongoing problem and issues of the underground mine and apply the latest gadgets, sensors, and networks for new technological alterations. A *turn-key project* on communication and automation, using IIoT, may be contracted to achieve a significant reduction in cost and complexity in a new or operational underground mine. Evidentially, the future of underground mining communications will be a mix of new and the latest.

5.7   Green mining – a new engineering concept

'Green mining' is a newly coined concept of engineering, and the word 'green' connotes the environment and eco-friendliness of the approach being adopted in the mineral excavation process. As long as the mineral-getting process, whether opencast or underground, is not causing negative impacts on our living environment, either biotic or abiotic, they are defined or termed as green or the *'Green Mining'*. Green technologies and methods, such as *bio-remediation* and *phyto-remediation*, are used to control environmental pollution in mines . Continuous transport system namely conveyor belts instead of road transport is green because it generates very less dust. Similarly, in a hill mine, *green mining* refers to dumping of waste in a designated dump yard rather than along the hill slopes, which would cause disturbances to the hill ecology.

Chemicals of different varieties used in the mines, day-to-day equipment for safety, and other ancillary activities required for the mine operation are 'green' if they are not causing adverse/ negative impacts on the environment we live with.

To describe the performance of industrial activity, we usually come across 'Greenhouse gas (GHG) emissions' as one major evaluating parameter. The GHG emissions include many gases like carbon dioxide (CO_2), methane (CH_4), and nitrous oxide (NOx) apart from the water vapour, ozone (O_3), and Chlorofluorocarbons (CFCs). The sources of CO_2 associated with mining are varied and specific to each mineral, method of extraction, and region. For example, carbonate minerals used to neutralize acidic mining waste also release CO_2 in the atmosphere; the direct emission of methane from the coal mine workings; and the unit operation of blasting in mines generates NOx. GHG emissions from diesel fuel burning in mines also contribute to the direct GHG emissions.

A practical question comes into foray that, for a mine, how to calculate GHG emissions (Ali Soofastaei, 2018)?

In mines, movable heavy earth-moving machines (HEMM) are generally deployed. They are operated by diesel engines, emitting a lot of greenhouse gases into the atmosphere. Diesel engines emit both GHGs and non-greenhouse gases (ANGA, 2013) into the environment. Total GHG emissions are calculated according to the Global Warming Potential (GWP) and expressed in CO_2 equivalent or $CO_{2\text{-e}}$ (Kecojevic and Komljenovic, 2010, 2011). The following equation can be used to determine the haul truck diesel engine GHG emissions (Kecojevic and Komljenovic, 2010; DCE, 2012):

$$\text{GHG Emissions} = (CO_{2\text{-e}}) = FC \times EF \tag{5.1}$$

where FC is the quantity of fuel consumed in Kilo Litre (KL) and EF is the 'Emission factor' for haul truck diesel engines $=2.7t$ $CO_{2\text{-}e}$ per KL.

GHG emission, by sources and removals by sinks, for the mining and quarrying sector of India in 2016 was as follows:

GHG Emission (2016):

- Total CO_2 equivalent $=4,095.79$ Gigagrams
- CO_2 emission $=4,082.09$ Gigagrams
- CH_4 emission $=0.16$ Gigagrams
- N_2O emission $=0.03$ Gigagrams (Source: MOEFCC, 2021; Page 211).

The total fugitive emissions in the year 2016 were 37,179 Gg CO2e, of which 46% was from coal mining and post-mining operations and 54% from oil and natural gas production and handling systems. Fugitive methane emissions registered a decrease of 2% between 2014 and 2016, mainly due to a relative reduction in underground mining activities. Fugitive emissions contributed to 1.8% of emissions from the energy sector (MOEFCC, 2021).

5.7.1 Carbon footprint

'Carbon footprint' describes the amount of GHG emissions caused by a particular activity or an industry. Here, we are discussing carbon footprints concerning mining which is an industrial activity. Mining is currently responsible for 4%–7% of GHG emissions globally. According to the USEPA, other GHG emissions in various sectors as a percentage of 2019 GHG emissions are as follows: Industry (23%); Agriculture (10%); Electricity production (25%); Land Use and Forestry (12%); Transportation (29%); and Commercial and Residential (13%).

These days, increasing interest in 'carbon footprints' has been observed, and discussions are held about its reduction across all industries of different types. As a result of growing public awareness of the effects of global warming (a result of increased GHG emissions, which are part and parcel of climate change and its impact), the global community recognizes the need to reduce GHG emissions to save our planet. Current contribution of mining industry towards GHG emissions can be brought down to 2% or less with a reduction in carbon footprints, especially through the coal mining industry. All across the globe, mineral-producing nations can do it by searching for alternate resources, e.g. United Kingdom/Great Britain (UK) is phasing out coal with other energy sources to reduce the carbon footprint of the British mining industry. Similarly, Germany is closing down its brown coal mines.

5.8 Clean technologies in mining

Could it be possible that in a dirty mining operation there is a clean technology option? They are not to be gauged with the literal meaning of the word 'clean'. In other words, if it is green, it is presumed as clean (sub-section 5.7). All 'best practices mining' or 'best mining practices' abbreviated as 'BMPs' (Soni, 2017a) may be considered as a part of clean technology options. Their elaborative description is very difficult to describe in limited pages; hence, the only definition of *Clean technologies in mining* is described here.

Clean technology in mining means those which cause less damage to the environment and ensures a clean and green environment around the operation, e.g. monitoring of GHG emissions and environmental impact assessment and abatement in a mining enterprise is a good example in the context of clean technology in mining because it helps to reduce the carbon footprints (Singh, 2019). Practically, clean technologies are like others yet green meaning their adverse effects on the population and environment are less. The challenges to handling these technologies are also quite demanding.

The environmental damage associated with mining is well-known in terms of land damage, ecosystem degradation, air pollution, water pollution, and noise pollution. Throughout the life cycle of a mine and from the exploration phase, exploitation phase to the closure phase, clean technologies can be put into practice. There are several sub-areas of mines and mining where clean technology options are applicable and possible in real practice. Broadly, they are:

a.   Eco-friendly mining of minerals and approaches.
b.   Coal washing, handling, and coal transportation.
c.   Environmental pollution in mines and its control.
d.   Ore/mineral processing (Extractive metallurgy).
e.   Quality control for ores and coal.

All these (a–e) areas have many sub-areas which embrace clean technology option in terms of equipment, plant, and processes of several varieties. Ensuring that the net effect of clean mining technology is beneficial and positive, these options require integrated strategy and careful assessment through an interdisciplinary approach. After evaluation, they can be put into practice for industrial use.

5.9   Climate change and mines

As of 2021, it is well established, without any valid disagreement, that climate change is a reality and happening. Climate scientists unanimously agree that the Earth is getting warmer and that the rise in average global temperature is predominantly due to human activity (Constantine and Travis, 2016). The global challenge of climate change is very big and responded to through international agreements such as the United Nations Framework Convention on Climate Change (UNFCCC), the Kyoto Protocol, IPCC (Inter-governmental Panel on Climate change), and Montreal Protocol and Paris Agreement, 2015. According to Odell et al. (2017), very few published articles exclusively on 'climate change and mining relations' are available. Only two have been searched, the first article was published in 1998, while the second article was published in 2005. This reflects the underdeveloped nature of this particular field requiring more attention from researchers (Odell et al., 2017). The following are the key points for climate change and mine:

•   Mining operations are vulnerable to climate change:
      Most infrastructure was built based on the presumption of a stable climate and is thus not adapted to climate change (Ford et al., 2010, 2011; Pearce et al.,

2011). Climate change will affect exploration, extraction, production, and shipping in the mining and quarrying industry (Pearce et al., 2011). An increase in climate-related hazards, such as forest fires, flooding, and windstorm, affects the viability of mining operations and potentially increases operating, transportation, and decommissioning costs, as observed at Australian and Canadian mine sites. Directly or indirectly, it impacts mineral production from the mines.

As climate change proceeds and intensifies, the implications for a range of economic sectors will become more apparent. The energy and water intensity of large-scale mining may make the mining sector particularly vulnerable (Odell et al., 2017).

- The impact of the intersection of climate change and mining will be visible on water (hydrological impacts, acid mine drainage) soil, ecosystem health, and biodiversity.
- Climate change could open new areas to mining, e.g. (a) an expansion of existing ones akin to changed climate conditions for real mining operations, (b) will open up access to natural mineral deposits (a resource) that were previously inaccessible, (c) alternatives/substitution of mined ore or resources with other, and (d) energy-efficient infrastructure. Thus, mineral resources, which are produced from mines and used by the industry for miscellaneous applications, are vulnerable to climate change and have linkages.
- Policy Driver: Climate change is an emerging driver of mining policy too both at the national and international levels. The *'Public policy'* and *'Industrial policy'* get affected by climate change and cause impacts on the ongoing practices in mines, hence climate change and mines/mining are interrelated.

The field of climate change and mining is still underdeveloped and more urgently in need of overall elaboration. More examples of how specifically climate change is impacting mining at different spatial scales and the particular ways in which governments and companies are responding either locally, regionally, or nationally are the need of the hour. More work on the societal implications of the interactions between mining and climate change and the implications on the biological and physical environment is required.

The present industrial age is fuelled by hydrocarbons and industrialization, leading to a universal rising concentration of GHGs in the atmosphere with unintended changes in climate and environment. The rise in average earth temperature and sea-level changes are some well-known and perceptible impacts of climate change well noticed and recorded in various parts of the World. According to the 2015 Paris agreement (An international treaty on climate change adopted by 196 countries in Paris, on 12 December 2015 and entered into force on 4 November 2016), various governments are now aiming to keep the average temperature increase to less than 2°C. At present, the sea level along the Indian coast is estimated to be rising at about 1.7 mm/year (at different rates in the various parts of the coast). It has the potential to exacerbate the inundation of low-lying areas during extreme events such as storm surges while leading to increased coastal erosion (MOEFCC, 2021).

In India at the national level, there exist a 'National Action Plan on Climate Change' (NAPCC) which was launched in 2008 to address climate change-related concerns and promote long-term sustainability. Through the involvement of many agencies, eight missions that form the core of the NAPCC have been formulated. This mission keeps on updating the progress of each of the missions periodically (MOEFCC, 2021).

5.10   Quality control, automation, and engineering aspects

The present age of cut-throat competition and large-scale production is survived by 'quality' only. For an established company, an expanding company, or a start-up of the modern time, only quality can survive. Therefore, quality control has become a major consideration for all industries whether it be a mine, a plant, or a mine mill for mineral processing. Mine, which supplies better quality coal, the best ore grade, and renders service to the consumers at a reasonable cost, can stand tall and survive in the business. Thus, 'quality control' is used to connote all those activities which are directed towards defining, controlling, and maintaining quality. It includes all those techniques and systems required for the achievement of the best in terms of superior products and good grades, i.e. quality.

In mining, quality control is needed at various levels be it a *quality control of coal for grading* or *for assessing mineral percentages in the mined-out metallic ore.* For Coal Handling Plants (CHP), which are installed at the pit head in some mines, quality control is a necessity. Besides this, the mine-to-mill and several mining operations are quality conscious. *For example, a limestone mine for captive cement plants has to maintain the cement-grade quality of limestone for their day-to-day requirement as raw material feed.* Similarly, manganese ore of an assigned quality is needed for the ferro-alloys industries. Thus, the importance of quality control is immense and its different practical aspects must be known to an engineer. Associated with quality control are the different aspects of engineering and automation, e.g. mechanization, sampling, monitoring assessment, and evaluation. Proper quality control ensures the most effective utilization of available resources and confers to the use of all the quality standards. If a quality control route is adopted, the chances of deviations from established standards are minimal.

Coal quality: Coal, which is produced from the mines and mainly consumed in thermal power plants to generate electricity, has one big industrial requirement regarding its quality assessment. Hence, coal quality assessment based on the parametric analysis assumes a major and very large industrial requirement for the coal mines in India as well as in other countries. Most of the collieries producing different grades of coal in India are operated by Coal India Limited (CIL) and its subsidiary companies. CIL as MDO has set a target of covering 100% of coal dispatch under the Third-Party Sampling (TPS) Project,[1] which will help to deliver the required coal grade to the thermal power stations (Table 5.1).

Such sampling arrangements and quality assessment of coal are meant to increase the operating efficiency of thermal power plants and lower the pollution level of electricity generating units thereby providing cheap electricity to consumers, i.e. households and industries. Hence, TPS is a practical quality control method that provides a practical solution to achieve the target of good quality coal for industrial units as per their requirement and specifications (Annexure 5.1).

Table 5.1 Coal Grade: Quality Assessment/Analysis Under the TPS Project

Date: 06.09.2022
Group: Solid Fuels
• Test Report No.: XYZ/2022-23/ 11074 (Discipline: Chemical, Lab testing)
• Name & Address of the Party: PBSM Pvt. Ltd., Madhya Pradesh, India
• Samples Received on: 01.09.2022
• Condition of the sample received: The sample was packed in plastic bags
• Test Method Used for Analysis: IS 1350 Part 1 for Proximate Analysis & IS
  1350 Part 2 for GCV
• Sample Size Tested: 212 microns (BS 72 mesh)
• Relative Humidity of laboratory: 67%
• Room temperature of the laboratory: 26.37 degrees Celsius (°C)

S. No.	Sample ID	Weight of sample received (In grams)	On 60% RH & 60°C basis					Total moisture in %	Grade
			Proximate analysis						
			Moisture in % (by mass)	Ash in % (by mass)	Volatile matter in % (by mass)	Fixed carbon in % (by mass)	GCV kCal /kg.		
1	Coal Sample – 01	2286	3.76	65.94	8.56	20.98	1987	7.31	Ungraded
2	Coal Sample – 02	2162	4.05	56.31	11.18	27.57	2812	8.88	G15
3	Coal Sample – 03	2174	3.76	59.73	8.92	26.84	2446	8.12	G17
4	Coal Sample – 04	2053	3.51	64.87	8.16	23.06	2122	7.22	Ungraded
5	Coal Sample – 05	930	4.98	43.32	13.05	37.65	3755	12.16	G12

Source :   CSIR-CIMFR, Nagpur.
Note :   (i) Coal samples are collected by the party and not by the testing agency, i.e. lab.
(ii) Dates on which the test was performed are: Proximate Analysis – 3.09.2022; GCV – 4.09.2022;
Moisture on 60% RH & 40°C – 5/9.09.2022; and Total Moisture – 3.09.22.
*COAL GRADES: These grades of coal apply to the coal mined in India or exported to the Indian
continent and as notified by the Ministry of Coal, Govt. of India and principally used for the sale of
coal in the market to industries and consumers.

Grades of coking coal (based on ash content)

Grade	Ash content
Steel Grade – I	Not exceeding 15%
Steel Grade – II	Exceeding 15% but not exceeding 18%
Washery Grade – I	Exceeding 18% but not exceeding 21%
Washery Grade – II	Exceeding 21% but not exceeding 24%
Washery Grade – III	Exceeding 24% but not exceeding 28%
Washery Grade – IV	Exceeding 28% but not exceeding 35%
Washery Grade – V	Exceeding 35% but not exceeding 42%
Washery Grade – VI	Exceeding 42% but not exceeding 49%

(Continued)

Grades of semi-coking and weakly coking coal (based on Gross Calorific Value, GCV)

Grade	Ash + Moisture
Semi-Coking Grade-I	Not exceeding 19%
Semi-Coking Grade-II	Exceeding 19% but not exceeding 24%

Grades of non-coking coal (based on ash plus moisture content)

GCV Band (kCal./kg.)	Grade
Exceeding 7,000	G-1
Exceeding 6,700 and not exceeding 7,000	G-2
Exceeding 6,400 and not exceeding 6,700	G-3
Exceeding 6,100 and not exceeding 6,400	G-4
Exceeding 5,800 and not exceeding 6,100	G-5
Exceeding 5,500 and not exceeding 5,800	G-6
Exceeding 5,200 and not exceeding 5,500	G-7
Exceeding 4,900 and not exceeding 5,200	G-8
Exceeding 4,600 and not exceeding 4,900	G-9
Exceeding 4,300 and not exceeding 4,600	G-10
Exceeding 4,000 and not exceeding 4,300	G-11
Exceeding 3,700 and not exceeding 4,000	G-12
Exceeding 3,400 and not exceeding 3,700	G-13
Exceeding 3,100 and not exceeding 3,400	G-14
Exceeding 2,800 and not exceeding 3,100	G-15
Exceeding 2,500 and not exceeding 2,800	G-16
Exceeding 2,200 and not exceeding 2,500	G-17
Below 2,200	Ungraded

Source: https://coal.gov.in/en/major-statistics/coal-grades

Coal, through production, sale, and dispatch process, reaches from mines to the consuming industry (power plant). To determine its quality, there are various parameters based on which the quality assessment can be done. Two of the primary quality parameters of a coal sample are:

- **Gross calorific value (GCV):** Gross calorific value of coal is the amount of heat liberated during the combustion of fuel completely. The GCV or Calorific value is a measure of the amount of energy produced from a unit weight of coal when it is combusted in oxygen. It is a measure of the heating ability of coal and is needed to estimate the amount of coal required to produce a desired amount of heat. Calorific values are used to define the coal grade/rank of different coal types. Calorific values can be measured in standard English units (Btu/lb) or metric units (kilojoules/kg or mega joules/kg) using the 'Bomb Calorimeter' in the laboratory.
- **Ash content:** Ash content of coal is defined as the percentage residue of coal after being burnt at a certain temperature.

As described above, the TPS of coal facilitates such assessment of coal quality by determining the above-mentioned parameters. In general, this coal quality analysis is

Figure 5.4 Online Ash Analyser (OAA).
Source: http://www.aricoindia.com/online-ash-analyzer.html

done in the laboratory. Sampling and analysing can also be done through automation, i.e. Online Ash Analyser (Figure 5.4), which is a mechanized means of analysis.

The *Online Ash Analyser (OAA)* uses a dual Gamma-ray penetration method to eliminate the influence of coal porosity, particle size, and other factors and to rapidly measure the coal ash and calorific value. This method is the most commonly used technique for fast online *coal quality analysis*. This analysis method uses a radioisotope that is safe and reliable (low radiation), has a low energy level, and can be used for a long time without requiring replacement. OAA finds extensive use in coal mining, coal washeries, coal blending plants, thermal power plants, coking plants, etc. In this way, quality control results are faster and instant for dissemination.

Monitoring the *calorific value* and the *ash content* in coal is an important technical indicator for the evaluation of coal quality and sales price of different coal grades. Similarly, the Moisture Content of coal (water percentage present in coal) is that factor of coal quality which has an impact on the operating efficiency of the thermal power plant and the cost of power generation because it has a cascading effect on energy consumption (for drying of coal used as an industrial fuel).

5.11   Hazardous mining operation – tools and solution

The word 'hazardous' concerning mining has a very wide coverage for description. It applies to humans, systems, operations, and the environment equally. If safe practices are not followed, the entire mineral-getting operations could be hazardous. From the viewpoint of practical applicability, we will be concentrating our description on a few operation tools and solutions that reduce the hazard to the minimum.

a.   **Robotics in mining**: To keep humans out of inhospitable and hazardous environments, 'robotics' provide an effective solution. More and more replicative unit operations of mining can be done automatically using the robot.

   Basics about What are *Robots* and *Robotics* can be explained as follows: *Robots* are programmable machines that are usually able to carry out a series of actions either autonomously or semi-autonomously. Robots are generally used in the manufacturing sector and address some of the modern-day problems of productivity, hazardous environments, and the scarcity of skilled workers through the innovation route.

Robotics is a branch of technology that deals with physical robots. Principally, there are four types of robots:

1. Articulated Robots
2. SCARA (Selective Compliance Articulated Robot Arm) Robots
3. Delta Robots (Spider Robots)
4. Cartesian Robots (Gantry Robots)

Robots are programmable and make use of AI in their functioning. Selecting the best machine and knowing its capabilities and costs associated with different robot types is a crucial first step. For mining applications and to make an informed decision, an in-depth understanding of this emerging field is an essential requirement.

Today's increased industrial challenges can be easily handled through automation, and 'robots' are the tools that provide a solution. Mining-related production operations namely drilling, blasting, bulk material handling, loading, and crushing are labour-intensive, repetitive, and time-consuming hazardous processes. Robotics is helpful to automate them making it an easy task for the organization in this competitive market. Automation and robotics could prevent injuries in the industries (WEF, 2017). Though considerable advanced research in the field of robot development and robotics has been done, its applications are available for industrial use. Still, this new field is in an early implementation stage in the mining industry. Industrial applications are not very costly these days and will become more economical and cheaper further in the coming decade.

b. **Unmanned Aerial Vehicles (UAVs) in Mining:** The use of aerial vehicles for industrial applications goes back to the early 19th century. Starting from 1860, balloons were used to take pictures. In 1903, pigeons carrying a breast-mounted aerial camera were used for photography. Around the beginning of World War I, 'aerial torpedoes' were developed, and since then, the origin of drones came into existence. These unmanned flying systems got their penetration in several areas and applications. Since Drones were quite successful and efficient in mission-mode projects, they became popular with time. These unmanned flying systems were able to carry different sensors based on the type of their missions, such as acoustic, visual, chemical, and biological sensors. They are also equipped with cameras, data and image analysis devices as per the requirement of the job to be done (Figure 5.5).

In recent years, attention towards UAVs has been growing among academic and industry communities worldwide. UAVs are called 'Drones'. Thus, Drone and its technology called 'Drone technology' is an emerging area that is relatively new to the mining engineering fraternity. UAVs/Drones have a variety of capabilities for civilian, space, and military utilization. They are generally classified based on their configurations and grouped into nine categories, such as *fixed-wing, flapping wing, rotary wing, tilt-rotor, ducted fan, helicopter, ornithopter, and unconventional types of drones.*

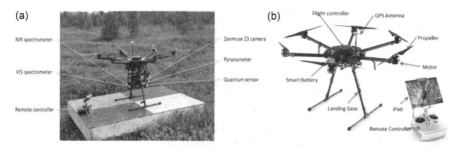

Figure 5.5 Drone for scientific application in mining.
Source: Francis et al. (2019).

UAV/Drones & mining: Generally, mines are located in far-flung and remote areas. Drones can be applied to monitor activities in the mines and the mining areas, thus preparing a guide for mine planning and safety. At the same time, monitoring mine is a challenging task which can be attended very well with UAVs/Drones. Therefore, time- and cost-consuming traditional methods of mine monitoring can be partly or fully replaced with Drones. Appropriately, drones can also be beneficial in surveying and mapping the mines.

Drone applications are in both surface mines as well as underground mines. In surface mines, weather conditions present a challenge by inducing deviations in the drone's pre-designated paths. In some cases, weather conditions can be damaging to the drones, leading to failure in their missions. Despite advancements in drone technology, the use of drones in underground mines has been limited. The application of drones in underground mines is challenging because access to unreachable and dangerous locations in underground mines is practically impossible for a drone operator to visualize. Harsh underground environments, confined space, reduced visibility, air velocity, dust concentration, and lack of wireless communication systems pose many obstacles to flying drones in underground working areas. Furthermore, and theoretically, Drones in underground mines have potential applications in gas detection, leakage monitoring, ventilation modelling, etc., all of which are essential for the workers' safety and health.

Limitations: Normally, a drone runs on a battery and consumes energy for hovering, wireless connection, data, and image processing. Due to the power restrictions as such, a decision needs to be made on whether data and image analysis should be performed onboard in real-time or offline to reduce energy consumption. This is the biggest limitation of Drone technology.

Application areas: Drones/UAVs have both commercial and scientific applications in the mining industry. Some of them are:

a. Mine operation (Continuous production monitoring)
b. Mapping of the accessible and non-accessible mining areas for the mine planning
c. 3D mapping for the mine environment
d. Ore control/Grade Control
e. Rock discontinuities mapping/Geotechnical Characterization

f. Post-blast rock fragmentation measurements
g. Slope monitoring and tailings stability monitoring
h. Construction monitoring
i. Subsidence monitoring
j. Landscape mapping
k. Aerial surveillance for safety and security
l. Drones in search and rescue operations
m. Facility management (for the mine /plant as a production pit)
n. Safety, health, and environment (SHE) monitoring

Advantages: Adopting drones in the mining industry can ease automation by providing visual and real-time primary data. Considering excellent manoeuvrability and low-cost maintenance, drones can make a huge benefit to the mine by surveying large areas in a short time period compared to the traditional /conventional methods used by the mining workforce. They can provide required data where there are health and safety hazards like abandoned areas and unsafe workplaces compared to the past. There are two main advantages of using drones in mining operations:

1. To conduct a quick inspection of an area in an emergency.
2. Quick inspection and identification of an area in a hazardous environment that is not easily approachable.

5.12   Industry 4.0 vs mining 4.0

Industry 4.0 is a vision and a concept for the development of future technologies needed for the economy of any country. It is a new *high-tech strategy*. It is a smart technology concept for a future industrial solution that includes mining too. It is recognized that 'labour' – a principal constituent/source of economic raw material production – is neither easily available nor cheap to hire, hence western high-wage countries have to route their highly productive mechanized industrial operations through automation. In this light and accordingly, *Industry 4.0 – A* Smart Industrial Solution became reality. The concept of Industry 4.0 revolves around a reliable, sustainable, and transparent raw material supply that has great importance to the industry. Economic strategic resources enable industrial production processes and innovations and are therefore indispensable for the raw material production centres. Newer and current technologies make necessary and essential changes in their system to keep pace with the latest trend(s) inevitably.

Industry 4.0 is the fourth industrial revolution driven by modern information and communication technologies which are becoming more and more popular in the industrial automation sector. Intelligent systems, virtual digital data systems, and physical real systems are all part of it. From Industry 4.0, we are heading towards *Industry 5.0* and the Green IoT with relevant future technologies. Taking into consideration the new and latest industry trend (Edge computing and Edge-AI), we have to get ourselves acquainted with *Society 5.0* and *Education 5.0* as well. *Edge computing, Edge-AI,* and *Green IoT systems* will make use of fewer raw materials and help in reducing energy consumption for a benefit to the manufacturing, operating, and recycling processes thus encouraging sustainable environmental practices. The Green IoT paradigm has emerged as a future research area to reduce carbon footprints.

Mining 4.0: Mining 4.0 means the advancement of automation in mines during extraction, transportation, and processing. Concerning the mines, *Mining 4.0* apply to all core equipment, machinery, and processes involved in mining. Mining 4.0 is not only a vision but, in parts, is already a reality. To maintain the technological leadership of the mining industry on the industrial front at the international level, it is extremely necessary to drive forward the R&D activities in the field of communication, sensor technology, automation. Technology interfaces between industrial enterprises (production companies) and other stakeholders must be healthy and progressive. The 'surface mining segment' is the biggest segment of the industry for implementation of Industry 4.0 /Mining 4.0. Industrial examples can be found in future mine projects or next-generation mining projects using established remote operations and heavy automation for improved productivity and effective mine operations.

Some of the most important aspects to meet the objectives of *Mining 4.0* are selective mining, autonomous production of raw materials, and low impact due to mining (Figure 5.6). Resource-sparing and sustainable mineral production are only possible with a high degree of selective mining which saves costs for transportation, treatment, and waste disposal. Another aspect of Mining 4.0 encourages the 'low-impact mine' – a mine with a minimal negative impact on the environment, people, and surroundings (Thomas Bartnitzki, 2017).

Industry 4.0 vs Mining 4.0: Industry 4.0 is a progressive upgradation of Industry 1.0 to 2.0 to 3.0 (Figure 5.7). With increasing computing power and communication capacity, the smart systems are networked, configured, and optimized and made autonomous to work for themselves either through machine-to-machine (M2M) communication or person to person (P2P) communication, and they can be extended with engineering support when required. Such smart machines and systems are networked with each another via the IoT also and react to internal and external events with learned behavioural patterns. The entire production and product life cycle, as well as the entire value chain, can be analysed smartly through Industry 4.0. An example of

The top three challenges for Mining 4.0.

Core-Technologies for Mining 4.0.

Figure 5.6 Mining 4.0.

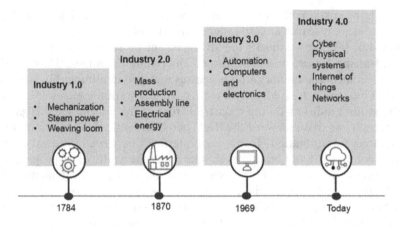

Figure 5.7 Progressive upgradation from Industry 1.0 to Industry 4.0.

'sensor technology' which is used in mines is described below for a better understanding of Mining 4.0/Industry 4.0.

An Example: *Environment sensors* or *Sensor networks* are used in the mining industry to record the state or properties of the environment within the mine workings (workplace environment). The environmental sensor system includes tasks such as (a) position determination e.g. positioning, collision avoidance, and mine model; (b) material determination, e.g. material identification, boundary layer detection, material quality, and mineral deposit model; (c) ventilation determination e.g. temperature, quality, flow rates, and ventilation models; (d) mine water determination e.g. water level, water quality, flow rate, and mine water models.

In mines, *small sensors* (handy and embedded) can be fixed quickly and in as many places as possible. Similarly, *Wireless radio-based sensors* (sensor networks) can be applied to each other without any major installation efforts and risk of cable damage. Other sensor types include *'energy self-sufficient sensors'* which are extremely energy-efficient and equipped with low-energy and energy-harvesting power packs.

A real-time trend can be established using 'sensor technology'. However, 'Sensor Networks' are not yet widely used in the mining environment and mines.

5.13   Emerging global trends in the mining sector

Some of the megatrends that will shape the world and industry over the next decade have been depicted in Figure 5.8. Industry consolidation, decarbonization, sustainability, and industry 4.0 are those broader areas that have a significant impact on the mineral sector globally.

Some advanced and mineral-rich countries are at par with the global trend and developing nations are gearing up. Comparing India with global trends, the Indian mining industry is heading towards self-reliance and infrastructure improvement in

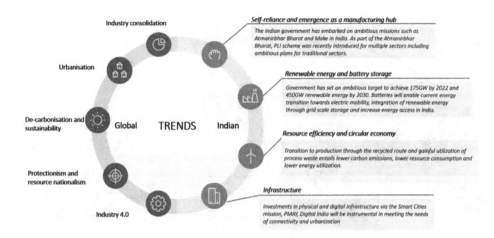

Figure 5.8 Emerging trends.

the mining sector. Its national contribution to the exchequer via industry route has a bright and good scope as mineral demand is on the rise. Hence, digital initiatives, smart mines, alternate energy sources replacing coal, and advanced skill and expertise will be the future for more application and learning in the mining sector of India.

For the mining industry and ancillary applications, the 'Internet of Things' (IoT) and 'Big Data' are current buzzwords and reflect a trend. While the term 'big data' is relatively new, the process of collecting and storing large amounts of information is old. A mining engineer can help the organization capitalize on the opportunities presented by IoT and Big Data. In a mining enterprise, capturing and analysing large amounts of data from a variety of data sources is an area towards which the evolution of technological advances is taking place quite rapidly, commonly, and usually referred to as a sort of digital transformation or the digital revolution.

5.13.1   Edge computing and edge controllers

'*Edge computing*' is a technique for handling large and voluminous data related to mining suitably, whereas 'Edge controllers' are a modern automation platform for enabling digital transformation and effectively applying IIoT concepts. An edge controller combines a real-time operating system (RTOS) with a general-purpose operating system (OS) like Linux. Their benefits can be realized by applying edge controllers to both new designs and retrofit applications. The ability to collect more data from different places has increased the volume, velocity, and variety of data.

Many aspects of mining operations require some level of automation, and nearly every area creates data useful for analysis and operation (Figure 5.9). Because mining operations are most often remotely located, any useful technologies must be suitable for these conditions and provide extensive communication options. Edge controllers are built for this operational technology (OT) environment, and they have the latest and most secure IT computing and networking features. Edge controllers are especially compelling for this service because they can gather and store data locally,

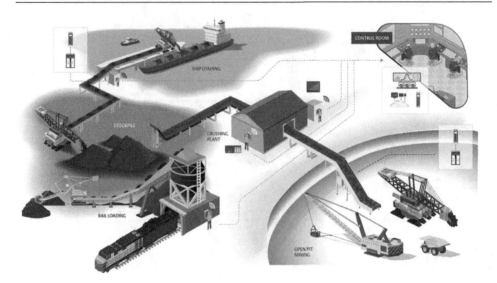

Figure 5.9 A pictorial representation of edge computing for the mining application.
Source: Emerson (2020).

process and analyse it, directly inform operational logic of optimal settings, and relay the most essential information to higher-level systems (Emerson, 2020). The result is more efficient, environmentally friendly, and safer operations for all types of mining and metals operations which are highly risky too.

For real-time, front-line operational technology automation platforms used in the mining operations throughout, some devices that can be named are:

- Traditional programmable logic controllers (PLCs)
- Variable frequency drives (VFDs)
- Distribution control system (DCS)
- Supervisory control and data acquisition system (SCADA)
- Local human–machine interfaces (HMIs)
- Remote terminal units (RTUs)

PLCs play an important role in almost every area and may include local HMIs. RTUs are also widely used, providing some functionality similar to PLCs and adding remote connectivity features. Increasingly, systems incorporate intelligent field devices like VFDs that can supply extensive operational and diagnostic data. Larger processing areas may rely on a DCS, and mine/plant-wide operations may be monitored by a SCADA system.

To tap the IIoT communications and analysis, these higher-level electronic gadgets can be adopted in the implementation schedule and the available data can be made easily accessible for processing and transmission. Because automation platforms, to be used with the mining equipment, have generally become more intelligent over the years, there are more opportunities than ever before for obtaining the right field data and acting upon it to overcome operational challenges.

Edge controllers, meant for edge computing, are physically built to withstand the harsh conditions found at remote mining sites, including extremes of temperature, contaminants, and vibration. The onboard general-purpose operating systems offer several advantages such as security, better internet connectivity, and flexibility for communication and network issues. Users can develop applications in IT-oriented languages like C, C++, Python, and Java.

Mining of minerals including the metal extraction from them is no more a low-tech digging venture. All these industrial and related ancillary operations will be driven by Industry 4.0/Industry 5.0 and the Green IoT concepts with relevant future technologies. Edge computing, Edge-AI, and Green IoT systems will help in efficient and cost-effective mine production including the sustainable development of mines in an environmentally friendly manner (Fraga Lamas, 2021).

5.14   Blockchain technology?[2]

Record-keeping of data and transactions is a crucial part of the business. Often, this information is handled by the mining company/business houses, or it is passed through a third party like manufacturers, equipment suppliers, financiers, brokers, bankers, or lawyers.

BT is a structure that stores transactional records, also known as the block, of the public in several databases, known as the 'chain', in a network connected through peer-to-peer nodes. Typically, this storage is referred to as a 'digital ledger'. Every transaction in this ledger is authorized by the digital signature of the owner, which authenticates the transaction and safeguards it from tampering. Hence, the information the digital ledger contains is highly secure. In simpler words, the digital ledger is like a spreadsheet shared among numerous computers in a network in which the transactional records are stored based on the actual. The fascinating angle is that anybody can see the data, but they can't corrupt it.

BT avoids the long process and facilitates the faster movement of the transaction, thereby saving both time and money. Technologically, Blockchain is a digital technology (a ledger). BT is gaining a lot of attention recently. In an increasingly digital world, the 'Blockchain' is an emerging technology with many advantages:

i. Retrievability of data, security and safety: Since BT uses digital data and signature features, these features are easily retrievable/recoverable with the click of a computer button. To conduct fraud-free transactions and change the data of an individual, by others without specific permission, is quite difficult in the BT making it a safe and secure system that is rather impossible to corrupt.
ii. Decentralized system: Conventionally, you need the approval of the concerned or the regulatory authorities like a government for official transactions, however, BT transactions are done with the mutual consensus of users, resulting in smoother, safer, and faster transactions.
iii. Automation capability: It is programmable and can generate systematic actions, events, and payments automatically when the laid down criteria or norms are fully met.

Blockchain applications are presently entering the mining sector slowly, while some industries have quickly embraced blockchain technology. The possibilities and gearing up of BT to transform the mining industry are immense, e.g. smart contracts and online/digital submission and processing with full transparency; a digital or virtual currency that works as a medium of exchange (cryptocurrency); tracking of goods' status in real-time, i.e. streamlining supply chain management); improved industrial insurance for the customer satisfaction and record management. As described above, the BT system has a decentralized approach that can influence our day-to-day life extensively even in mine and right from proving our digital identity to the way businesses function. Industry, financial institutions, governments, hospitals, insurance, media, print & publication, the entertainment industry, and legal institutions, all are exploring blockchain applications.

5.15   Technovations and applicability in the mining sector

Typical major operations concerning mining where technovations may be applied are:

i. **Mining and extraction**: Pit mine and underground tunnels, equipment and conveyors characterized by deployment across large areas, e.g. ore tracking, ore blending, vibration data and its leveraging, stockyard optimization, project and production planning and scheduling.
ii. **Bulk material handling**: Crushers, stackers, screeners, and autonomous vehicles; Waste and overburden handling in bulk.
iii. **Water management**: Systems supply, treat, store, recover, and reuse water to provide the right quantity and quality needed for processing.
iv. **Utilities**: Includes electrical power distribution and monitoring to support other systems, power optimization.
v. **Logistics**: Rail and port operations for handling mining production.
vi. **Metallurgical operations**: For metal extraction from mineral /ore namely:

   a. Comminution /grinding/ball mills and its operation
   b. Separation includes filtration cells, leaching, thickeners, solvent extraction, and filtration
   c. Refining, e.g. systems such as electro-winning, electrolysis, and smelting for extracting and purifying metals.

5.15.1   Creating sustainable solutions using technology

An opportunity to reshape the mining sector lies ahead through technology options for the day-to-day mining operation. As described earlier by leveraging technologies, such as satellite, blockchain, and digital payment, it is possible to modernize mineral mining and secure the mineral supply chain from earth to end user with traceability, transparency, and confidence. The faster the technology upgradation and implementation, the better will be the benefits to the industry. Building a green future and creating sustainable solutions using technology, especially for the mines and mining sectors, are as listed below. In the listed points, i.e. (a) to (h), the goals are kept outside and results are kept in the parenthesis.

a. **Narrow vein underground mining** (*selective mining using technology options*).
b. **Deep surface mining** (*enhanced production and productivity*).
c. **Better and more performing mining machinery** (*heavy earth-moving machines to safe underground machinery*).
d. **Digital transition** (*turning the conventional practices into new and modern*).
e. **Zero emission** (*battery-powered equipment over diesel-powered for reduced carbon footprint*).
f. **Low maintenance** (*innovative hydraulic and electric control architect, silent electrical components, corrosion-free material, oil and water cooling system, reduced mining machine vibrations, and quick replacement system*).
g. **Ergonomy and comfort for the mining machines and equipment** (*joystick control, digital display, ergonomic design for human comfort*).
h. **Safety in mining through technology** (*remote operation for hazardous environment and extreme working conditions*).

5.15.2 Future of mining jobs

Mining jobs are not white-collar jobs. This industrial work is labour oriented mostly despite technovations. The present/current trend of the labour market requires a combination of financial security, Flexi-work, medical benefits, and a sense of social security. From the future point of view, the coming generation, which is a pool of talent, will be attracted to the dirty mining profession only when social security is affordable both by the employer and the government. Lack of social security will make mining a less willing profession and that in turn impacts innovations and technical advances. Hence, for the new mines, the modernized labour market with social security in place is the future of mining jobs. From the production as well as corporate point of view, Flexicurity (*A welfare state model originated in Denmark that combines high mobility between jobs with an attractive income, active labour policy, and a comprehensive safety net*) is key to the talent pool of mining.

In the techno-savvy business world, it is common wisdom to move forward by learning from the past. Emerging technologies from advanced data analytics to AI have always had the potential to transform the mining industry by realizing improvements in operational efficiency, enhancing productivity as well as safety performance, and empowering employees to do more meaningful work. Mining companies and the likelihood of vectoring in the intelligent, integrated operations in mining in a comprehensive manner are likely to accelerate with the adoption of these new tools namely digital technologies, AI, ML, robotics, drone use, edge computing, and big data/data analytics in the mining industry as briefly highlighted in this chapter. While mining operations use many types of equipment and operating methods (traditional/ established), there are ample opportunities to realize benefits by embarking on a digital transformation.

Notes

1   *Third-party sampling project* is an arrangement of coal quality analysis to be done in a certified laboratory, where different coal quality parameters are determined and analyzed to categorize different coal grades. This industrial work is carried out by an independent organization other than the producer and consumer, hence the name third-party project.
2   https://www.simplilearn.com/tutorials/blockchain-tutorial/blockchain-technology#:~:text=Blockchain%20technology%20is%20a%20structure, to%20as%20a%20'digital%20ledger

References

Advance Research Instrument Company. (n.d.), Online Ash Analyzer, http://www.aricoindia.com/online-ash-analyzer.html; Accessed on 11/05/2021.

Ajay, K. Singh. (2019), Better Accounting of Greenhouse Gas Emissions from Indian Coal Mining Activities – A Field Perspective, *Environmental Practice*, Vol. 21, pp. 36–40., DOI: 10.1080/14660466.2019.1564428.

Ali, Soofastaei. (2018), *The Application of AI to Reduce Greenhouse Gas Emissions in the Mining Industry*, London: Intech Open, 18 p. DOI: 10.5772/intechopen.80878.

ANGA. (2013), Department of Industry, Climate Change, Science, Research and Tertiary Education, Ed – ANGA (National Greenhouse Accounts Factors, Australia), Govt. of Australia, pp. 327–341.

Barnewold, L. (2019), Digital Technology Trends and Their Implementation in the Mining Industry, In *Proceedings of the 39th International. Symposium on 'Application of Computers and Operations Research in the Mineral Industry (APCOM-2019)*, edited by Christoph Mueller et al. (Mining Goes Digital), June, Wroclaw, Poland, pp. 9–17.

Bartnitzki, T. (2017), Mining 4.0 – Importance of Industry 4.0 for the Raw Materials Sector, *Mining Report*, Vol. 153, Issue 1, pp. 25–31.

Boussalis, C. and Coan, T. G. (2017), Text-mining the Signals of Climate Change Doubt, *Global Environmental Change*, Vol. 37, pp. 89–100. DOI: 10.1017/j.gloenvcha.2015.12.001.

Canisius, F., Wang, S., Croft, H., Leblanc, S. G., Russell, H. A. J., Chen, J., and Wang, R. (2019), A UAV-Based Sensor System for Measuring Land Surface Albedo: Tested over a Boreal Peatland Ecosystem, *Drones*, Vol. 3, No. 27, pp. 22–41. DOI: 10.3390/drones3010027. www.mdpi.com/journal/drones

Chandra, Sagar. (2021), Meshing Mining with Technology, *Global Mining Review*, Vol. 4, Issue 1, pp. 15–17.

DCE. (2012), *Emission Estimation Technique Manual*, Canberra, Australia: The Department of Climate Change and Energy Efficiency (DCE), Australian Government, pp. 127–141.

Emerson. (2020), White Paper on Edge Computing Unearths New Value for Mining and Metals Applications, September 2020, pp. 1–10, www.emerson.com

Ford, J. D., Pearce, T., Prno, J., Duerden, F., Berrang Ford, L., Beaumier, M., and Smith, T. (2010), Perceptions of Climate Change Risks in Primary Resource Use Industries: A Survey of the Canadian Mining Sector, *Regional Environmental Change*, Vol. 10, Issue 1, pp. 65–81. DOI: 10.1007/s10113-009-0094-8.

Ford, J. D., Pearce, T., Prno, J., Duerden, F., Berrang Ford, L., Smith, T. R., and Beaumier, M. (2011), Canary in a Coal Mine: Perceptions of Climate Change Risks and Response Options among Canadian Mine Operations, *Climate Change*, Vol. 109, Issue 3–4, pp. 399–415. DOI: 10.1007/s10584-011-0029-5.

Fraga Lamas, P., Lopes, S. I., and Fernández Caramés, T. M. (2021), Green IoT and Edge AI as Key Technological Enablers for a Sustainable Digital Transition towards a Smart Circular Economy: An Industry 5.0 Use Case, *Sensors,* Vol. 21, p. 5745. DOI: 10.3390/s21175745. p. 36.

Gavin, M., and Mudd, G. M. (2010), The Environmental Sustainability of Mining in Australia: Key Mega-trends and Looming Constraints, *Resources Policy*, Vol. 35, Issue 2, pp. 98–115. DOI: 10.1016/j.resourpol.2009.12.001.

Hodgkinson, J. H., Littleboy, A., Howden, M., Moffat, K., and Loechel, B., (2010). Climate Adaptation in the Australian Mining and Exploration Industries. CSIRO Climate Adaptation Flagship Working Paper No. 5. http://www.csiro.au/resources/CAFworking-papers.html

Internet Society (IS). (2017), AI & ML: Policy Paper, p. 13.

Javad, S., Elaheh, T., Pedram, R., and Mostafa, H. (2020), A Comprehensive Review of Applications of Drone Technology in the Mining Industry, *Drones*, Vol. 4, Issue 34, pp. 1–25. DOI: 10.3390/drones4030034. http://www.mdpi.com/journal/drones

Kecojevic, V. and Komljenovic, D. (2010), Haul Truck Fuel Consumption and CO_2 Emission under Various Engine Load Conditions, *Mining Engineering*, Vol. 72, Issue 12, pp. 44–48.

Kecojevic, V. and Komljenovic D. (2011), Impact of Bulldozer's Engine Load Factor on Fuel Consumption, CO_2 Emission & Cost, *American Journal of Environmental Sciences*, Vol. 7, Issue 2, pp. 125–131.

MOC. (2021), Ministry of Coal's Agenda for the year 2021–22, A Document of Ministry of Coal (MOC), Govt. of India, p. 104.

MOEFCC. (2021), India – Third Biennial Update Report to the United Nations Framework Convention on Climate Change (UNFCCC), Ministry of Environment, Forest and Climate Change (MOEFCC), Government of India, p. 500.

Nagaraj, R. (2022), World's Biggest Hydrogen-Powered Mining Truck Begins Operations, Team BHP. https://www.team-bhp.com/news/worlds-biggest-hydrogen-powered-mining-truck-begins-operations

OdellScott, D., Anthony, B., and Karen, E. F. (2017), Mining and Climate Change: A Review and Framework for Analysis, The Extractive Industries and Society, p. 14. DOI: 10.1017/j.exis.2017.12.004

Paul, K. and Steve, W. (2017), *The Digital Mine – What Does It Mean for You?* London: Diggers & Dealers, Deloitte Touche Tohmatsu Limited, pp. 1–8.

Pearce, T. D., Ford, J. D., Prno, J., Duerden, F., Pittman, J., Beaumier, M., and Smit, B. (2011), Climate Change and Mining in Canada, *Mitigative Adapt Strategy Global Change*, Vol. 16, Issue 3, pp. 347–368. DOI: 10.1007/s11027-010-9269-3.

Singh, S. (2020), Application of AI in Mining - A Case Study, *Indian Mining and Engineering Journal (IM & EJ)*, Vol. 59, Issue 8–9, pp. 28–34.

Soni, A. K. (2017a), Process-Oriented Opencast Mine for Limestone, In: *International Conference & Expo on Mining Industry: Vision 2030 and Beyond*, Organized by MEAI Nagpur Chapter, Nagpur, India, 7–8 December 2017, pp. 188–193.

Soni, A. K. (2017b), *Mining in the Himalayas – An Integrated Strategy*, Boca Raton, FL: CRC Press/Taylor & Francis, p. 225. DOI: 10.1201/9781315367552.

UTTAM, Ministry of Coal. (n.d.), Sampling Process: Third Party Sampling. http://uttam.coal-india.in/sampling_process.html; Accessed on 11/05/2021.

WEF. (2017), White Paper on Digital Transformation Initiative Mining and Metals Industry, World Economic Forum (WEF), January, p. 35. http://reports.weforum.org/digital-transformation

Annexure 5.1: Quality control of coal by TPS method: SOP and detailed methodology

(Refer Section 5.10).

(Source: http://uttam.coalindia.in/sampling_process.html)

- **Sample collection**

    a. **Case of coal dispatch by rail**: Each rake of coal, a rail wagon, supplied to the purchaser from the delivery point will be considered as a lot for sampling.

    1. Each rake of coal can be further divided into several lots based on grades of coal or the number of wagons belonging to the customer.
    2. Each rake shall be divided into equal sub-lots as under:

No. of wagons in the rake	Number of sub-lots
Up to 30 wagons	4
>30 wagons up to 50 wagons	5
>50 wagons and above	6

    3. From each of the sub-lots, serial number of wagons will be selected as per a random table in BIS Standards (IS: 436-Part I/Section I, 1964) or its latest version for the collection of increments.
    4. In each wagon selected for sampling, the sample will be drawn from the spot in a manner in such a way that, if in one wagon the sample is collected at one end, the spot in the next wagon will be in the middle of the wagon, and the spot in the third wagon at the other end; this sampling procedure will be repeated for subsequent wagons.
    5. Before collecting the samples, the spot will be levelled and at least 25 cm of Coal surface will be removed/scraped from the top and the place will be levelled for an area of 50 cm by 50 cm.
    6. About 50 kg of the sample will be collected from each selected wagon in the rake by drawing 10 increments of approx. 5 kg each with the help of a shovel/scoop.
    7. Samples collected from all the selected wagons/sub-lots in a rake will be mixed to form a gross sample. Items (IV) to (VI) above shall be applicable for Coal supplied in all types of wagons.

8. In the case of the rake having a live overhead traction line, the power supply in the overhead traction shall be switched off to facilitate the collection of samples from the wagons according to points (IV) to (VI) above.

b. **In the case of coal dispatch by road:** Samples of coal will be collected at the delivery point/dispatch location during the day and night from the trucks of the purchasers who opted for TPS.

1. The first sample of coal shall be collected randomly from initial lot of 8 trucks at the dispatch point. Thereafter, subsequent samples of coal shall be collected from every 8th (eighth) truck from which the first sample has been collected. The same process shall be repeated for every 8th truck.
2. Before collecting the sample, the spot at the top of the truck will be levelled and at least 25 cm of coal surface shall be removed/scraped from the top; the place will be levelled for an area of 50 cm by 50 cm for collection of sample.
3. About 30 kg of the sample shall be collected from each truck by drawing 6 increments of approx. 5 kg each with the help of a shovel/scoop.
4. All the samples collected from selected truck will be mixed to form a gross sample.

• **Sample preparation**
The gross sample collected at the Delivery Point will be reduced to laboratory samples in the size of 12.5 mm for total moisture and 212 microns IS Sieve (Top Size) for Testing and Analysis. The preparation of lab samples is done at the Coal company preparation rooms located in the dispatch area.
The preparation consists of the following steps:

1. **Primary crushing:** The gross sample collected is fed to the primary crusher and the coal size is reduced to 12.5 mm size with the help of mechanical crushing.
2. After the primary crushing of the coal sample, one portion (one-fourth of the gross sample) called Part 1 will be used for the determination of total moisture and the other portion (three-fourth of the gross sample) called Part 2 will be used for testing and analysis.
3. **Secondary crushing:** After the primary crushing of coal, Part-2 of the coal sample is sent to the secondary crusher, coning and quartering of a coal sample is carried out at the secondary crusher, and the sample is further reduced to 3.35 mm in size.
4. **Pulverizer:** Coning and quartering of a coal sample is done, and the pulverizer will reduce the coal sample to powdered form and the top size of 212 micron is attained. Precaution will be taken so that further sieving and pulverising are not needed at the time of testing.

In case the Pulverizer is not in working condition in the field, sample packed and carried in 3.35 mm to lab and then pulverized for testing. The final pulverized sample will be divided into four equal parts viz. Set I, Set II, Set III, and Set IV.

a. Set I is handed over to third party for testing and analysis;
b. Set II of the sample shall be handed over to the coal company;
c. Set III of the sample shall be handed over to the purchaser; and

    d. Set IV of the sample called the referee sample is sealed jointly by the Third Party and representatives of each of the Parties and is kept in referee room with lock and key arrangement under CCTV surveillance in the custody of the third party. The referee sample is retained in double sealed condition (duly signed by the third party and the representatives of the parties) for 30 days from the date of sample collection.

• **Transportation of sample:**

Set I of the coal sample is transported to coal testing labs for testing and analysis of the coal sample in a sealed condition.

• **Testing and analysis:**

1. **Determination of total moisture**: Part 1 (one-fourth of the gross sample) of size 12.5 microns is used for the total moisture test. The total moisture test is carried out at coal company labs and witnessed by each of the representatives of the party.

2. **Determination of ash, moisture, and gross calorific value (GCV)**: The testing and analysis of coal samples are carried out at coal testing Labs, selected by a Third Party, as per *BIS Standards (IS: 1350 Part 1-1984)* or *BIS Standards (IS: 1350 Part-II-2022)*. The determination of ash, moisture, and GCV is done on equilibrated moisture basis (at 60% RH and 40°C).

• **Referee sample:**

In case of any dispute regarding the findings of the result, any of the parties may opt for an analysis of the referee sample. The communication of referee samples should be made within 7 days of communication of the report. The third party holds the right to choose one of the designated/empanelled referee labs.

Chapter 6

Case studies

Never give up on yourself.

(Bhagwad Gita)

Minerals and their extraction from Mother Earth are what we call 'mining'. It is well known that mining is a site-specific activity of business interest recognized as an industry. Several types of minerals from various categories, viz. fuel minerals, metallic minerals, non-metallic minerals and industrial minerals, are being mined for centuries, but we must know how they are mined in real. When we discuss the mining of minerals, the *conservation of minerals* also forms an extremely important and integral topic for their excavation. It should always be emphasized, be it an industrial mineral of high value or a commercial mineral of low value, that the mineral produced must be consumed/used judiciously. Since minerals are *wasting assets*, at once if they are unearthed, it takes millions of years to form naturally again. Considering this well-known fact, the given international case studies make us understand their mining importance as well as social relevance for people and society across national boundaries.

This chapter of the book describes seven *case studies* from different countries necessary for practical understanding. These are the real mines, where actual mining is done, and a variety of problems and solutions have emerged; hence, a full-length chapter is devoted to this purpose despite being lengthy. We will be explaining the case studies of importance only, which will give an insight into the actual mining depicting the field conditions or site details. Each case record, described here, is different from one other. The described case studies in this chapter give answers to the following: (a) How does the exploitation of minerals occur and method differ? (b) What are the topicalities of an operational mine and mine site in different parts of the world? and (c) What intricacies are involved at a specific mining site in the coal/metal extraction process? The reader can learn about the mining of *Coal, Lithium, Copper* (a base metal), *Gold* and *Uranium* and know how they are excavated from the Earth.

Most of these described studies are invited case studies from different countries and all contributors are field experts in mining operations. Their experiences are rich making each contribution valuable. One of the case studies of Bolivia is derived from the available web sources, whereas the Indian case study of Jayant Opencast Coal Mine, located in the Northern Coalfields of India (Singrauli, Madhya Pradesh), is prepared by the book's authors themselves. Concerning the coverage, attempts have been made to cover both the opencast and underground methods of coal and non-coal

DOI: 10.1201/9781003274346-6

minerals broadly. Special methods of mining (chemical extraction) in Canada and Bolivia are also touched for wider coverage. It should be made clear that only selected and important cases have been dealt with and not all minerals, as their delineation in limited pages will not be a focused attempt for this book.

Minerals are primarily natural deposits and variation in their exploitation characteristics (mining) makes the case study section interesting, charming, and an engineer's challenge. As a mining engineer and industry professional, one should be aware of some of the classical cases and their practical engineering problems. Through case studies as described in this sub-section of the book, it is explained that:

i.	How Potash is extracted in Canada?	(Solution mining, Canada)
ii.	How Lithium is extracted in Bolivia?	(Salt flat at Salar de Uyuni, Bolivia) – A Latin American mining case study)
iii.	How mining is practised in Poland?	(Underground Longwall Mining of Coal)
iv.	How coal is extracted in India?	(Jayant Opencast Coal Mine, India)
v.	How copper ore is extracted in Brazil?	(Salobo Copper Mine, Brazil)
vi.	How Uranium ore is mined/extracted?	(Mining of strategic minerals; Restricted)
vii.	How gold mining is done in America?	(A case study from the USA)

Undoubtedly, case studies in mining engineering have immense importance to develop sound professional judgement. Like other engineering fields, professional judgement in mining plays a key role in mine development, its smooth operation, and future prediction for effective planning of a mine or a mining project. With this in mind, we have organized the following sub-sections of the book that has wealth of knowledge and information and unique value. This can be encashed worldwide for professionals to use, for information exchange, and to train future engineers.

Case study 6.1: Southey potash project, Canada

Introduction

Solution mining is probably the only option available to achieve less destructive mining in the long term, especially as ore grades continue to drop off. It's the only way one can think of to avoid generating massive volumes of waste rock and tailings, although it will not apply to all mineral types. Wherever minerals to be extracted lie at a depth of 1,000 m or deeper and are easily soluble in water, *solution mining* provides the best economical solution. It is less damaging to the surface topography. However, it's going to be difficult to separate minerals from the leached-out brine (a type of mine water leached out from underground) through processing in the solution mining systems. It is really hard to manage water impacts. Handling both brine and process water in solution mining is a stupendous task because brine has corrosive characteristics. Ultimately, water management practices assume significant importance in this method and form a part of the valuable mining work along with the widening and continuity of the excavating cavern developed underground.

This case study is described with the help of the Audio-Visual Link given (see the last section of this case study). A short description of the region, company, and theoretical aspects has been described for an in-depth and better understanding of this special case study which is different from conventional mining.

The mine and the region

The Southey Potash Project, Canada, owned by Yancoal Canada, is located 60 km North of the city of Regina in the Saskatchewan province of Canada. Since nearly two-thirds of the world's recoverable potash reserves are in the *Saskatoon* region, a brief review about *Saskatoon (Saskatchewan)* has been given in the next paragraph for a detailed understanding of the region.

Saskatoon, the largest city in the Canadian province of *Saskatchewan*, is situated on the banks of the *South Saskatchewan River* which is crossed by seven bridges within the city limits. *Saskatoon* city (218 km^2 area and 481.5 m altitude above sea level) straddles a bend in the South Saskatchewan River in the central region of the province. *Saskatoon* is located along the Trans-Canada Yellowhead Highway and has served as the cultural and economic hub of central *Saskatchewan* since its founding in 1882 (Wikipedia). Further, the world's largest publicly traded uranium company, CAMECO, and the world's largest potash producer, NUTRIEN, have their corporate headquarters in Saskatoon (Canada). *Saskatoon* is also the new home of *BHP Billiton's Diamonds* and their specialty products business unit. The economy of *Saskatoon* has been associated with potash, oil, and agriculture (specifically wheat). Various other minerals like uranium, gold, diamond, and coal including their spin-off industries fuel the economy of *Saskatoon* besides potash (Nutrien Company).

The company

Yancoal Canada Resources Company Limited (Yancoal Canada) is a Saskatoon-based potash exploration and development company. The company was established in 2011 and is a wholly owned subsidiary of Yanzhou Coal Mining Company Ltd. (Yanzhou Coal), an international mining company listed in the New York, Hong Kong, Shanghai, and Sydney stock exchanges with over 40 years of mining experience and have businesses in China, Australia, and South America. Its core businesses include mineral resources exploration, coal-based chemicals production, power generation, and manufacturing.

Theoretical aspects of solution mining

Conventional mining involves extracting a lot volume of rock material to access the ore/mineral resource either by opencast method or by underground mines. Excess waste material mined along with the ore/coal is then dumped on the Earth's surface. Solution mining involves the dissolution of the potash ore with water underground itself and in in situ. With solution mining, a brine is heated and injected into the deposit to dissolve the natural potash ore occurring in underground rock formations.

A cavern-like structure is formed underground, which is enlarged slowly with the water circulation. The potash-rich brine is then pumped out of the cavern to the surface where the water is evaporated.

Initially, Potash as a commercial-grade ore is first established and determined through geological investigations and through a detailed explorations programme that included both drill holes (with core samples) and an advanced 3D seismic survey to determine the continuity of the deposit and its extension both length-wise and width-wise. Thereafter, the project feasibility is studied to know whether sufficient resources (reserves) are available for commercial exploitation that can demarcate a mine life of 50–100 years. Upon completion of the detailed exploration phase, the primary production phase is started. Well-drilling work is then started at this phase. The injection fluid (water) is arranged in sufficient quantity with a regulated supply. Slowly, the 'primary mining phase' and the 'secondary mining phase' are developed which include the approach from surface and cavern development. Secondary mining production is not possible until primary mining has been completed in the first developed caverns, which will be available for secondary recovery after startup, maybe 3–4 years after. A pillar of unmined rock material is required between caverns to maintain the isolation of the caverns and to support the overlying strata to control the surface subsidence. The cavern dimensions and pillar sizes have been decided based on the modelling work. These cavern dimensions are based on stress analysis and site-specific data.

Large size drill holes (wells) are drilled into the potash-bearing Sylvinite beds and water, later brine is pumped down the wells to dissolve the potassium chloride (KCl) (potash product) and sodium chloride (NaCl) (waste salt). The halites/Sylvinite bed is turned into a large hollow cavern. From this cavern, the brine is returned to the surface and is conveyed to the process plant through pipelines and separated by mechanical evaporation and crystallization. No underground workers are required, as the Potash ore (Sylvinite) is accessed by drill pads and production wells from the surface. When the potash to be extracted is at a depth of 1,000 m or deeper and/or the potash is located in sedimentary rocks, then solution mining provides a cost-effective, efficient, and safe way to extract the potash resource. Mining is planned to start from one section of the mine property and then migrate to another boundary.

Some added benefits of solution mining include lower up-front capital cost, no underground workforce, relatively lower volumes of waste salt, and more limited subsidence.

Processing

Potash processing includes the following:

- Injection and solution recovery
- Evaporation and crystallization
- Product drying, compaction, and screening
- Product storage and transport of potash ore including shipping (finished product)

In solution mining technology, a lot of state-of-art drilling, chemical processing work, rock mechanics investigations (cavern dimensions and pillar sizes, i.e. design part of the cavern, cavern stability, and widening rate of the cavern), subsidence prediction work and pipeline engineering (casing, leak detection, pipeline isolation, pipeline monitoring for safety) is involved. The Potash, being corrosive, also needs safe environmental practices to handle. A significant amount of metallurgical engineering, process engineering, and plant engineering aspects have been involved in the solution mining method. Potash ore processing is a trade secret mastered by private-sector mining companies. The solution mining exercise is time-consuming and capital-intensive but best suited to extract the valuable potash, which is a mineral of a special category, largely useful for agriculture.

Audio-visual links for the case study

A Video link of the Southey Potash Project, Canada, is provided (Yancoal Canada - Southey Potash Project, 2006) (Courtesy: Yancoal Canada Resources Company Limited). The reader should copy and paste (or click the web link given here) to get the required case study information which is self-explanatory. This unedited video provides captivating information about the Southey Potash Mining Project located in Canada and explains – how the Potash is mined or extracted by the 'solution mining' method. As Canada is being a major resource country for Potash, we found this video as the most suitable and educative for the discussed case study. The ensuing project is at the final stage of execution and development. Yancoal Canada aspires to successfully construct and operate the Southey Project. The project has created opportunities for local employment and businesses for a sustainable and positive economy and played a role in the social resurgence. Local society and surrounding communities have benefited from the project.

Case study 6.2: Lithium mining in Bolivia

Introduction

Lithium (Li) is a well-known solid element with three atomic numbers after hydrogen and helium in the periodic table (Figure 6.1). The element Li is the lightest metal on the periodic table that was discovered in 1817 by *Johan August Arfwedson* and was found on the Earth in the mineral form as lithium ore (lithium carbonate) and considered as a mineral of rare/strategic category. The word Lithium (Li) is derived from the Greek 'lithos' meaning *stone*. Lithium is one of the important metallic minerals of present day-to-day life because of the increased demand for high-grade batteries from mobiles to automobiles. Lithium is a *secondary mineral* in associated ores and is available at 0.01% or even below. Lithium makes up about 0.002% of Earth's crust, but in geologic terms, it isn't particularly rare; however, lithium is not concentrated enough to mine economically. Its distribution on Earth is quite uneven, e.g. Bolivia, Chile, Argentina, Australia, Canada and the USA has ample lithium reserves of commercial-grade, whereas India has no lithium reserves. Recently, in India, exploration has started in the search for lithium ore and the business atmosphere is conducive to lithium (Box 6.1).

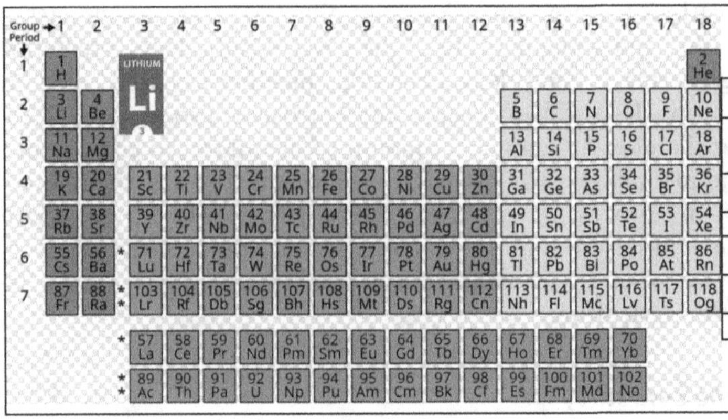

Name of parameters	Properties
Group, Period & Block in the periodical table	1, 2, & s
Relative atomic mass	6.94
Atomic number	3
Chemical formula and Key isotopes	Li
State (at room temperature)	Solid
Density (g cm⁻³)	0.534

Figure 6.1 Lithium in the periodic table.

Box 6.1: Lithium in India

Exploration

The Atomic Minerals Directorate for Exploration and Research (AMD), a constituent unit of the Department of Atomic Energy (DAE), is exploring lithium in potential geological domains in parts of Karnataka and Rajasthan. AMD is carrying out subsurface exploration in the Marlagalla area, Mandya district, Karnataka. According to a written reply by the Minister of Mines, Coal and Parliamentary Affairs, in the Indian Parliament – Reconnaissance surveys have also been carried out along the Saraswati River palaeo channel, in the Jodhpur and Barmer districts of Rajasthan, for locating lithium mineralization associated with brine (saline water in salt lakes). Preliminary surveys on the surface and limited subsurface exploration by AMD have shown the presence of lithium resources of 1,600 tonnes (inferred category) in the pegmatites of the Marlagalla – Allapatna area, Mandya district, Karnataka.

As per the approved annual field season programme (FSP) of the GSI (Geological Survey of India), different stages of mineral exploration have been taken up including reconnaissance surveys, preliminary exploration, and general exploration for augmenting mineral resource for various mineral commodities including lithium. During FSP 2016–17 to FSP 2020–21, GSI carried out 14 projects on lithium and associated elements in Bihar, Chhattisgarh, Himachal Pradesh, Jammu & Kashmir, Jharkhand, Madhya Pradesh, Meghalaya, Karnataka, and Rajasthan. During the current FSP 2021–22, GSI has taken up 7 projects on lithium in Arunachal Pradesh, Andhra Pradesh (Anantpur district), Chhattisgarh, Jharkhand, Jammu & Kashmir, and Rajasthan. However, the resource of lithium has not yet been augmented by the GSI (Geonesis, 2021).

Market potential

India is aggressively pushing for the faster adoption of Electric vehicles (EVs) and setting up infrastructure for recycling batteries. The infrastructure for battery-industry-related efforts is weak in India and the country as a whole lacks reserves of Lithium which is a crucial raw material for batteries. A recent study by the World Resources Institute (WRI) has said that India should make adequate arrangements for procuring such minerals from other countries to ensure a smooth path for electric vehicle growth in the country. Even though there is no national target for electric vehicles in the country, the government of India is increasingly pushing for policies to encourage a comprehensive eco-system that will encourage the use of electric vehicles. These policies and initiatives span various central ministries, including the Department of Heavy Industries (DHI), NITI Aayog, Ministry of Power (MoP), Ministry of Urban Development (MoUD), Ministry of Road Transportation and Highways (MoRTH), and the Department of Science and Technology (DST) (Geonesis, 2022).

Lithium does not naturally occur in its pure form due to its high reactivity. The top six countries with the largest lithium reserves in the world are Bolivia ($\approx$ 21 million tonnes), Argentina ($\approx$ 17 million tonnes), Chile ($\approx$ 9 million tonnes), the United States ($\approx$ 6.8 million tonnes), Australia ($\approx$ 6.3 million tonnes), and Canada. Russia, India, and China (in the Tibet region) have discovered lithium reserves but have no significant lithium mining activity.

The locations of the lithium reserves in Bolivia are at Salar De Uyuni; in Argentina at Salar del Hombre Muerto, in N-W Argentina.; in Chile at saline lagoons of Chile's Salar de Atacama; in the United States at Borate Hills, Nevada & at Thacker pass project, Nevada; in Australia at Greenbushes, Australia. (250 km south of Perth/Fremantle); and in China at lake Zabuye, China – A salt lake of the Tibet region, 1,050 km from Lhasa.

Concerning lithium, the word 'lithium triangle' is well known. It commonly refers to the lithium resources that have been found in three countries namely *Chile*, *Argentina*, and *Bolivia*. The lithium triangle holds more than 75% of the world's lithium reserves, mainly beneath the salt flats. Worldwide, there are three main sources of lithium: *Pegmatites*, *Brines*, and *Clays*. At first instance, the sources of lithium-bearing minerals and rocks become very important. Lithium-rich deposits are found in volcanic rocks, formed as a result of active tectonics, which contains a lot of water and heat and mixes well. Most pegmatites are a type of granite formed out of molten magma. The magma's chemical composition evolves with time. In the process of magma cooling, solid crystals of minerals are formed and liquid concentration drops out slowly. Elements, such as lithium, tend to be more concentrated in the rocks and lithium deposits are formed which contain more water or liquid.

Lithium: recovery and extraction

Lithium ore mining is not done by conventional mining methods. Retrieving the lithium ore requires a chemical process, usually. The open-pit mining of lithium ore is more expensive than pumping up the brine. Similarly, processing the clay to extract lithium carbonate or other industry-ready minerals is also pricey. Although American companies make claims for the extraction of lithium from clay, they haven't perfected it so far.

Besides lithium deposits, clay, tailings, pegmatite rocks, and old mud are the other principal sources of lithium. Clays or old mud, in the lake beds, which are the hard remnants, deposited by a slow sedimentation process containing fine sediments/grains are a rich source of lithium. The groundwater leaches the lithium from the rocks and transports it to a lake bottom where it becomes concentrated in the sediments. Getting lithium-enriched clay is a difficult task, e.g. Yellowstone caldera (in super-volcano craters that became lakes) in Wyoming (USA) has a potential treasure trove of lithium-rich clay. Lithium clays are not competitive with brines as they are costly to extract.

Lithium recovery from tailings and their upgradation for metallurgical purposes is equally necessary for the mining and mineral industry as lithium is in short supply. Given this, it is also essential to know the available resources of lithium-bearing tailings as it has strategic importance in the mineral and metal list. At present, most stress is given to extracting lithium from old used batteries. Considerable export savings are possible for the nations' exchequer if the technology of lithium enrichment, reuse, and recycling been well developed.

History of lithium extraction

Before the 1990s, pegmatites in the United States were the primary source of mined lithium. But extracting lithium, primarily from the ore mineral called 'spodumene' (a lithium-aluminium-silicate mineral from the pyroxene family having the chemical formula as $LiAlSi_2O_6$), was very costly. On top of the cost of actual mining, the rock has to be crushed and treated with acid and heat to extract the lithium in a commercially useful form. In the 1990s, a much cheaper source of lithium was discovered from the arid salt flats of the lithium triangle. In comparing the Li extraction from pegmatite, the process of extracting lithium from the brine is found extremely cheap. As a result, solution mining currently dominates the lithium market. But in the hunt for more lithium, the next generation of prospectors is looking to the third type of deposit, i.e. clay (Science News, n.d.).

Lithium mining in Bolivia: Salar De Uyuni, Bolivia

Bolivia – A Latin American country in South America with *La Paz* as its capital is a landlocked country. *Chile, Peru, Brazil, Paraguay, and Argentina* share their countries' borders with *Bolivia*. Bolivia is rich in Lithium resources and has a varied topography/terrain spanning the *Andes mountains*, the *Atacama desert*, and the *Amazon rainforest*. Lithium mines, their mining processes plants, and investment opportunities are the industrial offerings of Bolivia that earns a lot of foreign exchange for Bolivia. *Salar De Uyuni* is a site in Bolivia (Figure 6.2) where extensive lithium mining operations can be located.

In general, as the production of electric cars increases and mobile technologies take over the technological world, the lithium demand keeps on rising and an extensive

Figure 6.2 Location map of Salar De Uyuni, Bolivia.

Figure 6.3 Salar De Uyuni – a lithium mining site in Bolivia.
Source: http://www.lithiummine.com

look will be given to Lithium extraction. In Bolivia, large and broad strips or areas (swaths) of salts are found called 'Salt flats'. Just beneath the arid *salt flats* circulates the salty, lithium-enriched groundwater. At *Salar De Uyuni*, the brine method is used to produce Lithium metal because it is the most efficient and cost-effective process.

Salar De Uyuni is the world's largest salt flat (Figure 6.3). It is an expansive desert of over ten billion tons of salt covering nearly 5,000 miles^2 (12,950 km^2). It is located in the *Potosí and Oruro* in SW (southwest) Bolivia, near the crest of the Andes, and is elevated at 3,656 m (11,995 ft.) above mean sea level. It is covered by a few metres of salt crust, which has an extraordinary flatness across the entire area of the *Salar* and contains 50%–70% of the world's lithium reserves, which are in the process of being extracted. Harvesting of salt is continuing here for hundreds of years. These salt plains are one of the most remote and inaccessible plateaus in the world. It is estimated that over 650 million euros have been invested in the *Salar De Uyuni* Lithium mining project. Lithium, the white gold (nearly 5.5 million tons), is making Bolivia rich in terms of foreign exchange and employment for the locals.

Mining: We first imagine a lithium mining site as a conventional mine of the shaft, incline, or cave-like tunnel in the Earth where workers excavate rock and soil to unearth ores with pickaxes and heavy diggers. But this is not true, because a different process is followed in lithium mining. As described earlier, lithium has historically been produced from two sources either from *'brines'* or *'hard rock mining'*.

In the brine method, adopted at Salar De Uyuni, the miners have to drill holes in the salt flats to pump salty, mineral-rich brine to the surface. Miners pump the salty water to the surface, sequestering it into ponds and letting it evaporate in the sun. What's left behind after the evaporation is a sludgy, yellowish liquid called 'brine', which is a concentrated solution containing Lithium. Mining salt is difficult and it is back-breaking work (extremely arduous, exhausting, or demoralizing) done almost entirely by hand. The water evaporation takes months together. After 12–18 months, the evaporation process is complete, forming a mixture of potassium, manganese, borax, and lithium salts. The re-filtering process is done once more and the solution is left to evaporate. After filtering and re-filtering in 12 to 18 months, lithium carbonate can be extracted.

Since mother nature does most of the work of mining/extraction, lithium production becomes cheap. The cost-effectiveness of brine operations attracts even large producers to develop their brine sources in Bolivia. This brine contains lithium, which is derived mainly from the leaching process of the volcanic rocks and varies greatly in lithium content, largely as a result of the extent to which they have been subject to solar evaporation. The lower-concentration brines have modest evaporation rates and dilution is constant due to a large volume of freshwater inflow and small lithium concentrations varying between 30 and 60 ppm. Salt evaporation ponds of lithium mines are similar to the 'salt pans' meant to produce salt from seawater. The water or brine contained in large evaporation ponds has vivid colours – from pale green to bright red due to the variable concentrations of algae. The colour indicates the salinity of the ponds. Microorganisms change their hues as the salinity of the pond increases. Once the lithium is recovered from brine (Figure 6.4), other by-products can also be extracted from the concentrate which includes potash, boron, and other saleable

Figure 6.4 A view of lithium ore extracted at the mine site in Bolivia.

chemical compounds. Lithium recovery from brines in general leads to a significant carbon footprint reduction because of nearly zero-waste mining. The evaporation ponds also provide a productive resting and feeding ground for many species of water birds, which may include endangered species.

The mining of lithium is though simple yet challenging from an environmental angle. Investing in Lithium is like investing in future. Tesla – An electric car-making company is getting into the mining business, shortly by buying a lithium mining lease of 10,000 acres in Nevada (Lithium Americas, n.d.). With simple basic principles of mining and mineral processing (metallurgical processes), lithium ore can be extracted, but lithium mining by conventional open pit mining methods is uneconomical.

Ore processing

To extract lithium from the concentrate in commercially useful forms, particularly lithium carbonate and lithium hydroxide (battery-grade Li), the miners add different minerals to the brine, such as Na_2CO_3 and $Ca(OH)_2$ (sodium carbonate and calcium hydroxide). Reactions of concentrated brine with these added constituents cause different types of salts to precipitate out of the solution, ultimately producing mineral lithium. The faster process of lithium extraction from the mined ore is a bit complicated, requiring a chemical plant; however, most simply, it has been explained above. Private companies in the lithium production and marketing business keep the whole ore process a trade secret. As such, lithium extraction is one of the driest processes. Today, industrial companies involved in ore processing take sustainable development very seriously and responsibly and focus on environmental management more than ever.

Environmental impacts of lithium mining

Lithium mining is hazardous, dominated by chemicals and processes causing environmental contamination (Figure 6.5). Like any other mining operation, lithium mining has an impact on the environment. In lithium mining, associated pollution forms

Figure 6.5 Contaminated land and water at a lithium mining site.
Source: http://www.lithiummine.com

include water pollution, land pollution as well as health hazards. Their magnitude may vary depending on the involved area and chemical reactions. The Earth system and the engaged workforce get affected directly as well as indirectly. In brief, it leads to the following severe environmental impacts:

1. The cheapest extraction method of evaporating salt brines in the solar ponds deploys the usage of cheap and toxic PVC which causes health problems in the workforce.
2. The lithium-rich mining regions make use of fresh water in large quantities leading to groundwater depletion and contamination of groundwater through leaching.
3. A large land area where extraction of the metal is done is inflicted with chemicals causing severe corrosion of metals. Invariably, the large magnitude of the brine present makes the land become barren. Thus, land degradation is one important impact result of lithium mining.
4. Lithium metal, due to its alkaline properties, is corrosive and reacts with water. A large area becomes devoid of fresh drinking water, and good water quality becomes precious and a distant dream in the lithium mining areas.
5. Breathing lithium dust or alkaline lithium compounds irritates respiratory tracts. Prolonged exposure to lithium can cause fluid to build up in the lungs, leading to pulmonary diseases.
6. Lithium metal itself is a handling hazard because of the caustic hydroxide produced when it is in contact with water causing an explosion.
7. Prolonged exposure to lithium can cause nervous system disorders.

In brief, lithium mining carries high environmental protection costs and has great environmental concerns.

Conclusions

Lithium mineral deposits are mined from rocks underground. Wells are drilled into the aquifers below the salt flats to bring the brine to the surface which is a concentrate containing lithium as the principal constituent. The chemical process separates the lithium from the brine through a series of evaporation pools. The whole process of lithium extraction from brine is water-intensive and energy-intensive (Upton et al., 2021).

This case study has highlighted the importance of the lithium ore mineral and its exploitation aspects, which will inevitably be more in near future for shaping the market for a green future (Upton et al., 2021). It is quite clear that lithium deposits are deep-seated and their mining is hazardous and environmentally challenging. Hence, the mine developers cum operators (MDOs) should be well-equipped to deal with this. The field personnel should mitigate environmental challenges as they arise.

Case study 6.3: Underground coal mining in Poland (longwall mining)

Zbigniew Burtan and Piotr Małkowski

Introduction

Energy generation in Poland relies on fossil fuels, of which hard coal is predominating, and its share in energy production in the year 2020 was 46%. Currently, hard coal excavation continues in 20 coal mines, where 19 of them (most of them multi-operations plants) are in the Upper Silesia Coal Basin (GZW) and one mine in the Lublin Coal Basin (LZW) (Figures 6.6 and 6.7). The documented resources of hard coal in 2019 were 64,330 MT, while the recoverable reserves were 4 779.20 MT. In the year 2020, the hard coal production was 54.4 MT while employing 101.2 K workers (Kabiesz, 2021).

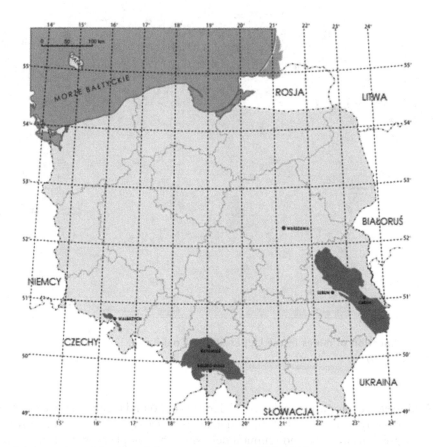

Figure 6.6 Location of hard coal deposits in Poland (PIG 2019).

Figure 6.7 Location of mines in the Upper Silesia Coal Basin.
Source: PIG (2019).

Multi-seam exploitation of hard coal deposits in Poland is carried out at increasing depths (currently exceeding 1,000 m), and in areas of geological disturbances, zones affected by previous mining, and remnant deposits. The complex geological and mining conditions are the cause of a high level of natural hazards in the mines, i.e. methane exhales, fire hazards, rockburst hazards, methane and rock outburst hazard, and their concurrent presence and interaction lead to a high number of dangerous occurrences.

Currently, 16 out of the total 20 mines are operational and have been identified with either methane hazard or rockburst hazard, while the production from these mines is 79% and 57% of the total production, respectively (Kabiesz, 2021).

At present, more than 99.9% of coal output is extracted using a *longwall system with roof caving*. There were 137 longwalls with roof caving active in 2020 (the longwall with backfill system is generally not in use; SMA, 2021). The coal production from longwall varies from several to over a dozen tons per day, while in the mine *Bogdanka* in the LZW, it reaches 2.0 MT/day. Two private mines that use the remnant reserves and infrastructure of the already closed state-owned enterprises apply the roadway with a hydraulic backfill system for excavation of the remnant thick coal seam and achieve production of several hundred tons per day.

Geological and mining conditions of excavation of hard coal deposits in Poland

Hard coal mining in Poland over the decades extended downward currently reaching a depth of 1,260 m, and six mines operating below 1,000 m. The average depth of mining operations is approx. 800 m and increased around 10 m per year in recent years (Kabiesz, 2021).

The thickness of the coal seams in Poland varies from several dozen centimetres to above 20 m, and most of them are dipping at less than 35° angle. The average thickness of the currently exploited seams is 3.05 m, and their dip varies from 6° to 19°. Coal seams thinner than 1 m are not a subject of exploitation, while seams thicker than 4.5 m are excavated by layers.

Coal mines in Poland are operative in the productive part of the Carboniferous, typically developed as a series of claystone, mudstone, and sandstone with coal beds. The compressive strength of hard coal is on average approx. 20 MPa, claystone – approx. 35 MPa, and sandstone – approx. 65 MPa. However, very weak types of coal can be found of UCS (uniaxial compressive strength) approx. 5–6 MPa, and thick beds of sandstone of a strength 120–140 MPa. The strength parameters of most of the coal types show a decrease with the depth of occurrence. Whereas the rocks surrounding the coal beds show strength increase with the depth, they are uneven; more apparent in sandstones than in claystone. A series of dozen to several dozen metres thick strata of strong rocks are recognized as prone to shock generation.

Since coal beds were subject to washing out, they show varying thickness and vertical distance between beds, and they may even disappear locally. Tectonic faults are also common in both coal mining basins Lublin Coal Basin and Upper Silesian Coal Basin. Faults of a very high drop ranging from 40 to 300 m determine the boundaries of mining areas, sections of coal beds, and exploitation fields, therefore implying the shape and parameters of longwall lots, as well as often are the cause for abandoning parts of the resources. The mines are compelled to abandon large parts of coal seams on both sides of a big fault, often of significant reserves. Exploitation in the vicinity of faults of a large drop often leads to their reactivation, which in turn causes high-energy shocks and even rock bursts. Driving mining openings through faults creates numerous technical difficulties and may cause sudden exhales of methane, or methane and coal outbursts. Despite the gradual drop in coal production in Poland, the share of it from seams with methane hazard is 79% and from seams with rockburst hazard is 56.6%.

The multi-seam character of coal deposits in Poland causes exploitation difficulties related to the presence of earlier excavations, goaves, excavation boundaries, and remnant pillars. This legacy may either reduce or escalate the geo-mechanical hazards, i.e., rock bursts and rock and gas outbursts. The significant problem for most Polish mines of hard coal, and in particular for those that exploited their reserves to a high degree, is the need to operate within irregular remnant parts of the deposit, within support pillars and protective pillars around shafts and main galleries, and within parts of coal seams excluded from exploitation by a previous operations' decision.

Gradual deepening of the exploitation, and at the same time, reduced spending on access works (e.g. sinking new shafts) causes the use of sublevel exploitation in most of the mines. Presently, out of 20 mines (28 operations), 16 (22 operations) carry exploitation below the initial access level, and the production from sublevels amounts to 40% of total coal production (SMA, 2021).

Longwall exploitation system

Parameters, components, and equipment of a longwall system: The commonly used exploitation system in the mines in Poland is the system of longwall with roof caving (Figure 6.8). Most often it is used as a dip-oriented option (Figure 6.8a). However, in most cases, the longwall direction vs strike of the seam is determined by the presence of tectonic disturbances and abandoned workings in its vicinity, which also determine the length and run of the wall. The option of strike-oriented longwall with hydraulic backfill was used in the past in several mines due to the requirement of surface protection (Figure 6.8b).

The salient components of a longwall lot in the dip-oriented option are (Figure 6.9):

- undisturbed coal seam and goaves (either caved-in roof or backfilled);
- startup drift, longwall drift, recovery drift;
- the main gate (transport gallery); and
- tailgate (ventilation gallery).

While the parameters of the longwall panel are:

- length of the wall W_L,
- longwall run L_L,
- longwall height h_L, and
- longwall opening width FR.

The length of the longwall in the GZW is typically less than 350 m (maximum 400 m) and their run ranges from several hundred to more than 2,500 m. In this aspect, the exception is the mine *Bogdanka* where the longwall runs reached more than 7,000 m in recent years, which was feasible due to the lower intensity of faulting. The longwall systems with standard machinery and equipment can be used up to 35° dip of the coal seam, and to seams from 1.0 to 4.5 m thick.

Thicker seams are excavated by two or three layers in a top-down sequence leaving a thin beam of coal between the excavated layers.

A less often used strike-oriented option with roof caving is applicable for seams inclined up to 20° for longwall progressing upward, and up to 15° for longwall

(a)   (b)

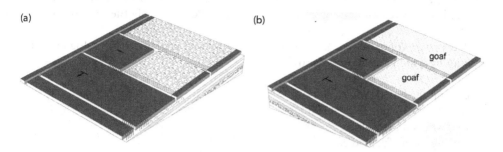

Figure 6.8 Longwall system with the roof caving: (a) dip-face and (b) strike-face.

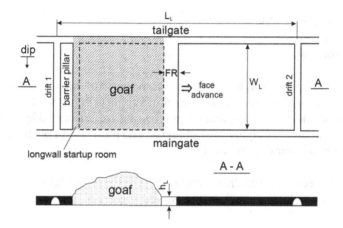

Figure 6.9 Components and parameters of dip-oriented longwall system with the roof caving.

progressing downward. The strike-oriented option is suitable for the excavation of thick seams by layers and with re-consolidation of goaf. It is also advantageous where the hydraulically placed ash backfill is used to aid the re-consolidation and to reduce the endogenic fire hazard.

Longwall progress may be directed:

- from the access roadways towards the outer boundary of the exploitation lot (advancing longwall); the output haulage distance extends progressively.
- from the exploitation lot boundary towards the access roadways (retreating long-wall); the output haulage distance shortens progressively.

In the case of the direction towards the boundary, the longwall face closes to the gates' faces which can be progressed simultaneously with the longwall. That way the exploitation of the panel can start earlier; there is no need to wait for the main gate and tailgate fully completed. However, the disadvantage of this solution is the difficulty in maintaining the gates behind the longwall face. Also, the ventilation requirements become stringent because the air flows along the goaves which increases the endogenic fire hazard. In the case of the retreating longwall, the preparation time is longer, yet the advantage is that the ventilation flow bypasses the goaves which reduce the fire hazard. For this reason, all longwalls nowadays in Poland apply the retreating direction option.

The process of exploitation by a longwall consists of repetitive technological activities among which the main ones are:

- cutting the hard coal,
- loading and haulage of the muck,
- longwall support, and
- closing of the formed cavity.

All these activities are carried out simultaneously or successively at a short time spacing, and they are performed by the longwall system machinery (Figures 6.10 and 6.11):

- longwall shearer or plough
- longwall scraper conveyor
- powered roof support

The longwall shearer or plough, the scraper conveyor, and the longwall crew operate under the shield of sections of the powered roof support (Figure 6.12).
In most longwall, the coal cutting and output loading are performed by longwall shearers, and in several cases, only by coal ploughs. The most used Polish shearer loaders manufactured by Famur company are:

- longwall shearer-loader hydraulically powered (Figure 6.13a) and
- longwall shearer-loader electrically powered (Figure 6.13b),

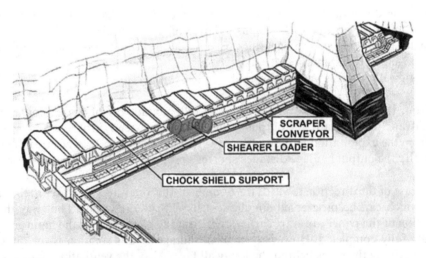

Figure 6.10 Longwall machinery set components.

Figure 6.11 The longwall machinery set.
Source: famur.com

Figure 6.12 Photograph of a longwall with machinery.
Source: nettg. pl

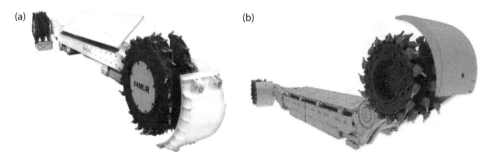

(a)   (b)

Figure 6.13 Longwall shearer-loader: (a) hydraulically powered and (b) electrically powered.
Source: famur.com

designed for two-way cutting, without starting niche, and for 1.4–4.8 m thick coal seams. Their total power output is from 475 to 1705 kW, driving speed up to 36 m/min, and the cutting depth from 0.63 to 1.0 m.

In several mines, for the coal seams from 1.0 to 1.7 m thick, the applied machines are longwall ploughs manufactured by the company Bucyrus (at present – Caterpillar), with 2 × 210/630 kW power output, driving speed up to 100 m/min, and cutting depth up to 0.21 m. The *Bogdanka* mine in the LZW basin, by using the longwall plough method, achieved production of 33,612 MT at an average daily drive of 28.3 m (Famur, n.d.).

Hauling out of the coal output is performed by scraper conveyors located along the wall and in the gate entries. Scraper conveyors manufactured by Famur company (Figure 6.14a), i.e. type FFC Famur, Nowamag, Glinik, and Rybnik, of a total length from 400 to 450 m, and capacity from 1,050 to 3,000 ton/h, enable uninterrupted production. Belt conveyors for gate entries (Figure 6.14b) manufactured by companies JZR and Famur (type FSL Famur, Nowamag, and Ryfama Grot), which have a total length from 80 to 120 m and capacity from 1,500 to 3,000 ton/h, enable longwall production without the need of frequent shortening of the belt. For further hauling, belt conveyors of 1,000–1,200 mm width and capacity up to 3,000 ton/h are used (Famur, n.d.; Jastrzębskie Zakłady Remontowe w liczbach, n.d.).

Support of the longwall opening is provided by powered roof support, consisting of 1.5–1.75 m long linear sections and endings. Currently, most of the supports used are manufactured in Poland, strut and shield types, i.e. Fazos (Figure 6.15a), Glinik

(a)   (b)

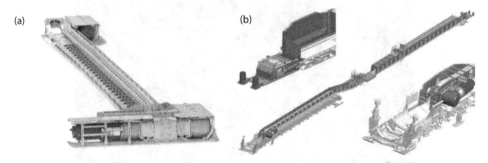

Figure 6.14 Scraper conveyor: (a) for longwall and (b) for gate entries.
Sources: (a) www.famur.com and (b) jzr.pl

(a)   (b)

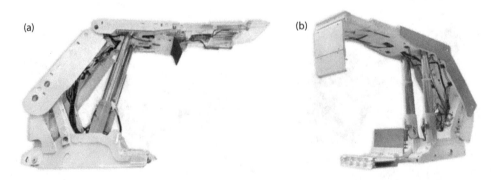

Figure 6.15 Section of linear powered roof support types: (a) Fazos and (b) Glinik.
Source: famur.com

(Figure 6.15b), and Tagor. Each section has two crops extending from 0.6 to 6 m with a bearing capacity of 0.55–1.5 MN/m². The complete longwall set-up includes also support sections designed for junctions of longwall opening and gate entries. These are the Fazos type, with working height from 1.7 to 4.3 m and bearing capacity from 0.37 to 0.77 MN/m² (Figure 6.16).

An essential part of longwall excavation is its ventilation, whose objective is, apart from supplying a sufficient quantity of air to the longwall opening, to prevent concentrations of methane, dissipate the excess heat from cutting, cool the coal faces, and reduce the endogenic fire hazard. In presently active coal mines in Poland, the ventilation systems used are "U" and "Y" types in the retreating longwall option (progressing from the lot boundary) (Figure 6.17a and b), with the "U" type being more prevalent (approx. 81% of all longwall in 2022). The "U" type is advantageous for fire hazard prevention because of the fresh air and the return airflow along the solid coal walls. The "Y" system is used at high methane and climatic hazard.

In the presence of high methane and fire hazards, the ventilation solution is the so-called "short Y" (Figure 6.18). This system, however, requires an additional, parallel ventilation drift.

The main *advantages* of the longwall system are:

- a small number of preparatory works,
- low exploitation losses,
- possibility of the entire mechanization of the process, and
- high production rate achieved from the small mining area.

Figure 6.16 An end section of the powered roof support Fazos type.
Source: famur.com

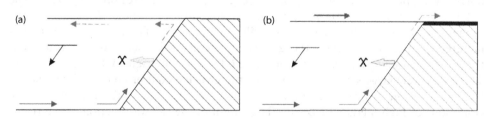

Figure 6.17 Longwall ventilation system: (a) "U" type for retreating wall and (b) "Y" type for retreating wall.

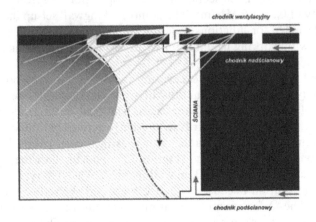

Figure 6.18 So-called "short Y" ventilation system.

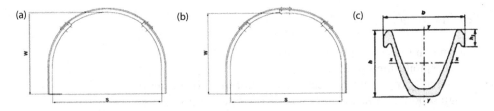

Figure 6.19 Yielding steel arch support ŁP type: (a) 3-section arch and (b) 4-section arch, profile V.

while its *disadvantages* include:

- high cost of machinery and equipment
- a long time of installation and dismantling works in the longwall workings

Longwall gate entries

In Polish hard coal mines, the used support types are standing support, combined standing & rock anchor support, and sporadically rock anchor support depending on the geological and mining conditions encountered. The longwall gate entries are normally supported with yielding steel arches type ŁP (Figures 6.19 and 6.20). The ŁP type support is a steel frame consisting of three to four sections (Figure 6.19a and b) made with a specially designed V-profile (Figure 6.19c). The ŁP support is manufactured in various sizes where the larger frames consist of 4 sections. The width S of the presently used frames ranges from 5.0 to 6.1 m and the height W from 3.5 to 4.2 m. These relatively large sizes are dictated by the growing coal production rate, and related to it, the growing volume of air needed to ventilate the workings, especially in conditions of high methane hazards. The length of the sidewall section of the frame is 3.1–3.5 m, while the roof section is 2.7–3.0 m (Figure 6.20). The V-profile is of three

Figure 6.20 Photograph of a longwall gate entry.

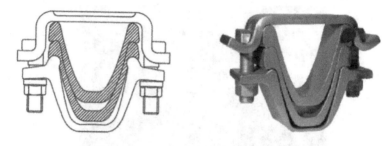

Figure 6.21 SD type clamp: (a) schematic and (b) view.

types V29, V32, and V36, where the number states the weight of one meter of the section in kilograms and is normally made with high tensile strength steel, e.g. S550W.

Sections of the steel arch are joined by a two-bolt clamp to enable slip along the joint, therefore yielding the whole arch. The commonly used clamp is the SD type (Figure 6.21), whose design has practically confirmed very good slip properties between V-profile sections. The clamp is approx.150 mm long and is tightened by two bolts M27.

Spacing of the support frames at the depth of 800–900 m is most often 0.8 m, and depending on geological and mining conditions, it can be lowered to 0.5–0.75 m or sporadically increased to 1.0–1.2 m.

An open profile of the standing support may not be sufficient at higher depths of the headings due to high geostatic stress in the rock mass. Therefore, at depths of ≈ 1,000 m and with weak rocks on the floor, the support used is of the shape of a closed frame (Figure 6.22).

In these conditions, additional rock anchor support is also applied, and the frame is stabilized perpendicularly to its plane direction by joists. The joist is a steel beam of a V-profile that is attached to the neighbouring frames with screw clamps (Figure 6.23).

Steel rods up to three metres long are used for rock bolting. In a situation where the calculated fracturing zone is wider than three metres, strand bolts are used. The strand bolt is a cable bolt built as a bundle of straightened wires. The front end is cut in such a way as to enable penetration and mixing of glue cartridges in the hole.

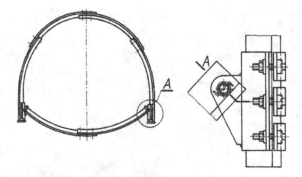

Figure 6.22 Closed profile of yielding standing support.

Figure 6.23 Two steel joists, first anchored to the roof with strand bolts and the second – with clamps.

The back end is welded and shaped into a hexagonal pin to enable rotating by the bolting machine when inserted into the drill hole (Figure 6.24). Strand bolts are 4.5–8 m long and their loading capacity is 280–450 kN.

Bolting the roof arches to the rock and installing rock bolts between the standing support frames allows for bigger spacing of the frames, i.e. fewer frames on the given length of the heading (Figure 6.25). This installation time of this type of support appears shorter and its cost is significantly lower compared to standing support. Since rock bolting is performed most often manually (with the use of hand-held roof bolters), the process is relatively slow. Therefore, to avoid disruption of the heading progress, this activity often starts behind the manoeuvring zone of the roadheader, yet in advance to the approaching longwall face.

Figure 6.24 shows the reinforcement of the arch support with three lines of joists (1), including short joists (2) bolted with strand bolts (3), and three rows of rod steel bolts (4) installed between frames. *Wood wedges* are used occasionally to provide tight contact between joists and arches (5).

A separate issue is ensuring good contact between the rock mass and the support. This can be achieved by filling up the void between the support arches and the

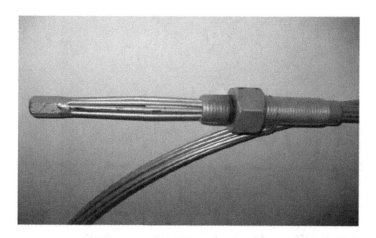

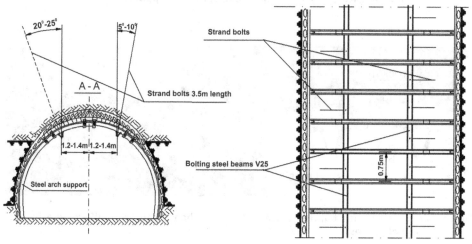

Figure 6.24 (a) Welded back end of the strand bolt. (b) Steel arch with asymmetrical rock
bolts support system – reinforced by steel beams.

rock face with bags filled with the mineral binder (Figure 6.26) – a method which
become popular recently.

Another method used in the mines for the protection of longwall gate entries is the
stabilization of the rock mass by injections of cohesive binders. The areas close to junctions of a gate entry with longwall openings are subject to particular treatments. Also,
maintenance of the junctions requires significant efforts mainly due to the need of dismantling the sidewall arches at the longwall side, as a part of the longwall excavation
cycle. The size of the drives of the longwall machinery situated at the junction causes
the strut supports, typically used for junctions to be difficult, or even impossible to
apply. Therefore, additional stabilization of the structure is often used in the form of
steel or wooden joists fixed to the earlier installed roof anchors. An example of a support structure on a junction heading with a longwall opening is shown in Figure 6.27.

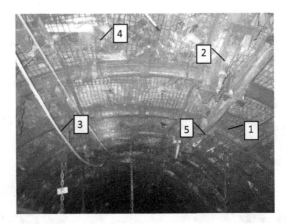

Figure 6.25 View on gate entry with reinforced standing support.

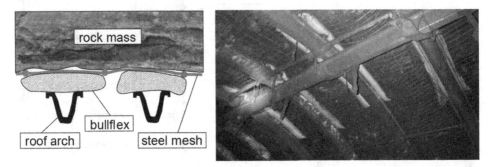

Figure 6.26 Lagging with bullflex system: (a) the scheme and (b) bullflex-type bags on roof arches.

This is practised in several mines of Poland longwall system with a single (shared) gate entry which necessitates maintaining the gate entry along the goaves for the excavation of the next longwall plot that requires additional reinforcement of the support and other geotechnical works. The standard types of this kind of work are *bearing beams* supported by backfill strips, concrete columns of high bearing capacity, or crib boxes made with prefabricated, hardwood elements (beech or oak; link 'n' lock type). Currently, the most often used link'n'lock cribs are four-point type, 90 × 90 cm or 120 × 90 cm in size, and load capacity of 200–250 tonnes.

The concrete pillars are usually 60–90 cm in diameter, have a bearing capacity of 560–1,270 tons, and are used in particularly difficult geological and mining conditions. All these stabilization measures are used in addition to the earlier mentioned anchor support, either short or long bolting, and traditional joists resting on steel or wooden props (Figure 6.28). Due to the more and more difficult geological and mining conditions, the mines are compelled to use several support reinforcement methods simultaneously.

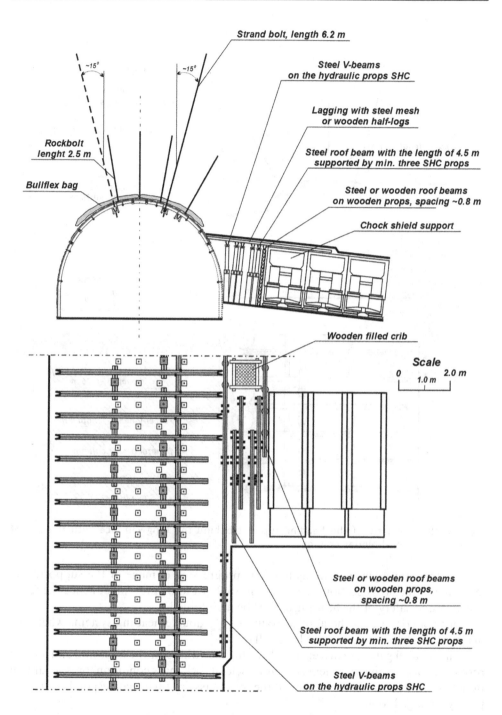

Strand bolt, length 6.2 m

Steel V-beams
on the hydraulic props SHC

Lagging with steel mesh
or wooden half-logs

Steel roof beam with the length of 4.5 m
supported by min. three SHC props

Steel or wooden roof beams
on wooden props, spacing ~0.8 m

Chock shield support

~15°   ~15°

Rockbolt
lenght 2.5 m

Bullflex bag

Wooden filled crib

Scale
0   2.0 m
        1.0 m

Steel or wooden roof beams
on wooden props,
spacing ~0.8 m

Steel roof beam with the length of 4.5 m
supported by min. three SHC props

Steel V-beams
on the hydraulic props SHC

Figure 6.27 An example support structure at junction longwall – gate entry.

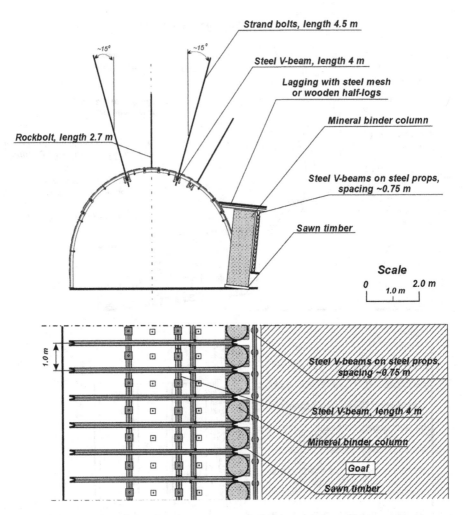

Figure 6.28 Schematic of the heading support behind the longwall face with the use of mineral binder columns.

Often, the convergence of the gate entry, progressing damage of the support, and lessening of the cross-sectional dimensions are hard to be prevented, hence the functionality and safety of the heading cannot be maintained.

In such conditions, the applied solutions are double-heading systems where the main gate and tailgate serve one longwall only.

The parallel gates are driven at a close distance wherever the fractured and jointed roof rock mass retains sufficient strength (yielded rock mass condition) or is separated by a 4–5 m wide pillar between them if the rock mass fracturing corresponds to a completely damaged roof.

Case study 6.4: Jayant open cast coal mining project, NCL India

Introduction

Jayant Opencast Coal Mining Project is one of the several greenfield projects of Coal India. It is located in the Singrauli district of Madhya Pradesh and is considered one of the biggest coal mines in India. The Jayant coal mine is a large opencast mechanized mine that has a planned annual coal production capacity of 25 million tonnes and was commissioned in the year 1975–1976. The mine is planned for overburden removal of @75.45 million m³/year over the lease area of≈3,200 hectares (CMPDIL, 2017). The total minable reserve available is 223.77 million tonnes.

The Jayant mine and its surrounding areas geographically lie between Latitudes 24° 09′50″ N to 24°11′ 25″ N and longitude 82°38′ 25″ E to 82° 41′ 42″ E (Survey of India topo sheet No 63L/12). The project is situated on a high plateau ranging from 375 to 442 m above MSL The planned ultimate depth of the Jayant pit is 225 m and the average stripping ratio is 2.99 (≈ 3). Currently, the pit has reached an approximate depth of 180–200 m (year 2021). The Jayant is a greenfield project of Northern Coalfields Ltd. (NCL). The Jayant mine is located in the south-central part of the Moher Sub-basin of Singrauli Coalfield and is bounded by Dudhichua and Nigahi Opencast mines on the east and west, respectively (Figure 6.29). The coal produced from the project is fed to NTPC's Singrauli Super Thermal Power Station

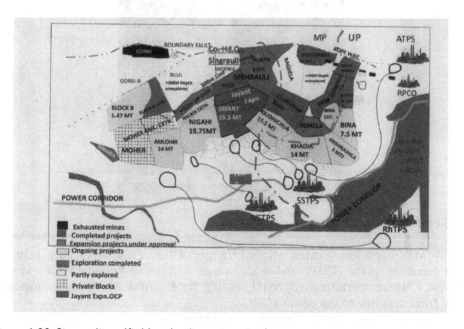

Figure 6.29 Singrauli coalfield and other mines in the area.

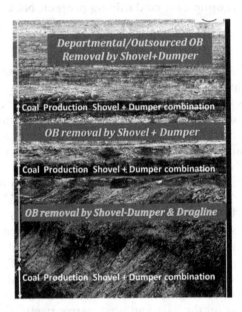

Figure 6.30 Pictorial lithological pattern with equipment deployment at NCL.

Figure 6.31 CHP and siding at Singrauli (NCL).

(2,000 MW capacity) located at Shaktinagar, Uttar Pradesh (Figure 6.30). A coal handling plant (CHP) and a dedicated Merry-Go-Round (MGR) system (Figure 6.31) are operational for coal handling, loading/unloading, and transportation, from the mine to the power plant.

Jayant mine site is well connected by both road and rail. The nearest Railway Station Shaktinagar is about 5 km towards the east and Singrauli Station is 12 km from the project office. Ranchi-Rewa Highway passes E-W through the Southern part of the Project. Jayant mine is also well connected to Singrauli Township on the north and

Waidhan, HQ of Singrauli district on the south at 12 and 8 km respectively. The road to Singrauli passes over the coal-bearing area of the project between the Nigahi and Jayant mines.

About coalfield

Singrauli Coalfields (latitude N 23°45′–24°38′ and longitude E 82°30′–83°32′) have an area of about 2202 km² and are one of the well-known coalfields in India with huge coal reserves of power grade (Figure 6.32). The coalfield is mainly mined by the Northern Coalfields Limited (NCL), formed in November 1985 as a subsidiary company of Coal India Limited (CIL) with its headquarters located at Singrauli, Madhya Pradesh (M.P.), India.

The area of the Singrauli Coalfield can be divided into two main basins, namely (a) Singrauli main basin (1890 km²) and (b) Moher sub-basin (312 km² area). The major part of the Moher sub-basin lies in the Singrauli district of Madhya Pradesh and a small part lies in the *Sonebhadra* district of Uttar Pradesh. Singrauli's main basin lies in the western part of the coalfield and is largely unexplored. The present coal mining activities are concentrated in the *Moher sub-basin* and future mine/ coal blocks will be developed in this basin only.

Singrauli Coalfield has five coal seams of varying thickness namely the Kota seam, Turra A seam Turra B seam, Purewa top seam and Purewa bottom seam. Thin coal seams (<4.5 m) and thick seams (>4.5 m) exist in the coalfield, however, the Turra seam is the thickest seam of the Singrauli Coalfield. Turra and Purewa (top & bottom) coal seams are being worked on at the Jayant project. The strike direction of the coal seams is towards E–W and the coal seams are dipping in a northerly direction at a dip angle of 1°–3°. The coal measuring rocks are lower Gondwana rocks, belonging to the *Barakar* formations. The stratigraphy of the rock formation seams considered for mining is shown in Figure 6.30 and its sequence is given below:

Lithology	Thickness of the beds (m)	
	From	To
Soil/Alluvium	0.00	12.20
Sandstone with 2–3 m clay bands	16.40	204.90
Coal seam: Purewa Top	4.00	13.07
Sandstones with thin shale bands	2.80	33.04
Coal seam: Purewa Bottom	8.69	18.54
Sandstone with thin carb shale	46.61	66.40
Coal Seam: Turra	13.90	23.61

Climate and drainage pattern

The climate of this area is tropical with severe summer. The temperature in summer rises as high as 48°C in May and June. In winter, the temperature comes down to 4°C and varies up to 21°C. The rainy season is generally from July to September with an average rainfall of around 1,218 mm. The percentage of days is the highest for the wind blowing in the ESE direction and lowest in the south direction. The average wind velocity is around 5 km/hour.

Figure 6.32 A view of the open cast mine of Singrauli coalfield.

Bijul Nalla, a tributary of the Sone River, traverses the north-eastern part of the block. Many seasonal nallas, flowing from north to south and south to north drain through this area and meet the master drain, the Rihand Dam (Govind Ballabh Pant Sagar) which is located south of this area and the Sone river located North of this area. Bijul Nalla, Motwani Nalla, Balia Nalla, Amjhar Nalla, and Tippa Jharia Nalla drain this area.

Mining at Jayant

'Draglines' are the giant excavating machine deployed to remove coal (mineral) and overburden in large open mines. Depending on the encountered site conditions (i.e. excavation material, height, width and slope of the working faces, seam thickness) and limitations of the machine (reach, boom length, machine capacity), they are deployed in commercial mining. The mega Jayant project which has an average extension of 4.8 km [4.8 km -along strike and 3.5 km wide along dip side] has a 'dragline mining' face for overburden between the Purewa bottom and Turra seam.

Coal production at Jayant coal mine is done by deploying large capacity Heavy Earth Moving Machines (HEMM) namely Dragline, Shovel, Dumper since its inception year 1977. The coal extraction method adopted at Jayant opencast mine is dragline-cum-shovel dumper combination with four draglines each in the eastern and western wings of the mine. The dragline excavates the rock parting between the coal seams of various sequences occurring at the mine site, i.e. *Purewa bottom* and *Turra* coal seam.

The coal extraction method adopted at Jayant opencast mine is dragline over casting-cum-shovel dumper combination. For this mining combination of Dragline and Shovel-Dumper, two flanks namely the east and west of the mine are covered. The longitudinal area of 2.8 km on the east side and 2.0 km on the west side from across the total mine length of 4.8 km of mine is excavated to produce coal at the mine pithead. The dragline excavates the rock parting between the coal seams of various sequences occurring at the mine site, i.e. *Purewa* and *Turra* coal seam (side casting /overcasting method). The average working height of the dragline bench is 32–35 m with a cut width of 80 m. An average of 80 m width is extracted through the 'dragline cut' with an average working height of 40 m and the rest of the parting, i.e. 15–20 m, is extracted by shovel–dumper combination.

Two Draglines (15/90 and 24/96) on the west side and two Draglines on the eastern side (24/96) of the mine are deployed for coal production. There are two mid-entries, i.e. east and west mid-entry, apart from the central entry to mining out Turra seam. These entries facilitate the draglines to operate continuously without any idle time while taking the new seating in the next cut. Also, the hauling distance of dumpers gets reduced with the provision of mid-entries. The OB cover varies from 10 m at the outcrop to about 180 m at the highest. Shovel and Dumper combination is deployed for Turra, Purewa (top & bottom) coal seam and all OB benches beyond the dragline benches. A total of 13 shovels with different capacities ranging from 10 Cum to 20 Cum are deployed for coal and OB excavation, while 84 dumpers with various capacities ranging from 85 to 190 tonnes are engaged for the transportation of coal and OB. Two surface miners are also deployed on a coal bench having a drum width of 4 m. The overall capacity utilization of the machines is recorded as very close to 100%.

Two types of drilling patterns are used: (a) For dragline bench: 259 & 311 mm dia drill holes at 10–12 m spacing; (b) For shovel–dumper bench: 259 mm diameter drill holes at 9–11 m spacing. Blast design and firing pattern, using electronic detonators and detonating fuse, is different for dragline benches as well as shovel-benches and a typical one. Both blast designs yield good fragmentation; however, the vibrations and flyrocks are kept under control through scientific experimentation. The results of a dragline blast at the Jayant mine are depicted in Figure 6.33. The mine sump for dewatering is located at the intersection point of both flanks.

Since this opencast mine has a stripping ratio of 2.99 and the mined-out land covers a huge area, the generation of overburden (OB) will also be more. Due to this reason, degraded dump areas containing waste rocks/material are visible at the Jayant mine site. OB removal to the tune of 75 million m^3/year will require a very large land area and this is also the reason why the backfilling practice is adopted at Jayant OC. Furthermore, it is analysed that the Jayant opencast mine has a slushy floor layer of crushed rocks with coal dust and water (1–2 m) at the mine floor level and the dragline dump was formed over it. This slushy layer (not a soft floor) is causing the failure of dragline dumps, many a time, though the slope angle is maintained at 45° which is safe. Dragline dump has a height of > 85 m, and periodically, the slope stability investigation has been carried out (Sharma and Roy, 2015). The surface road transport system for coal production has been in use at the Jayant mine, and the standard practice of height and width of benches as well as haul roads is practised in this surface mine because Draglines have been used in combination with shovel–dumper mining.

Backfilling to the de-coaled area is a common practice at Jayant coal mine because the volume of waste generated is very high and requires re-handling, The berm height and width are more than in conventional open-pit mine because of dragline operation at the mine. After backfilling, reclamation through tree plantation is an ongoing practice to protect the greenery of the area and maintain the environmental balance. The green cover mission of massive plantation being carried out every year in and around the Jayant project with the help of Madhya Pradesh Rajya Van Vikas Nigam Limited (MPRVVNL) is the current practice. It covers the mined-out areas and dumps areas, containing waste or overburden (OB) material. Saplings planted include species

Name of mine	Jayant, NCL
Face location	East dragline bench
Strata	Medium hard Strata (OB)
Face condition	Face choked one sidewith blasted material
Bench height	28.0 m
Depth of drilled holes	29.0 m
Dia. of drilled holes	269 mm
No. of rows	8
Pattern of holes	Square pattern
No. of holes	25 + 2 (Pilot holes)
Average spacing	10 m
Average burden	10 m
Sub grade drilling (if any)	1 m
Name and type of explosive used	IOCL (IBP) – SME – 40149 kgs Cast booster- Solar – 72.5 kgs
Type of initiation used	Electronic detonator—solar
Explosive charge per hole	1550 kgs
Maximum charge per delay	1550 kgs
Percentage of booster	0.2%
No. of decks (if any)	One deck – 3 m
Stemming material used	Drill cuttings
Water column in hole (if any)	7–9 m
Length of stemming (top)	5.5
Volume of rock blasted	72000 cu.m
Powder factor	1.79 cu.m / kg
Blast results	
1 Fragmentation	Very Good
2. Throw	10–12 mtrs from face on previous blasted side
3. Percentage of boulders	1 to 2%
4. Vibration	2.76 mm/sec at 3 km, 3.03 mm/sec at 2 km distance
5. Noise	Very low
6. Muck pile profile	Power trough of 4 m at back

Figure 6.33 Dragline blast results at Jayant O/C mine.
Source: Pandey et al. (2017).

such as Jamun, Jungle Jalebi, Seesam, Sirus, Mahua, Subabul, Bel, Amla, Kachnar, Karanj, Neem, Amaltas, Bamboo, Bougainvillea, Cassia, Gulmohar, Khamer, and Peltophorum. According to the land reclamation report based on satellite data for the year 2020, the green cover has increased from around 1,180 ha (pre-mining forest cover) to 1,419 ha which is about 45% of the total leasehold area of the Project. The target is to have over 2,600 ha of the area covered under green cover after the closure of the mine, which will be more than double the pre-mining stage (MOC, 2021). Comprehensive environment protection measures adopted at NCL's Jayant project have helped in increasing the carbon offset and reducing the impact of various types of pollution to a considerable extent. By meeting Corporate Social Responsibility (CSR) and maintaining the area green, the mine is achieving the coal and OB production targets and recording profit.

Lessons learnt from the case study

The Jayant coal mine case study makes us learn that a giant machine-like Dragline is suitable for soft, loose coal-like strata. It can yield high production results provided the mineral property (deposit) is large enough. Coal seams having a thickness of more than 4.5 m (thick seams) and which have horizontal mineral deposition are easily minable with a dragline. 'Side Casting' or 'Over Casting' of bulk material namely coal and overburden can be handled economically, but Draglines are not suited for 'selective mining' as contamination of coal/mineral with overburden is more.

Case study 6.5: Mining for copper ore in Brazil (base metal)

Introduction

The Carajás mineral province of Brazil is rich in several mineral resources. It has iron ore mines, manganese deposits, low-cost gold mines, copper sulphide, and nickel laterite resources. *Sossego, Antas and Pedra Branca, Sequerinho, Jatoba*, and *Serra Dourad* are some other copper mining projects in this mineral-rich province. *Sossego* project alone has an estimated 313 MT of copper ore reserves discovered way back in 1997. The Carajás area has excellent infrastructure including all-weather roads, a commercial airport at Carajás, Tucuruí Dam, abundant water, good roads, and many industrial and social institutions. The area is well-served by railroads and highways that connect the villages and cities.

Salobo Copper Mine, Brazil

Salobo open-pit mining operations (Salobo operations) are located in northern Brazil in the southeastern portion of Pará State of Brazil (Figure 6.34). Geographic coordinates for the operation are 5°47′25″ S latitude and 50°32′5″ W longitude. The Salobo mine is a surface mine located 159 km SW from Maraba, Brazil and is owned by the company **Vale SA** which is a Brazilian multinational corporation engaged in the metals and mining business in Brazil.

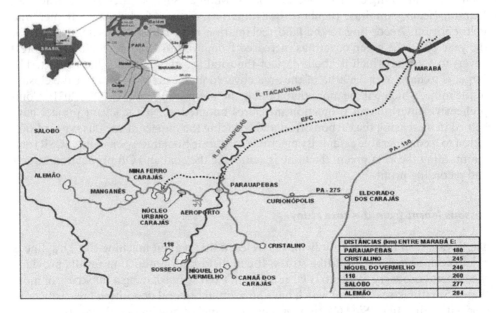

Figure 6.34 Location map of the Salobo copper mine – project.

The Salobo project operations consist of an operating copper-gold open-pit mine and an ore processing plant. The mine is currently producing at a rate of 24 MT per annum through a conventional mining method described later and mainly crushing–grinding–floatation is the metallurgical process involved in producing the copper concentrates.

Salobo is the largest copper deposit ever found in Brazil and the deposit contains copper as well as gold in an association. Salobo has reserves of over *1 billion tonnes* with gold and copper grades of 0.43 g/t and 0.692% respectively. This newly constructed, low-cost copper-gold mine began operating in May 2012 with a designed capacity of 12 MTPA and subsequently began the second phase of construction to expand the mine to 24 MTPA with additional resources. The company has produced an estimated 172.7 thousand tonnes of copper ore in 2020 and the mine has an operative life until 2053 at the current production rate.

The ore mineralization and geological setting

The Salobo deposit extends over an area of approximately 4 km along strike (W-NW), is 100–600 m wide, and has been recognized to depths of 750 m below the surface. The copper mineralization typically consists of assemblages of magnetite–chalcopyrite–bornite and magnetite–bornite–chalcocite. Accessory minerals include hematite, molybdenite, ilmenite, uraninite, graphite, digenite, covellite, and sulphur salts. The mineral assemblages can be found in several styles (Figure 6.35).

Geologically, the Carajás mining district, located in the SE of Pará State, lies between the Xingu and Tocantins/Araguaia Rivers and covers an area of about 300 km × 100 km. It is hosted in the Carajás Province, forming a sigmoidal-shaped, west-north-west to the east-southeast-trending late Archean basin (Figure 6.36). The Archean

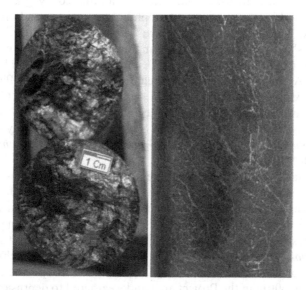

Figure 6.35 Bornite, chalcocite, and chalcopyrite, Salobo mine. Cu mineralization occurs in veins.

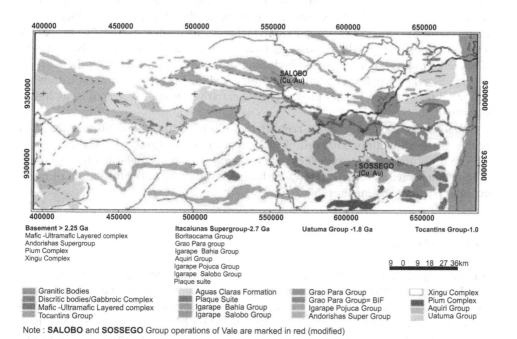

Basement > 2.25 Ga	Itacaiunas Supergroup-2.7 Ga	Uatuma Group -1.8 Ga	Tocantins Group-1.0
Mafic -Ultramafic Layered complex	Boritaocama Group		
Andorishas Supergroup	Grao Para group		
Pium Complex	Igarape Bahia Group		
Xingu Complex	Aquiri Group		
	Igarape Pojuca Group		9 0 9 18 27 36km
	Igarape Salobo Group		
	Plaque suite		

Granitic Bodies
Discritic bodies/Gabbroic Complex
Mafic -Ultramafic Layered complex
Tocantins Group

Aguas Claras Formation
Plaque Suite
Igarape Bahia Group
Igarape Salobo Group

Grao Para Group
Grao Para Group= BIF
Igarape Pojuca Group
Andorishas Super Group

Xingu Complex
Pium Complex
Aquiri Group
Uatuma Group

Note : **SALOBO** and **SOSSEGO** Group operations of Vale are marked in red (modified)

Figure 6.36 Regional geology of the Carajás Province, Brazil.

basin contains a basement assemblage that is dominated by granite–tonalitic or-tho-gneisses of the Pium Complex, and amphibolite, gneisses and migmatites of the Xingu Complex. The basement assemblage defines a broad, steeply dipping, E-W trending ductile shear zone (Itacaiúnas shear zone) that experienced multiple episodes of reactivation during the Archean and Paleoproterozoic periods. The metamorphic rocks are cut by Archean-age intrusions, including the calc-alkaline Plaquê Suite (2.73 Ga[l]), and the alkaline Salobo and Estrela granites (2.57 and 2.76 Ga respectively). The basement rocks are overlain by volcanic and sedimentary rocks of the Itacaiúnas Supergroup (2.56–2.77 Ga). The Itacaiúnas Supergroup is informally subdivided into the *Igarapé Salobo Group, Igarapé Pojuca Group, Grão Pará Group, and Igarapé Bahia Group,* from oldest to youngest.

The Itacaiúnas Supergroup hosts all the Carajás iron ore-copper-gold (IOCG) de-posits, including *Salobo* and *Sossego,* and it is thought to have been deposited in a ma-rine rift environment. The metamorphism and deformation have been attributed to the development of a sinistral strike-slip ductile shear zone (the 2.7 billion years Itacaiúnas Shear Zone) and sinistral, ductile-brittle to brittle trans-current fault systems.

Mineralization at the Salobo deposit is hosted by upper greenschist to lower am-phibolite metamorphosed rocks of the Igarapé Salobo Group. The group thickness varies from 300 to 600 m in the Project area and weathered to depths of 30–100 m. The rocks strike approximately N70°W and have a sub-vertical dip. The major host units are biotite (BDX) and magnetite schists (XMT). Granitic intrusions (GR) occur adja-cent to the north and south sides of the BDX and XMT, and a series of much younger diorite dykes (DB) cross-cut the mineralization forming barren zones.

There appear to be two classes of copper-gold deposits in the Carajás region. The first group includes Cu–Au–(W–Bi–Sn) deposits which contain quartz veins, may or may not have associated iron oxides, and are genetically related to the cooling of Pal-aeoproterozoic granites. The second group includes iron oxide Cu–Au ($\pm$U – rare earth elements) deposits (e.g. Salobo, Sossego, Cristalino, 118, and Igarapé Bahia) that may be related to more alkaline rocks, including the alkaline complexes of the Carajás belt, e.g. Estrela Complex, Old Salobo granite,. and the base metal mineralization-as-sociated intrusive. The second group of deposits are commonly referred to as IOCG deposits.

Weathered rocks in the upper regolith (30–100 m) of the Salobo project area also contain traces of metallic minerals. These chemically weathered rock traces contain iron ores, saprolitic gold, supergene copper, uranium, bauxite, and other heavy miner-als in residual accumulations. Deep weathering is the cause of the formation of many secondary and supergene ores in it.

The mineral tenure

Brazilian legislation separates the ownership of surface rights from mineral owner-ship. In Brazil, a mining company can operate a mine even if does not own the surface, provided it owns the minerals. In this case, it is necessary to pay a royalty to the sur-face owner. The royalty is calculated as 50% of the CFEM (Compensation for Finan-cial Exploitation of Mineral Resources), which is paid to the government. The mining concessions are updated every year on the presentation by Vale of the annual report

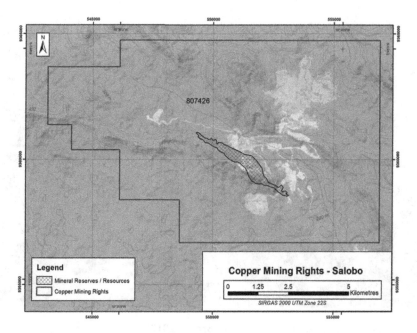

Figure 6.37 Mineral tenure layout plan.

of mining production to the DNPM (National Department of Mineral Production or in Brazilian language – Departamento Nacional deproducao mineral). The mining rights are managed by DNPM. The Salobo operations tenement title is 100% owned by Vale S.A. (copper ore, DNPM 807.426/74; Exploration Permit no. 1121 dated July 14, 1987, and 2017) and defined as a polygon of 9,180.61 ha (Figure 6.37).

A marginal cutoff grade of 0.253% was calculated for copper and is applied in equivalent amounts of copper. This cutoff is used for the determination of mineral reserves and mining sequencing.

Mining and method of ore excavation

Salobo mine (Figure 6.38) utilizes standard open pit methods, developed in 15 m benches, with trucks and shovels. After drilling and blasting the material, cable shovels, large front-end loaders, and hydraulic excavators are used to load this material. A fleet of 240 t and 360 t trucks are used to haul the waste material to waste dumps proximal to the pit or ore material to the primary crusher. Lower-grade ore is stockpiled for later processing.

The mine planning objective is to mine the ore sequentially in different mining phases, considering the largest possible vertical spacing between phases. The plan is to provide an approximately steady annual production of 24.0 million tonnes to the mill. In 2017, the ultimate pit was redesigned based on the 2016 pit optimization results and incorporated the revised pit wall designs. The internal phases were also redesigned in 2017 with an eighth phase plan added to improve the overall mine productivity. The overall phase-wise site layout is shown in Figure 6.39. A practical and executable

Figure 6.38 A view of Salobo copper mine, Brazil (2018).

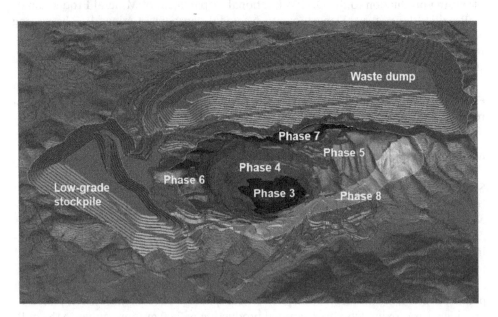

Figure 6.39 Salobo phases of the mining.

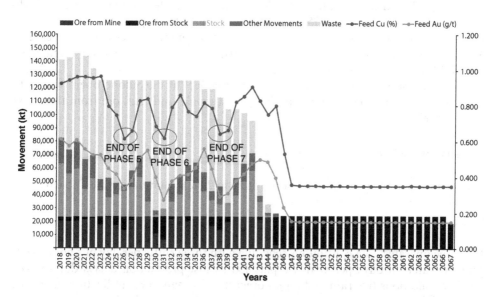

Figure 6.40 Life of mine plan (mineral reserve only).

production schedule is developed by short- and long-term mine planning teams. The ultimate pit has been subdivided into eight phases two of which have been mined out the remaining six phases form the basis of the life of mine plan.

In general, the phases have been sequentially scheduled with a maximum ore plus waste production rate of 126 million tonnes per year feeding 24.0 million tonnes of ore to the processing plant (Figure 6.40). The mine plan and the entire schedule are as per the detailed planning done by Vale SA.

The open-pit mine life is approximately 28 years, ending in 2045. However, the process plant will continue to operate by reclaiming stockpiled material until 2067. Phasing of the open pit development and application of the cutoff grade strategy allows higher grade ore (above 0.90% Cu) to be processed in the initial years of the operation.

The Salobo bulk mining operations primarily utilize large electric (rope) shovels for ore and waste/overburden (Saprolite) production. Hydraulic shovels are used for the oxide Saprolite and transition material where a lower ground pressure is required. Wheel loaders are used for miscellaneous cleanup jobs and backup of the shovels when needed. A fleet of off-road haul trucks is used to transport the material to either the waste dump or the primary crusher stockpiles. Low- and medium-grade ore is stockpiled near the open pit. Cycle times for haulage calculations are determined for each mining period using the Mine Haul software. The track dozers are assigned to maintain the production areas, waste dumps, and cleanup of the benches. Wheeled dozers, road graders, and water trucks complete the remainder of the auxiliary equipment fleet. The drilling equipment consists of electrical and diesel-powered drills of 12 1/4″, 10″, and 6 1/2″ diameters, cable shovels of 42 yd^3 and 63 yd^3, hydraulic excavators of 38 yd^3, and wheel loaders of 33 yd^3 capacity. The diesel drills and the wheel loaders are mainly used for ore exploitation that needs more mobility, whereas electrical drills and cable shovels are used for waste removal and bulk ore portions. The wheel loaders were considered in the fleet to support the larger units in narrow areas and in opening new accesses. Komatsu 830EAC (240 t), Caterpillar 793 (240 t) and 797 (360 t) trucks are selected to haul all materials within the pit. The auxiliary units are bull-dozers (D475A-2, Cat D11, D375A, Cat D10 and D6R); this equipment is necessary to maintain the production areas, waste piles, and cleaning the material on the benches. Wheeled tractors, motor graders, and water trucks complete the auxiliary equipment fleet. Table 6.1 details the current fleet.

Table 6.1 provides a summary of the mining fleet at the Salobo operations. The equipment listed is used to develop, drill, blast/muck, and haul ore from the active mining levels.

Production at Salobo mine

The ore production since the start of the mine in 2012 is summarized in Table 6.2.

Manpower

The mine operates on a continuous schedule with three shifts per day of 8 hours each. Approximately 10 days per year are planned as lost production delays due to poor weather conditions (i.e. rain and fog). Forecast mine manpower utilization takes into account delays for training, blast moves, and other operational delays.

Ore stockpiles and waste disposal

Low-medium-grade ore and waste rock from the mine are stored in three locations along the perimeter of the pit at a designated place and as per the mine plan and

Table 6.1 Production Mining Fleet at Salobo Mine

S. No.	Type of equipment	Fleet details	Nos of equipment deployed (quantity)
1	Drilling Machine		21
		Pit Vipper 351 (121/4")	6
		BE 49 HR (12 1/4") 6	6
		Pit Vipper 235 (9 7/8")	3
		ROC-L8 (6 3/4") 5	5
		ROC-D7 (5 1/2") 1	1
2	Loading Shovels		12
		BE 495 HD (42 yd3)	4
		BE 495 HR (63 yd3)	1
		PC 5500 (38 yd3)	4
		L 1850 (33 yd3)	3
3	Hauling Machine		51
		Komatsu 830 (240 t)	18
		Komatsu 930 (320 t)	3
		CAT 793 (240 t)	14
		CAT 797 (360 t)	16

Table 6.2 Production at Salobo Mine

Year	Feed			Concentrate		
	Tonnage (Kt)	Cu %	Au (g/ton)	Tonnage (t)	Cu (%)	Au (g/ton)
2019	22,486	0.97	0.68	509,778	37.2	22.47
2018	23,657	0.95	0.66	509,811	37.8	22.05
2017	23,650	0.95	0.67	498,172	38.8	21.63
2016	21,401	0.94	0.67	445,238	39.5	22.18
2015	20,290	0.88	0.57	402,592	38.6	19.41
2014	12,474	0.97	0.62	255,511	38.5	19.51
2013	7,366	1.09	0.76	1,65,471	39.4	21.92
2012	1.823	1.13	0.74	32,231	40.8	20.42

Table 6.3 Waste Dump and Stockpile Design Parameters

Parameter	Units	Waste dump	Ore stockpile
Lift Height	M	20	10
Angle of Repose	degrees	32–35	34
Berm Width	M	10	15
Overall Slope	degrees	20	25

schedule. The main waste rock dump is to the west of the pit and contains both oxidized and fresh rock. Geotechnical investigations were conducted to develop the dump design parameters. A 35% swell factor is used to compute the required storage volumes in the stockpiles and waste dumps. Material is end-dumped in 20 m high lifts with 10 m berms between lifts. The bench face angles range from 32° to 35°, depending on the angle of repose for the material (Table 6.3). Including the berms, the overall slope of the dumps ranges from 2H:1V to 2.5H:1V. The resulting slopes have an estimated 1.5 factor of safety against large-scale circular slip failures.

The waste materials and the low-medium-grade ore have been characterized as having low acid rock drainage potential. The long-term storage of medium- and low-grade material in a tropical environment may lead to some oxidation of contained sulphide minerals, impacting the recovery of metals during the eventual processing of the stockpiles. To control the infiltration of surface water and minimize resultant leaching, the water management requires additional attention, particularly during the rainy season, that includes pit dewatering, control and runoff from the surrounding area to the open pit. With over 1,920 mm of rainfall each year, sumps and pumps need to be well managed to maintain the roads and pit working surfaces. The operation recognizes this and has allocated appropriate resources to this task. During the dry season, dust control is maintained through the use of water trucks. Evaluations are ongoing to determine the effectiveness of additives, such as calcium chloride for dust control.

Ore processing

Mining and ore processing are interlinked operations. Accordingly, three phases for the Salobo mine and ore processing plant were designed and commissioned since the start of the project in 2012 as given below:

- Salobo I (2012): 12 MTPA of ore – 100-kilo ton of copper concentrate
- Salobo II (2014): 24 MTPA of ore – 200-kilo ton of copper Concentrate
- Salobo III (2019–21): 36 MTPA of ore – At the commissioning stage

Salobo I and II plants were almost similar, but the Salobo III has better and more advanced metallurgical processing features. The Salobo III Project includes all equipment and unit operations necessary for the processing of copper ore, from the receipt of ROM in primary crushing to the storage of concentrate at the plant, including all utilities, infrastructure, and operational and administrative support functions.

The process flow sheet (Figure 6.41) consists of the basic stages of comminution (crushing and grinding), classification (wet screening and cycloning), concentration (floatation), regrinding and solid–liquid separation (thickening and filtering). The process route is evolved through the various practical study phases of the project incorporating the additional knowledge gained from metallurgical test work and the relative importance of the identified lithologies in the mineral resource and mineral reserve estimates.

The mining operating costs and processing costs are updated periodically based on the current costs and exchange rates. For the Salobo mine, the average overall mining cost comes out as ≈ $3.37 per ton, whereas the process cost is $ 7.91 per ton for the year 2019–2020. The processing cost and the mining cost indicated above have been worked out separately and this has the effect of not increasing the marginal cutoff grade. While trading the copper metal (for buying/selling), taxes and international exchange rates are considered in addition.

Vale SA produces copper in Brazil and Canada and its copper mining operations in Brazil are a profitable venture because of the presence of gold in the ore. The mine is well-positioned relative to the infrastructure and has substantial exploration and future expansion potential Burns et al. (2017, 2019).

Note: This case study is summarized from the technical reports referenced here. The original author of the technical report is approached and the case study content is duly checked and verified.

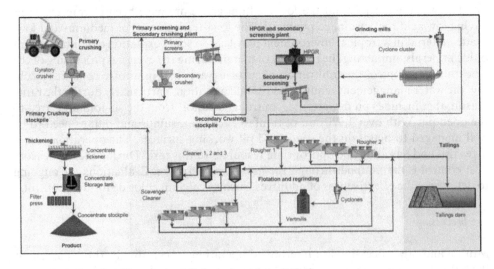

Figure 6.41 Process flowsheet, Salobo mine plant (I & II).

Case study 6.6: Tummalapalle uranium mine, Andhra Pradesh, India

Ajay Ghade and K.K. Rao

Uranium is an ore mineral produced by mines. It is found in small amounts in most rocks and even in seawater, but nature has created Uranium ore in the rock formations containing complex ores, e.g. Uraninite, brannerite, monazite, coffinite, xenotime, gummite, and *pitchblende*. All these ores contain radioactive minerals (Uranium and Thorium) and have geological-grade Uranium with basic common elemental properties (Table 6.4). The vast majority of uranium is used for power, usually in controlled nuclear reactions. The leftover waste, i.e. *depleted uranium*, can be recycled to harness other types of power, such as the Sun-power.

Uranium mines operate in many countries, but more than 85% of uranium is produced in six countries: Kazakhstan, Canada, Australia, Namibia, Nigeria, and Russia. India is among those countries which have limited Uranium reserves for their domestic needs only (Table 6.4). Uranium ore, being a mineral of strategic importance, is essentially required for nuclear power generation in nuclear reactors as the raw material feed. Its various other applications for industries and peaceful purposes have enhanced its importance tremendously, e.g. the pharmaceuticals industry for Cancer drugs and electronic/semiconductor applications for human welfare are some known applications of Uranium. Weapon-grade uranium for military application is useful for warfare. Uranium has explosive potential too because it can sustain a nuclear chain reaction through fission meaning that its nucleus can be split into several neutrons with the same energy as its ambient surroundings (through chain reaction). This *fission* and *fusion* phenomena of the Uranium ore generate heat that is then used to boil water, creating steam that turns a turbine to generate a power called 'nuclear power'.

It is well known that Uranium is associated with radioactivity. In ore form, its rate of decay is low and handling is rather easy but with great care. It is possible to enhance the Uranium percentage through processing making the ore Rich-Uranium. For example, U-238 and U-235 have a half-life of an incredible 4.5 billion years and

Table 6.4 Some Basic Elemental Properties of Uranium and Thorium

S. No.	Parameter/Properties	Values	Values
1.	Atomic symbol (on the Periodic Table of Elements)	U (Uranium) (Solid phase at room temp}	Th (Solid phase at room temp)
2.	Most common isotopes	U-234; U-235 ; U-238	Th-232
3.	Number of isotopes	16	9
4.	Atomic number	92	90
5.	Atomic weight /mass	238.02891	232.04 g/mol
6.	Density	18.95 g/cm^3	11.72 g/cm^3
7.	Melting point	2,075°F or 1,135°C	1750°C
8.	Boiling point	7,468°F	4,500°C or 8,100°F

Note: While thorium is a source of nuclear power, it is not usable as a fuel directly; instead, it is a fertile nucleus that can be converted to uranium in a reactor. Only after uranium conversion does thorium become useful as a nuclear fuel.

over 700 million years respectively. With this background and in consideration of the fission bomb like the one that destroyed Hiroshima in Japan, we will put our focus on the mining aspects of Uranium in this case study.

The mine, project, and location details

Tummalapalle Uranium mine (14°18′36.6″N 14°20′20″N: 78°15′16.57″E 78°18″3.33″E) of Uranium Corporation India Limited (UCIL) is situated in the Cuddapah district of Andhra Pradesh and falls in the Survey of India topo sheet No. 57 J/3 & 7 and is about 12 km NNW of Pulivendula town (A.P), India (Figure 6.42). The mine is operational since 2012 and is named after the *Tummalapalle* village of Cuddapah district (alternatively Cuddapah is named as YSR district also). The neighbouring villages of the mine area are Tummalapalle, Mabbuchintalapalle, Bumayigaripalle, and *Rachakuntapalle* of *Velpula* and Medipentla Mandals (districts are locally referred to as Mandals). Nearly 60 ha in the Kottala village of Vemula Mandal were acquired by UCIL (Uranium Corporation of India Limited – A government enterprise) for 'tailing disposal'. Only 1,305 g of uranium can be extracted out of the 2,350 tonnes (First Post, 2018). The Tummalapalle project has an underground mine and a mineral processing plant.

According to the Dept. of Atomic Energy (DAE), India has one of the largest reserves of uranium in the country. Though the production, grade, and reserves of Uranium ore are classified information according to the Atomic Minerals Act, Rules and Notifications, according to the sources, at Tummalapalle mine, approximately 3,000 tonnes/day of uranium is being produced every day.

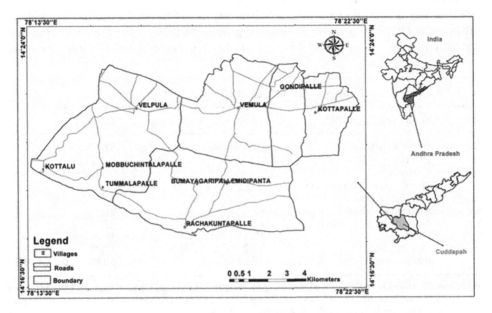

Figure 6.42 Location map – Tummalapalle uranium mine.

The processing unit of the Tummalapalle project processes 2,350 tonnes of ore per day though it has a higher installed capacity of 3,000 tonnes/day. This commercial processing plant has a slated capacity to treat using state-of-the-art alkaline pressure leach process technology and is operational on a similar pattern to that of the UCIL plant of the Jaduguda complex made jointly by BARC, AMD, UCIL, and NPCIL (Nuclear Power Corporation of India Limited). The plant is meant to extract Uranium from its ore (Deccan Herald, 2011).

Geology and geo-mining conditions

Geologically, the area falls in the SW part of the 'Cuddapah' basin. Uranium mineralization at Tummalapalle uranium mine is hosted by *Vempalle dolomites* of the *Papaghni* group of the Cuddapah supergroup. Mineralization is of strata-bound type, confined between two lithological units, viz. red shale and massive limestone. The radioactive minerals present in the ore are *pitchblende, coffinite,* and *uranium–titanium* complex. The average geological grade of the ore in the hanging wall is 0.038% U_3O_8, whereas the footwall has 0.041% U_3O_8. Uranium mineralization is occurring as two lodes in dolomite (impure siliceous, phosphatic, stromatolitic dolostone) and is designated as Hang wall lode (width 3.2 m) and footwall lode (width 2.5 m) separated by 1–3 m lean zone. The strike direction of the ore body is N68°E- S68°W, dipping towards N22°E with an average dip of 15°. The dimension of the ore body is 5.6 km along the strike and 1 km along with the dip with overburden depth ranging from 15 to 275 m. The hanging wall lode is more persistent in its thickness and grade than the footwall lode in the entire strike length. The stratigraphy of the Tummalapalle mine as derived from the borehole study of the area is given in Table 6.5. Uranium mineralization has taken place in Dolostone and the overlying rock is purple/red shale. On the top of the shale, there is cherty limestone. Dolostone is quite strong having RQD 83%–95% and can withstand more loads. However, the layer of purple shale having RQD 67%–88% is not very strong and requires support during underground excavation. In case the shale is exposed, the shale layer becomes problematic from a support angle and may collapse.

Table 6.5 Stratigraphy of Tummalapalle Uranium Mine

Pulivendla Quartzite	
	Disconformity
	Cherty limestone
	Purple shale (20 m)
Vempalle formation (1,900 m)	Dolostone (Uraniferous)
	Intraformational Conglomerate
	Massive limestone
Gulcheru Quartzite	Quartzite / conglomerate
	Eparchean Unconformity
	Archean and Dharwars

The geological reserves of hang wall (HW) and footwall (FW) lodes are 28.64 MT and 16.84 MT respectively, whereas taking 70% extraction of HW lode and 40% extraction of FW lode, the total mineable quantity of Tummalapalle mine is about 26.79 MT. Considering a planned production capacity of 3000 TPD (0.9 MT/annum), the Tummalapalle mine has a life of about 30 years (MECON, 2006).

Method of mining

At the Tummalapalle uranium mine, the *room-and-pillar method of mining* abbreviated as R&P is followed (Figure 6.43). The physico-mechanical properties of the ore are such to make it suitable for extraction by the inclined room-and-pillar stoping method. Ore and host rocks at the Tummalapalle mine have an average RMR (rock mass rating) of 50–60 (Fair) and an average Q of 3.23 (poor category) (Ghade et al., 2016). This mechanized underground mine is developed with three sets of declines (5 m × 3 m) at 15 m intervals and driven at a 9-degree gradient in the apparent dip direction of the ore body. The advanced strike drives (ASDs) of size 4.5 m (W) × 3 m (H) are driven in strike direction from both services declines at vertical intervals of 10 m. Levels are connected with the ramps developed at 9° apparent dip with an interval of 120 m. Mining of HW lode is generally undertaken only after the complete filling of mined-out areas of the footwall lode. Stope blocks of dimension (120 m × 39 m, length: width) have been developed by driving ramps at every 120 m interval between the rib pillars of 7 m wide, which are left at 120 m interval for supporting the roof. After developing the ramps, stope drives have been developed on either side of the ramp-up to the limit of the rib pillars. These stope drives will be interconnected by forming a room and pillar of dimension 5.5 m × 4.5 m. Several levels (more than 18 levels) have been developed in the footwall lode on either side of the decline and are confined to the footwall lode only.

Low-height drill jumbos for drilling, rock breaker, low-profile dozers, belt conveyor system, and dump trucks have been deployed. The blasting operation is carried out using emulsion explosives with millisecond delay detonators. Muck loading and

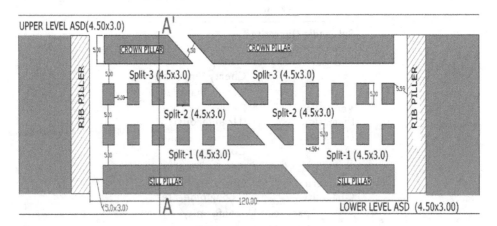

Figure 6.43 A schematic diagram for the room-and-pillar method of mining.

hauling work, for the mine development as well as stoping, is done by diesel-driven loaders. The ore transportation is done principally by mechanized means using low-height LHDs. Vital aspects of day-to-day production and related mining operations, e.g., drilling, blasting, mucking, transportation, and support, for this mine are done as per the approved mine plan & design and executed by M/s SMS Limited as Mine Developer and Operator (MDO). The mine development work and other construction of inclines are carried out according to the contractual practices. However, government stake and interference, as per the AMD directive, remains applicable to the principal custodian of the mineral properties, UCIL.

Analysis and discussions

The Uranium mineralization occurring at the Tummalapalle mine in two lodes, i.e., FW lode and HW lode, has an average width ranging from 2.5 to 3.2 m respectively and is separated. This is a good thickness of the ore body for an underground metallic deposit. Since it has a uniform thickness, is inclined, and has a good strike length, the stope formations are rather easy. The host rock for uranium mineralization is dolomitic limestone and is overlined by shale which creates some support problems and water problems while excavating.

The depth of the mine workings is below 100 m with the presence of the weathering zone (W2), extending from 40 to 50 m below the surface. The in situ stresses acting on the rock mass are not a big problem to tackle. However, periodical rock mechanics investigations are necessary for this mine to devise optimum solutions for better safety and extraction ratio.

Mining issues and problems

Some mining issues and problems of the Tummalapalle mine that are likely to occur or encountered during mining are listed below:

1. Room-and-pillar dimension optimization for stopes and stoping operation.
2. Roof stability problems due to the presence of roof shale in the mines underground.
3. Support design at deeper levels is a problem for underground workings.
4. Problems of mining in HW lode.
5. Solution cavities and problems of shale as the immediate roof of host rock, i.e. Dolostone.
6. Water seepage problem from the roof (presence of the 'Red Shale').

(As per the hydrogeological investigation report of the Central Mining Research Institute (CMRI), India, the overall average water table of the Tummalapalle region is 4.19 m (pre-monsoon season = 5.80 m and post-monsoon season = 2.58 m for the year 2005) and the estimated water quantity would be 1,230 m³/day approximately which varies from year to year and season to season).

Room-and-pillar (R&P) mining is a very old successful method applied to inclined, horizontal or nearly horizontal veins/deposits that have been refined over the years. The method is used in both coal and metalliferous mining. In general, the R&P

method at Tummalapalle has facilitated multiple working places/stopes for safe ore extraction with maximum ore recovery. In addition, the incline R&P method provided better ventilation and transportation routes at the working faces. However, depending on the mining conditions encountered, the size of rooms and the supporting pillars, i.e. their dimensions, should be optimized after due research. However, for more and more extraction percentages, continued research is always suggested and advisable as well at the Tummalapalle uranium mine.

In brief, this case study educates us that mining strategic minerals like Uranium is economically feasible with safety **(Box 6.2)**. Optimum stope design with proper planning has led to mineral conservation as well as optimization (highest ore extraction in percentage) at the Tummalapalle underground mine which has inclined R&P as its method of mining. Both production and efficiency are the keys to underground mining at the Tummalapalle uranium mine.

Box 6.2: Uranium Facts

(i) Some facts about uranium ore found in India

Mines and Minerals (Development and Regulation) Act, 1957 (MMDR Act, 1957) is an act of the Parliament of India enacted for the development and regulation of mines and minerals that extends to the whole of India. The First Schedule of the MMDR Act of 1957 deals with specified minerals, i.e. Hydrocarbons, Energy Minerals, Atomic Minerals and Metallic and Non-Metallic Minerals listed under Part A, Part B, and Part C respectively. Atomic minerals are specified under Part B of the First Schedule of the MMDR Act, 1957. Uranium ore is covered in Part B under the 'Atomic Minerals' category, similar to 'lithium ore' which is also listed in the same mineral category. The Central Government promulgated Atomic Minerals Concession Rules (AMCR), 2016 for the regulation of mineral concessions in respect of Atomic Minerals.

The state of Andhra Pradesh in India is the largest producer of uranium in India, besides the state of Bihar (Singbhum district), Jharkhand, Chhattisgarh, and Rajasthan state. According to the Atomic Minerals Directorate for Exploration and Research (AMDER), a unit of DAE, **Beach Sand** also contains minerals like 'Monazite' and other heavy minerals of the atomic mineral category. More than 0.75% monazite and total heavy minerals are found in it.

In India, the Uranium ore of the Himalayan region in Jammu and Kashmir state, particularly the Ladakh region, has immense strategic significance. The Uranium ore of the Ladakh in the 'Ladakh batholith', all along Kohistan (Ladakh) and Southern Tibet (from east to west), has an exceptionally high concentration of Uranium (0.31%–5.36%) and thorium (0.76%–1.43%) and said to be of recent origin. Uranium-rich rocks in India, such as in Andhra Pradesh, are very old geological formations of the Precambrian period about 2,500–3,000 million years old, whereas uranium-bearing magmatic rocks located in the Himalayas are of very young origin, between 25 million years and 100 million years old.

Uranium-rich zircons from young magmatic intrusions of the 'Shyok suture zone and associated sequences' also contain Uranium ore. Similarly, the state of

Chhattisgarh and Rajasthan has some notable Uranium deposits of commercial grades but contains very less Uranium percentage.

'Narwapahar Mine' of Uranium Corporation of India Ltd. (UCIL) with good geological reserves at depth is the first fully mechanized underground mine operating since April 1995 where trackless mining with decline access is practised for the Uranium ore excavation in India.

(ii) Sustainable mining options: In situ Recovery (ISR) technique for uranium ore mining

The ISR technique includes injection, extraction, and well monitoring with lixiviant in the mineralized zone. The lixiviant is a 100% non-toxic, bio-degradable solution developed by research in which the ore gets dissolved. The ISR process captures ore in solution which is brought to the surface for processing without any ground disturbances. The feasibility of the ISR technique mainly depends on the solubility of ore in the lixiviant. Before confirming that the orebody is suitable for mining, several 'tracer tests', 'hydrogeological test', and 'push–pull tests' (pumping test) need to be conducted to know the hydraulic connectivity between the wells which are dug out in the form of drill holes to extract the solvents from the underground. The ISR technique resembles somewhat that of *'solution mining'*. It eliminates the development of an expensive traditional mine on the surface and hence an environment-friendly, sustainable, clean mining option with low carbon emissions.

ISR technique, though not new, was tried for Uranium mining where rich grades are available, e.g. in the country namely Kazakhstan, Canada, but this technique is currently in use for copper-gold extraction at Alford West & Alford East Project, Australia (150 km S-W of Adelaide) by Thor Mining Inc. and Enviro Copper Company; Excelsior's Gunnison Project, Arizona USA; and Van Dyke Project, ARIZONA USA by Copper-Fox Metals Inc. (Warland, 2022).

Case study 6.7: Turquoise Ridge underground gold mine, Nevada, USA

(A narrow vein gold mining in weak rock mass)

Arun Kumar Rai

Location and description

The Turquoise Ridge Joint Venture (TRJV) is located in north central Nevada, within the Basin and Range province (Figure 6.44). Newmont has a 25% stake in the joint

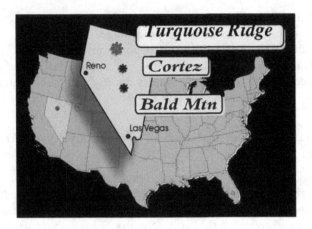

Figure 6.44 TRJV mine location plan map.

venture with Barrick controlling the remaining 75%. Mineralization in the area is typical of other Carlin-type gold deposits, with gold in the lattice of arsenic-rich iron sulphides that formed chiefly within carbonate-bearing rocks that have undergone significant decalcification and marginalization.

The Turquoise Ridge (TR) underground mine is accessed by two shafts used for ventilation, men and materials transportation, and ore hoisting. Current production is approximately 1000 tons/day and utilizes the underhand cut-and-fill mining method or box stopping due to the relatively low rock quality in the ore zones and the relatively shallow dipping 25°–45° ore body geometry. The primary ground control uses bolting, mesh, and shotcrete. Methods of supplemental support and rehab may utilize spiling, shotcrete arches, and cemented rock fill (CRF) arches. The geology and support strategies used at TR are described in an earlier paper (Sandbak et al. 2012). It includes the use of CRF and a quick mining sequence to minimize ground exposure time and unravel ground.

Due to the high clay content and inherent weakness of the ore, the ore bodies must be dewatered as much as possible before the commencement of mining individual zones. Dewatering is done as necessary, using up holes drilled from beneath ore zones from the access ramp systems, and collected locally as new areas are accessed.

Mining methods

Undercut drift and fill (UDAF)

The majority of our production is based on underhand cut-and-fill or box stoping methods. In the underhand cut-and-fill method (UDAF), the ore is initially mined out in 3m × 3m (10′W × 10′H) or 3m × 3.7m (10′W ×12′H) panels called top cuts. Drifting is normally completed using jackleg-drilling techniques with the excavation by conventional drill and blast or by underground LHD. Small mechanized jumbos are being tested to eliminate the use of jacklegs and to minimize workers at the drift face. Usually, the drill pattern is smaller than the excavation size, and the amount of explosives is minimized. Mucking is performed with a 1.5–2.3 m³ (2–3 yd³) LHD. In the very weak ground, excavation is often completed by mucking in advance without drilling and blasting.

After the excavation of one panel, it is completely backfilled tight to the back with CRF. The next panel is advanced adjacent to the backfill. This cycle is repeated in adjacent drifts, with one rib following the backfill of the adjacent drift until the desired width of the stoping area is reached.

The backfill creates a pillar of strength that reduces the possibility of back failure. The ideal sequencing of a top-cut panel will involve driving an initial access panel down the centre of the ore zone and retreating back towards the access. In areas where this is not feasible, panels will be started at the edge of the ore zone closest to the access with subsequent panels being mined in an advanced fashion. This procedure is used throughout the entire ore zone of the top cut level. Top cuts are occasionally backfilled earlier than planned due to the instability of the ground. Figure 6.45 is a schematic of an actual UDAF design configuration.

Upon the completion of the top cut of a level, the next level is advanced 3.7 m (12′) or 4.6 m (15′) lower in elevation. The undercuts are mined at a height of 3.7 or 4.6 m (12 or 15 ft) using the backfill from the top cut as the back. The width of an undercut varies from 3.7 to 6 m (12′–20′) depending on ground and water conditions. It has been demonstrated that a span of at least 9 m (30 ft) in width can be safely excavated beneath the consolidated backfill for the new undercuts.

Tertiary access is usually advanced first to the undercut area. An undercut is mined typically from the end of this access to the far edge of the undercut area, and the undercuts are mined and backfilled in retreat (mining and backfilling in retreat are proven to be more productive for operations). The ore body is accessed off of a secondary development heading, utilizing a heading with a cross-section of 4.3 m wide × 4.3 m high (14′W × 14′H), 4.6 mW × 4.9 m H(15′W × 16′H), or 4.9 m W × 5.2 m H (16′W × 17′H) configuration. In the wider and more massive zones, two headings in a production panel may be worked at any one time. Drilling is done with either an electric over

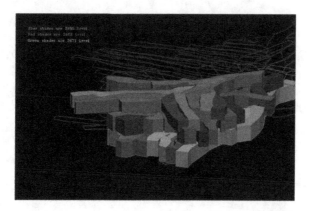

Figure 6.45 Underhand drift and fill design as-builts showing 3495 level top cuts (in black) and subsequent undercuts below; 3483 (in dark grey) and 3471 (in light grey).

hydraulic jumbo or a hand-held jackleg drill. Rib bolting is installed with a bolter or hand-held jackleg drill. Drift round lengths are determined by the type of drilling machine being used. Loader sizes vary from 1.5 to 4.6 m³ (2–6 yd³) LHD.

Box stoping method

The box stoping method is used in areas of longer and steeper dipping ore segments/ panels. In box stoping, the top cuts are mined using the same procedure as in the undercut drift and fill method. The access being driven in the undercut (box stope top cut) is advanced at the edge of the ore body. A second access is developed on the opposite side of the ore body. One of the accesses is backfilled, providing a backfill cap above the next level's access. Every other panel is driven 14′ × 12′, connecting from access to backfilled access (leaving 20′ in between panels). The box stope top cut panels are backfilled immediately after they are complete. Figure 6.46 is a typical box stoping design illustration showing the general layout and sequencing.

The box stope undercut is then advanced directly under the backfilled panel. After the undercut is completed, the bottom is taken out to the desired elevation, followed by backfilling. When a box stope top cut and undercut are backfilled on both sides of a non-mined panel, they create continuous strength pillars, allowing for advancement without having to backfill between levels. The box stope top cut is taken first at 20′ × 12′. The undercut is then taken, also at 20′ × 12′, followed by taking a bottom-up. After the bench, the bottom 20′ minimum is backfilled. The upper portion of the stope or bench top is filled with uncemented waste fill or gob.

The chief advantage of the UDAF method is the ability to adjust the mining to the specific ore zone encountered. Drifts are mapped and sampled for assays as they are driven, and the design can be adjusted to fit the ore zone or adjusted for the actual ground conditions. Because of the variability of the ore zones, the UCAF is the most

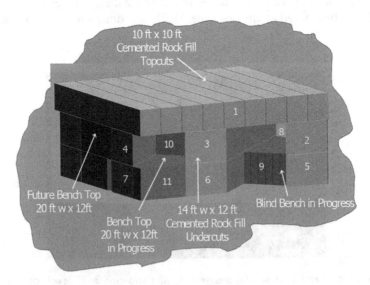

Figure 6.46 Typical box stope design.
Source: Holcomb and Dorf (2006).

selective method and can be tailored to smaller and shallower dipping ore zone geometries. It is also much quicker to develop than the box stoping method. The chief disadvantages to the UCAF method are the relatively high cost of mining and the increased use of cemented backfill needed.

The advantages of the box stope method are in the efficiency in blasting, mucking, and backfilling, as well as the ability in developing higher spans. The use of waste gob in the bench top stopes eliminates the need for cemented backfill and the use of a Jammer for the waste. This results in at least 25% in reduced cement and hoisting costs, as waste rock developed on the level is used in place.

One of the drawbacks of the box stope method is that it takes longer to develop the access drifts and infrastructure to start mining. It also requires a larger horizontal and vertical ore thickness. The sequencing of the mining is more restrained and must be developed in pieces. Because of the sharp breaks between the very high-grade and low-grade material, there is more risk of making mistakes in the early placement of the drifts that could mean more waste and dilution as the box stoping progresses.

Support elements

Bolts, mesh, and shotcrete

The support elements used at the Turquoise Ridge Mine include inflatable bolts, friction bolts, spiling, and groutable anchor bolts (Table 6.6).

Mesh and shotcrete control the loose and broken rock from unravelling from drifting and mining activities. When used in conjunction with shotcrete, the mesh also provides some tensile strength to provide more stable ground support. The shotcrete retains loose material, protects the mesh from oxidation, and prevents small loose material behind the mesh from falling out. The primary welded mesh used here is panel

Table 6.6 TRJV Bolt Support Elements (Sandbak et al., 2012)

Bolt type	Lengths (m)	Primary use
Split sets: SS39, plates; mats Coated (galvanized) or uncoated	0.46, 0.9, 1.2, 1.5, 1.8	0.9 m at the face, 1.5 and 1.8 m in ribs; formerly used 2.4 m in back
Inflatable bolts; Swellex or Super Swellex; plates; coated or uncoated	12 ton rated bolts (1.5, 1.8, or 2.3 m); 24-ton rated (3.7 m, 3–6 m connectable)	1.5 or 1.8 m in ribs 2.4 m in backs 3.7 m or 6 m in intersections; back or special applications Higher loads in weak RMR<35 ground (4.5 t/m); 9–15 t/m in RMR>35; better than split sets in clay zones
Spiling: #10 rebar	4.6 m	Backs with weak or expected difficult ground Nominal 25 cm close spacing; good coverage in sandy and clay
IBO Self-drilling anchor bolt	3 m length standard; couplers for extra length	Substitute for #10 rebar spiling, self-drilling into place or grout option

wire type 7.6 cm × 7.6 cm (3″ × 3″) 6 gauge diameters in 2.4 × 3 m (8′ × 10′) lengths. The mesh is overlapped to prevent rock from falling out between mats. In addition, the flexible chain link type of mesh support is available in galvanized 1.8 m × 7.6 m (6′ by 25′) and 1.8 m × 7.6 m (8′ × 25′) lengths in rolled forms.

Shotcrete strengths are approximately 28–34 MPa (4,000–5,000 psi) for a 28-day test. Shotcrete is applied to all primary and long-term development headings and may be applied in other headings where the primary bolts and wire-mesh support are needed to supplement. Ore headings are only supported with shotcrete if the ground is so poor that it will not stand long enough to be mined through. Portable shotcrete equipment that will work in the smaller dimension top cuts is being tested. The shot-crete uses a wet-mix technique, where the remixed shotcrete mix is delivered to the pump, and an accelerator is added to the spray nozzle to quicken the setting time. The initial mix is loaded into a trans-mixer truck from the concrete batch plant and deliv-ered to the underground site. There it is sprayed from a predesigned robotic boom. Shotcrete is delivered as a dry mix by a supplier. Approximately a 75 mm (3-inch) layer of shotcrete is applied to all permanent headings in the mine. The shotcrete used is a blend of coarse sand and fine gravel with cement, water, and small amounts of perfor-mance additives. All of the existing workings have ground support in place, which is usually a chain link screen, split set or inflatable friction bolts, or grouted resin rebar bolts. On average, just less than one cubic yard of shotcrete is used for each foot of waste development advanced.

Cemented rock fill (CRF)

The backfill system at the Turquoise Ridge Mine is a typically CRF system that has been proven to be effective during production. CRF consists of a cement/fly ash mix-ture with crushed and sized aggregate. The material is batched in a mixing plant, dumped down transfer raises from 1,250 to the 1,715 level, and then hauled to the stope to be backfilled and jammed into place using a modified LHD with a push plate as the jamming tool. In the top cut ore headings, the backfill may have to be trammed to the backfill face using an LHD. A backfill replacement factor of 01 ton for each ore ton removed is accurate for estimating the quantities required (Pakalnis, 2009). Backfill is an important part of the total ground support in active mining panels. The major benefits of backfill are as follows:

* Provides lateral support to pillars to stop spalling and prevent collapse
* Reduces stope wall closure
* Creates stable mining blocks
* Solid, uniform backfill is usually better than jointed rock it replaced

Support criteria

Design guidelines that were developed were based on historic conditions evidenced at the Getchell and Turquoise Ridge mines. Ground conditions at TRJV were subdivided into three categories for mine design and cost estimating. The support was originally based on a study by Barton et al. [1] guide to support and then modified to encompass ground conditions at TRJV (Figure 6.47). At least 50 mm (2″) of shotcrete were recommended for any rock with an RMR value less than 23. The minimum ground support was based on studies on RMR values encountered in drifting and the history of failures in different rock types and alterations (Bieniawski, 1976). A series of minimum ground control standards have been compiled according to the size of the drift or mine opening present at TRJV (Table 6.7).

The ground support is custom designed for each type and size of the opening and is issued with all design drawings. The development headings are designed with arched backs for the designated drift widths.

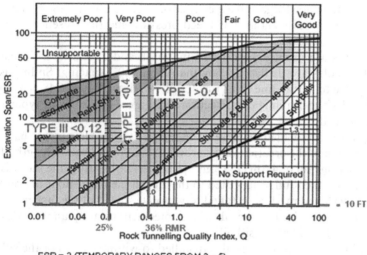

ESR = 3 (TEMPORARY RANGES FROM 3 – 5)

SPAN = 10ft ~ 3m THEREFORE SPAN/ESR = 3/3 = 1

NOTE REQUIRE SHOTCRETE WHEN Q<0.1 OR RMR < 23%

Figure 6.47 TRJV top cut design chart based on Q and RMR ratings based on the chart by Barton, Lien and Lunde (1974).

Table 6.7 Typical Minimum Ground Support Standards

Heading width	Fully supported ground	Nominal spacing
3 m × 3 m (10′ × 10′) Development headings up to 5.5 m (18′) Headings >5.5 m (18)′ or intersections >7.3 m (24′) (Additional)	2.4 m (8′) uncoated inflatable bolt (12-ton) 2.4 m (8′) coated inflatable bolt (12-ton) minimum 3.7 m (12′) coated inflatable bolt (24-ton) 6 m (20′) inflatable bolt; (24 ton) or 4.6—6 m (15–20′) inflatable bolt	1.2 m × 0.9 m (4′ × 3′) pattern 1.2 m × 1.2 m (4′ × 4′) pattern 1.8 m × 1.8 m (6′ × 6′) pattern 1.5 m × 1.5 m (6′ × 6′) pattern or as engineer warrants

Extra support in weak ground

While drifting through those areas of low RMR<20 or Type IV ground, or in faulted and altered ground in development or tertiary drifts, extra support is usually added to prevent unravelling or potential caving. TR has tried several methods and types of support including spiling, groutable bolting type of spiling, shotcrete arches with lattice girders, and even mining around and through mineralized weak areas using CRF arch.

Spiling

The decision to spile is based on changing ground conditions and qualitative assessment based on the knowledge and experience of the lead miner, miner, and supervisor. Factored into the assessment are drill feed pressure and the presence of slips, joints, and voids.

Spiling bars are usually 4.6 m (15 ft) in length and 32 mm (1–1/4″) diameter #10 rebar. Spiling holes are drilled on nominal 25 cm (10″) spacing. Spiling is installed from spring line to spring line or as the design requires. Figure 6.48 is a hypothetical development drift and spiling cross-section and Figure 6.49 shows a typical spiling in a development drift.

Spiling holes are drilled as close to the back as possible s to get back on grade (drilled to eliminate the creation of a "brow". A maximum round length of 2.4 m (8′) is drilled under one set of 4.6 m (15′) long spiling. After blasting, the muck is removed, and the ground is then fully supported to the face in the back. Bolts are on a 1.2 m (4′) staggered pattern and mesh is installed to within 0.3 m (1′) of the face in the back and ribs. Mesh is not required in the rib where intersections are planned. Mesh (wire) is installed within 1 m (3′) of the sill pillar.

An additional row of bolts is also installed to within 1 m (3′) of the sill. The tails of the spiling bars are pinned to the back using steel mats and a minimum of 1.8 m (6′) long split sets. Split sets will be installed at right angles perpendicular to the back. All intersections are supported with 3.7 m (12′) Swellex inflatable bolts. All spiling is then shotcreted with a 25–75 mm (1–3″) layer.

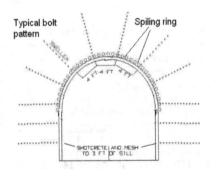

Figure 6.48 Design cross-section of proposed spiling pattern in a typical 4.9 m W × 5.2 m H (16′W X 17′H) development heading.

Figure 6.49 Photo showing spiling bars and steel mast holding spiling ends. Extra bolts through shotcrete for rehab support.

IBO groutable anchor bolt

The IBO anchor bolts have been used instead of using #10 rebar as spiling in development or larger development. It is only 3 m long, but has connectors to extend the length as needed. The bolts can be used with throw-away bits that allow the ISO bolt to be drilled into the perimeter and then left in place. This option can be used where the ground is more of a gravel formation. This allows the IBO bolt to stay installed without the drill hole collapsing. It also has a hollow core, which allows it to be grouted. Depending on ground conditions, the installation can be in two different forms: the throw-away bits with the IBO bolt or pre-drill holes without bits and hand-installed IBO bolts.

Recently, IBO bolts were tested in top cuts in the very sandy and gravelly ground as an alternative to shotcreting. Figure 6.50 is a design cross-section of a top cut showing the pattern of 15 bolts running parallel to drift on nominal 0.3 m (1′) spacing, and Figure 6.51 is a photo showing a typical installation. Spiling holes or grouted bolt spiling are drilled as close to the back as possible and drilled to get back on a grade, 15°–18° from horizontal in the back and upper spring line to achieve good protection and to eliminate the creation of a brow. The purpose of the ISO bolt is to produce a grout curtain in the back and portions of the upper ribs ahead of the active working face.

Shotcrete arches (w/lattice girders and cementitious grout)

The shotcrete arch support system was developed to be used in development drifts or tertiary drifts encountering very weak ground such as mineralized or faulted zones or in areas that are currently taking weight and needing supplemental support or rehab.

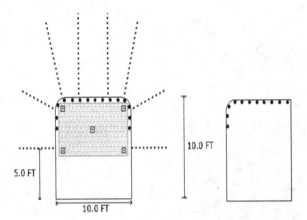

Figure 6.50 Typical cross-section of 3 m × 3 m (10 × 10 ft) top cut design showing ISO anchor bolt pattern (black dots). Dashed lines are approximate traces of normal support bolting.

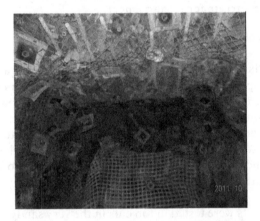

Figure 6.51 Photo showing the application of groutable ISO bolt in a top cut.

Steel arch shotcrete sets and shotcrete support are in addition to the normal development ground support of bolts, wire mesh, and spiling.

The shotcrete arch sets can be installed in a fully supported development drift with bolts, mesh, and shotcrete. The ring sets are made of four sections of triangular lattice girders that bolt together and are designed to fit inside the currently existing development drift. Figure 6.52 is a schematic cross-section of the lattice girder for a 4.9 m × 4.9 m (16′ × 16′) finished product.

The system can also be used under spiling as part of the drift cycle. Between two completed ring sets set on 1.2 m (4′) spacing, 75 mm (3″) thick wire 'bedspring' panels are installed to provide tensile support for subsequent coats of shotcrete to cover the sets, which is at least 1.8 m (6″) thick to cover the sets (Rai and Westhoff, 2013).

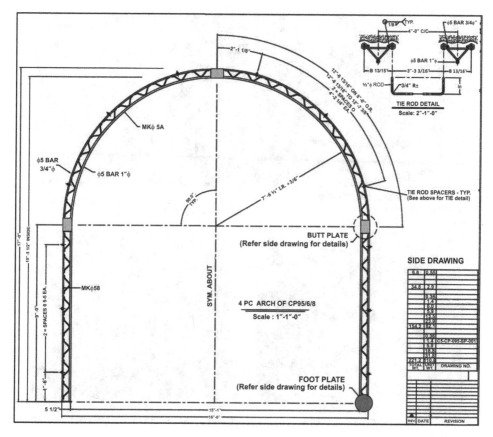

Figure 6.52 Cross-section of typical lattice girder set-up for 4.9 m × 4.9 m (16'WX16'H) drift.

The gap between the arch sets and the back is filled with a cellular cementitious grout of about 1.4 MPa (200 psi) strength (Figure 6.53).

Shotcrete arch sets were used to rehab an area of 33 m (110') long that was previously moving at a 0.5–0.8 mm (0.02–0.03/inch) per day rate, with extensive slabbing and drift shelling. Since the installation in June of 2010, there has been no sign of cracking in the covered drift, and the drift looks as pristine as when it was installed. There has been over 10 mm (0.4") of movement as measured by cross-drift tape extensometers, but the subsequent installation of two wireless 3.7 m (12') MPBXs installed at each rib has shown no evidence of movement in the rock. This suggests that the cellular cementitious grout may be deforming without putting any undue stress on the steel sets.

The arch design with expandable grout in the back thus provides excellent yielding support around the more rigid lattice girders. This provides an alternative to areas that have already experienced damage needing rehab but are expensive and labour-intensive. It does however provide one more way for supplementing support to keep drifts accessible.

EXCAVATE DAMAGED SHOTCRETE; PUT
IN GIRDER ARCHES BOLTED TO
CONCRETE FLOOR

BEDSPRING 3" DEEP. 4ft X 8ft X 3 in. SHEETS
BETWEEN SETS, SHOTCRETED

SPRAY BOTTOM ZONE 6 INCHES
SHOTCRETED TO COVER GIRDERS

TOP PORTION FILLED WITH BEDSPRINGS;
INSTALL POUR PIPE FOR EXPANDABLE GROUT

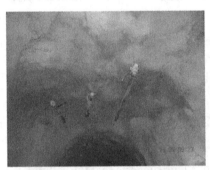

CEMENTITIOUS GROUT – EXPANDABLE GROUT.
200psi. NOTE ~1-2ft BETWEEN ARCH AND BACK
(GAP).

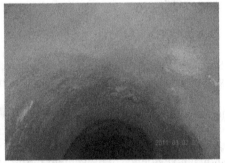

FINAL PRODUCT 15ft W X 16ft. H. ORIGINALLY WAS 16'W
X 17'H.

Figure 6.53 Photos showing stages of construction of shotcrete arch from girders to
shotcrete, and cementitious fill in the back.

Source: Pakalnis (2010).

CRF arches and MBD ramp

Another optional support method for areas of weak ground support is using CRF itself as a support when we encounter isolated pods of mineralized lenses sometimes encountered in driving development drifts or when backfilling weak existing drifts that are taking excessive weight or convergence.

The MBD Ramp is an important East-West drift providing ventilation and an escape way. The drift was a typical 4.6 mH × 4.9 mW (15′H × 16′W) development drift designed to be in the competent ground of RMR of 40–60. Upon excavation, it was found to parallel and cross a very weak and altered mineralized zone in mudstones and thin limestone with an RMR of 20 (Type IV) ground. The Ramp Drift had been open for three years, and mining close to the area initiated increasing horizontal drift closure. Drift damage continued despite two separate rehab campaigns installing 3.7 m (12′) and 6 m (20′) Swellex bolts through the deforming shotcrete.

Figure 6.54 is a photo of the MDT ramp that had squeezed down to 3.7 m (12 ft) from the original 4.9 m (16′) width, and it shows the supplemental rehab bolting to try to slow shelling and shotcrete spalling in the back. In the North rib sill, the rock un-ravelled behind the shotcrete–mesh interface (Figure 6.55).

Because bolting was not slowing the movement and because removing the existing shotcrete and mesh might leave a too-wide drift which could converge, it was decided to backfill the original MDT drift and create a new drift surrounded by a CRF arch to reconnect with the undisturbed drift further east. Figure 6.56 is a plan map of the con-figuration of the backfill arch in the MDT area and Figure 6.57 is an N-S cross-section showing the resulting geometry of the new MBD created by the CRF arch.

Figure 6.54 Looking East down MBD Ramp. Note supplemental bolting and spalling in the back and ribs. The drift profile was reduced to 3.7 m (12 ft) from the original 4.9 m (16 ft) width.

Figure 6.55 Damage in North rib and sill of more than 0.6 m (2′) with unravelling behind shotcrete and mesh. Longer 3.7 m (12′) and 6 m (20 ft) bolts were an attempt to tie drift to the deeper undisturbed ground.

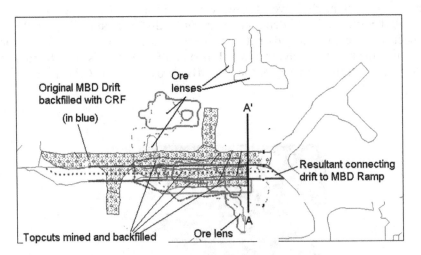

Figure 6.56 Plan map of MBD ramp area utilizing CRF arch.

Firstly, the weakened original MBD drift and intersecting N-S bays were back-filled. The next step was to ramp up, excavate, and backfill four separate top cuts in the mineralized zone above and beside the original drift. This created a new MBD drift surrounded by the strong CRF (Figure 6.57). The CRF arch is a more viable option if the mineralization can help pay for the mining. Other options such as the shotcrete arches can be a supplemental rehab, or emplaced as early support when encountering very weak RMR<25 sandy or gravelly ground or used in combination with the spiling or grouted ISO bolts.

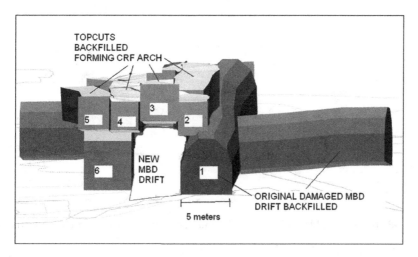

TOPCUTS
BACKFILLED
FORMING CRF ARCH

3

5   4   2

NEW
MBD
DRIFT

6   1

ORIGINAL DAMAGED MBD
DRIFT BACKFILLED

5 meters

Figure 6.57 Cross-section A-A′ looking west illustrating CFR arch sequence and resultant new MBD drift reconnecting to old MBD ramp.

Movement monitoring and stress modelling

Instruments are installed at TRJV to understand the movements in the rock mass associated with stresses after excavation, to determine ground control cause and effect, and to build a base of information for stress modelling and movement prediction. The ultimate goals of such instrumentation were safe and efficient excavation and minimization of movement through proper ground support.

The results of the movement monitoring (displacement) and stress modelling have been utilized in the design of mining method, rib design, backfill design, and detailed stress analysis quite successfully and implemented into practice largely for the safety of men and workplaces (Pakalnis, 2010).

Summary and conclusions

The very weak and low RMR rock of the gold-bearing lenses consists of weak and fractured siltstones, clays, mudstones, and highly altered limestones.

The effective management of the challenges contributed to a history of significantly increased production (Figure 6.58).

The *underhand cut-and-fill* or *box stoping* methods are ideal for the low dipping nature of the orebody pods and lenses. The basis of the ground control design in the weak ground at TR is the quick installation of reinforcement and retention elements, quality control and sequencing of backfill or CRF, and the minimization of rock damage from blasting. The emphasis is on safety first and production next.

The support elements of inflatable bolts and wire mesh in top cuts, as well as shotcrete in tertiary and development headings, are necessary to keep the rock from unravelling. In top cuts, the width of the openings is made as small as possible, with mesh on the ribs and back. This maximizes support and limits movement in the yielding low-strength rock masses. The mining cycle also induces an element of mining-induced stresses and changing strain or movement.

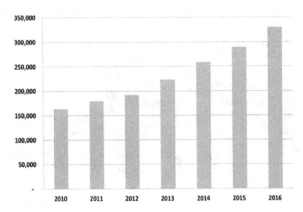

Figure 6.58 Trend of gold production at Turquoise Ridge underground mine (2010–2016).

Bolt testing in the ore bodies has measured bolt bold strength of only 3–6 t/m (1–2 t/ft), limiting the safety factor. The safety factor also depends on the shape of the opening, with arched backs potentially doubling safety factors over flat backs. Shotcrete may also be needed as we increase the top cut size beyond the 3 m × 3 m dimensions currently in use in the weak RMR type IV ground.

The high strength CRF or backfill strength is typically 2–3 times the design strength criteria of 600 psi for a 14-day test. Quality control of the aggregate sieve size and placement is important to maintain the integrity of the design.

Supplementary support in the weak ground is supplied by spiling or groutable spiling. Rehab can also be implemented with shotcrete arches or using "CRF arches" that incorporate the inherent backfill strength.

Instrumentation monitoring in the drift and the rock are necessary to calibrate the stress modelling to help understand the failure mechanisms in and around the orebody. We typically do not see stress buildup in the rock, but rather a yielding or unravelling rock mass. Movement is initially confined to the first metre (1–3 ft) in the rock, and most damage is observed in the back and spring line in the shotcrete–mesh shell of only 0.6″ thick. The 'shelling' effect of shotcrete and mesh moving over itself is due to horizontal convergence, and the differential movement induced from mining in the vicinity of the drift or at wider excavation intersections. Stress models provide estimates of predicted movement in different elastic moduli for different rock mass ratings.

Acknowledgements (for Case Study 6.6 only)

I would like to thank the Engineering Department at TRJV including Sandbak L.A. I would also like to extend special thanks to consultants Rimas Pakalnis and Rad Langston, and Shelia Ballantyne for their computer modelling insights.

Note

1 In geology, **Ga** is a common scientific abbreviation for *Giga-annum* derived from the Latin 'Giga' plus 'annum'. Commonly Ga signify the time before the present; e.g. 1 giga-annum = 1 billion years = 10^9 years.

Bibliography

Case Study 6.1

Saskatchewan. (2023). Wikipedia. https://en.wikipedia.org/wiki/Saskatchewan#Economy
Yancoal Canada - Southey Potash Project. (2006). https://youtu.be/EHz_iRjWGPU
Yancoal Canada Resource Co. Ltd. (Yancoal Canada). https://www.yancoal.ca/
Nutrien Company (https://www.nutrien.com/)

Case Study 6.2

Geonesis, Vol. 8, Issue 10, September 2021. p. 7. (http://www.geonesis.in)
Geonesis, Vol. 9, Issue 04, March 2022. pp. 5 and 6. (http://www.geonesis.in)
Global Mining Review. Lithium. (n.d.). https://www.globalminingreview.com/search/?q=lithium.
Haynes, W. M. ed. (2021), CRC *Handbook of Chemistry and Physics*, Boca Raton, FL: CRC Press/Taylor and Francis, 95th Edition; Internet Version 2015, Accessed on 29/09/2021.
http://www.lithiummine.com/lithium-mining; Accessed on 29/09/2021 and 30/09/2021.
Kaye, G. W. C. and Laby, T. H. (1995), Tables of Physical & Chemical Constants, 16th Edition, Version 1.0, Middlessex, England: National Physical Laboratory; Accessed on 29/09/2021.
Lithium Americas. (n.d.), https://www.lithiumamericas.com; Accessed on 30/09/2021.
Science News. (n.d.), https://www.sciencenews.org/; Accessed on 29/09/2021.
Upton, M., Longstaff, R., Sekar, S., and Figueiredo, J. (2021), Lithium: Shaping Market for a Green Future, *Global Mining Review*, March, pp. 10–14.

Case Study 6.3

Famur. (n.d.), www.famur.com; A company dealing with equipment for underground mining, Accessed on 12/2021.
Jastrzębskie Zakłady Remontowe w liczbach. www.jzr.pl; A Company: For powered supports used in underground mining, Accessed on 01/2022.
Kabiesz, J. (ed). (2021), Raport roczny o stanie podstawowych zagrożeń naturalnych i technicznych w gornictwie węgla kamiennego, Annual Report of Natural and Technical Hazards in Coal Mining. Katowice: Central Mining Institute (in Polish).
Wyższy Urząd Górniczy. (2021), Ocena stanu bezpieczeństwa pracy, ratownictwa gorniczego oraz bezpieczeństwa powszechnego w związku z działalnością gorniczo-geologiczną w 2020 roku [The Evaluation of Occupational Health, Mining Rescue and Common Security Regarding to Mining and Geological Activity in 2020]. Katowice: State Mining Authority (in Polish).

Case Study 6.4

CMPDIL. (2017), Summary of EIA for Public Hearing of Jayant (Expn.) Opencast – From 15.50 MTPA to 25.00 MTPA, Northern Coalfields Limited (NCL), September, p. 15; Accessed on 14/02/2022
MOC. (2021), Ministry of Coal Press Release of Press Information Bureau (PIB), New Delhi, Posted on: 13 Sept 2021. https://pib.gov.in/PressReleasePage.aspx?PRID=1754523; Accessed on 15/02/2022

Pandey, G., Choudhary, A. K., Singh, C. P., Kumar, M., Singh, P. K., and Mishra, A. K. (2017), Optimization of Blast Design Parameters for Reducing the Social and Economic Impacts during Operation in Large Opencast Mine: Case Study, In *Proceedings of the International Conference on NexGen Technologies for Mining and Fuel Industries*, Eds. Pradeep K. Singh et al., New Delhi, India: Allied Publishers Pvt. Ltd., Feb 15–17, 2017, Vol. 1, pp. 227–234.

Sharma, S., and Roy, I. (2015), Slope Failure of Waste Rock Dump at Jayant Opencast Mine, India: A Case Study, *International Journal of Applied Engineering Research*, Vol. 10, Issue 13, pp. 33006–33012.

Case Study 6.5

Burns, N., Davis, C., Diedrich, C., and Tagami, M. (2017), Salobo Copper-Gold Mine Carajas, Pará State, Brazil, A Technical Report Prepared by Wheaton Precious Metals, Rio de Janeiro, Brazil: Companhia Vale do Rio Doce (Vale SA), December, p. 126.

Burns, N., Davis, C., Diedrich, C., and Tagami, M. (2019), Salobo Copper-Gold Mine Carajas, Pará State, Brazil, A Technical Report – Salobo III Expansion, Prepared by Wheaton Precious Metals, Rio de Janeiro, Brazil: Companhia Vale do Rio Doce (Vale SA), December, p. 160.

Case Study 6.6

Deccan Herald. (2011), New Method to Recover Low-Grade Uranium from Tummalapalle Ore. https://www.deccanherald.com/content/150550/method-recover-low-grade-uranium.html

First Post. (2018), The Real Cost of Uranium Mining: The Case of Tummalapalle. https://www.firstpost.com/long-reads/the-real-cost-of-uranium-mining-the-Case-of-tummalapalle-4749521.html

Ghade, A., Rao, K. K., Vijay Kumar, K. S., Raju, G. D., Venkateswarlu, V., and Bharath, A. Y. Kumar. (2016), Rock Mechanics Investigations at Tummalapalle Uranium Mine of UCIL, In *International Conference on Recent Advances in Rock Engineering (RARE-2016)*, Bengaluru: Atlantis Press, pp. 501–504, DOI: 10.2991/rare-16.2016.80. https://www.researchgate.net/publication /311365083; Accessed on 04/10/2021.

MECON. (2006), A Project Document of Tummalapalle Uranium Mine of Uranium Corporation India Limited (UCIL), Volume – II, Chapter 3: Mining, Prepared by Metallurgical & Engineering Consultants Limited (MECON) - A government-owned engineering consultancy service provider, pp. 1–31.

Warland, Nicole G. (2022), Sustainable Sourcing to Meet Demand, *Global Mining Review (GMR)*, Vol. 5, Issue 3, April, pp. 33–35.

Case Study 6.7

Barton, M., Lien, R., and Lunde, J. (1974), Engineering Classification of Rock Masses for the Design of Tunnel Support, *Rock Mechanics*, Vol. 6, Issue 4, pp. 183–236.

Bieniawski, Z. T. (1976), Rock Mass Classification in Rock Engineering. In *Proceedings of the Symposium on Exploration for Rock Engineering*, Johannesburg, South Africa, pp. 97–106.

Holcomb, T. and Dorf, T. (2006), Geology and Mineral Reserves at Turquoise Ridge, SME Technical Session Slide Presentation, AIME-SME Annual Meeting, March 26–29, St. Louis, MI.

Langston, R. (2010), Analysis of Deformation in the HGB/Ozy Ramp Area-Turquoise Ridge Joint Venture Mine, Golconda, Nevada.

Pakalnis, R. (2009), Report on Mining under CRF; Turquoise Ridge Mine No. TQR-7/09.

Pakalnis, R. (2010), Report on Site Visit to Turquoise Ridge Mine No. TQR-9-11w Mining under CRF; Turquoise Ridge Mine No, TQR-7/09.

Rai, A. R., and Westhoff, D. (2013), Controlling Drift Profile in Mining Small Top Cut Panels-TRJV Case Study, In *Proceedings of the 47th US Rock / Geomechanics Symposium*, San Francisco, CA, June 23–26, 2013, Paper No. ARMA 13-136, 6 pp.

Sandbak, L. A., Rai, A. R., Howell, R. S., and Bain, N. G. (2012), Ground Support Strategies for Weak Ground Masses at the Turquoise Ridge Joint Venture, Nevada, In *Proceedings of the 46th US Rock Mechanics / Geomechanics Symposium,* Chicago, IL, June 24–27, 2012, Paper No. ARMA 12-288, 17 pp.

Chapter 7

Training and skill development for mining engineering

Selflessness is the only way to progress & prosperity.

(Bhagwad Gita)

Mining engineering is applied engineering wherein the importance of training, skill development, and human resource management has never lessened even in this information era of the internet (Figure 7.1). All technical studies have become part and parcel of the World wide web whether a piece of company information or a detailed research paper. It is also true that soft copy formats of both *data* and *information* are preferred over hard copy formats because they are easily accessible and retrievable. But the field knowledge of a real mine can't be accomplished without going to a mine site physically. Hence, hands-on learning at mine, periodical field visits to a mine for updation, and book reading can't be simply ignored. Concerning applied science, namely mining and geology, the mine is like a laboratory where practical field knowledge is available at a very low cost and with immense ideas for good growth potential.

Figure 7.1 Training and skill.

DOI: 10.1201/9781003274346-7

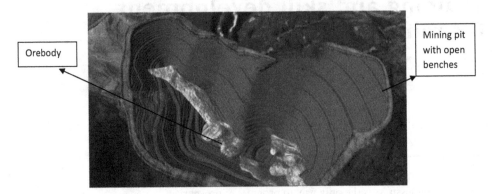

Orebody

Mining pit
with open
benches

Figure 7.2 A virtual mine with the ore body (orebody in the ground is highlighted with different colours for different minerals).

PART I: TRAINING AND SKILL DEVELOPMENT

7.1   Virtual mine for field training

Hands-on practical training in mines is a current practice to learn. To gain practical knowledge of mine, mining equipment, machinery, and instruments, the concept of *"Virtual Mine* for field training" has emerged. A *Virtual Mine* is a definitive master class for learners and beginners to help develop their professional skills (Figure 7.2). It is the future of mining teaching and learning. The blending of video-based technology and virtual communication can be leveraged for better, more productive, and more powerful learning. Since 'virtual learning' is today's reality, a *Virtual Mine* for field training becomes important for a mining engineer. Simulative environment conditions, as found in mines, are created in the virtual mine that represents the exact scenario as that of the actual mine. Both operative and non-operative conditions can be created through audiovisuals, models, and so on. This enables a win-win negotiable condition to be done in less time and at a lower cost.

In every mining engineering course, practical training and real exposure to the industry are compulsory be it at the undergraduate degree level or a post-graduate diploma level. More mines and case studies mean a better understanding of the practical concepts. Undoubtedly, training has its special importance in the technical-oriented engineering field and mining is no exception to it. Mining is such a subject where without field knowledge or mine practical visits, learning by rote is a very difficult task. All such practical pieces of training are participative involvement that stimulates our subject knowledge and makes us acquainted with the geographical area, people, environment, and society. Working through mining problems despite hardship or failure is essential for becoming a true mining professional.

Virtual mine is meant for the development of skills and teaching mining students besides acting as a tool or model for the wider society learning. It helps popularize mining among the general public and how minerals are extracted from Mother Earth. It helps in increasing the awareness among the local community about the importance of raw materials and their possible uses for better development of society or regions. Planned mining activities and practical applications of mineral extraction methodology can be taught easily and conveniently with the help of a Virtual mine.

One should not get confused by the terms 'Model mine' and 'Experimental mine', which we also come across in the field of mining. A virtual mine is a physical form of a real mine, whereas the experimental mine is meant for carrying out mining-related technical experiments in simulative conditions. By using advanced tools, underground mines, surface mines, mining machines, HEMM (heavy earth moving machine), mining equipment, and different mining and excavation operations can be simulated and training imparted irrespective of the mine sites (**https://5dt.com/**). For skill development, creativity, and entrepreneurship skill development, the Virtual mines are helpful. Since they apply mining in everyday life as the model of sustainable development, for vocational education/training and scientific informal education, the Virtual mines are a good facility. Thus, a Virtual mine is a public or exclusive entity that helps in skill development. Using new computer-based techniques and modern electronic gadgets, sensors, and artificial intelligence, the 21st-century Virtual mine can be made extremely attractive.

7.1.1   Mock rehearsal

The learning process of rescue, fire, gas-testing, and first-aid training for mines traverses through the mock rehearsal route. A replica of the mining equipment, machinery, and instruments can give an idea when not seen in actuality. *Virtual mode* and *mock rehearsal* together provide a real and convenient learning platform for the trainee. 'Learning by training and doing', through interactive sessions, thus has an important role in the human resource development and generation of a skill.

7.2   Skill development

Skill development is a continuous process. The past in the present continuous can be learnt easily and very well with skill development. The sectoral approach to skill development paves the way for improvement in the performance of the industry. It is recognized that the mismatch between theory and practice, between demand and supply can be easily filled with a skilled labour force. Thus, the role of skill development in improving quality of life, women empowerment, education, and labour reforms is possible through the skill development approach. Skill development is a niche area that not only imparts training but is also helpful for the government authorities in policy formulation. A focus on Integrating skill development with technology, newly researched methods, and economic considerations all lead to the development of institutional and regulatory mechanisms in any industrial sector, be it mining or otherwise. Therefore, skill development is a necessity for the mining industry globally, and in this, the role of the mine as a production unit and organization as an institution are extremely important constituents. To get an insight into the concept of skill development, specifically for the student readers, the following may be adopted as an antecedent.

7.2.1   Skill Council in Mining (SCM)

Skill development is the process of identification of the skills gap in youth and the workforce engaged in an industry. Providing skills, imparting training, and generating employment benefits for them is the basic goal of the *Skill Council*. In the mining sector, to provide skill development at the national/state level, setting up a government organization is one of the best ways for continual improvement. The core activity of the council can be set up to formulate the skill sets for occupational safety, health, environment, and industry-specific standards at the national level for different job roles aligned to the mining industry.

One such set-up is existing in India by the Ministry of Skill Development and Entrepreneurship (MSDE), Govt. of India, vide its notification dated 17 March 2015 which has authorized SCM as a non-statutory agency for certifying the mining workforce of the Indian mining industry. This Skill Council for Mining Sector (SCMS) is promoted by the Federation of Indian Mineral Industry (FIMI) and supported by the Ministry of Mines, Govt. of India, to train and meet the requirement of a skilled workforce for the mining industry in PPP mode.

The practical knowledge of mining methods and procedures, mining equipment, machinery, and instruments, all have been imparted under the guidance of experienced industry professionals. Different levels of professionals, from face workers to supervisors and managerial functionaries, can be trained under the skill council umbrella.

7.2.2   First aid for mines

A mine's vulnerability to accidents calls for training in First Aid. The First Aid is an essential element for the mine and mining premises. Everyone, right from the workforce to the top-level management, must be trained and equipped. Because of this vulnerability, the well-recognized 'St John Ambulance First Aid examination' has been made compulsory for all in the mining industry. This has served its very purpose of safety and care in the mining industry as a whole. This need and idea have impeccable results as well.

7.2.3   Gas-testing certification

In the case of gassy coal mines, where hazardous gases, namely, carbon monoxide (CO), methane, firedamp, and blackdamp, are commonly occurring, gas-testing certification has been made mandatory for workers as well as managers. Statuary provisions do exist in the coal mines' rules and regulations of many countries.

7.2.4   Fire and rescue training

Mining is a hazardous operation and rescue is a part of it. The unwarranted and accidental situations caused by mine accidents, mine fires, inundation, gas explosions, and mine disasters are likely to occur without any notice. Hence, skill development, training, and fire and rescue are a must from the industry perspective. For an underground mine, its gravity is more severe, and hence, precautionary measures and proactive approaches are always desirable and needed.

PART II: ADDS-ON USEFUL IN MINING ENGINEERING

The literal meaning of the word *'adds-on'* illustrates that one can keep on adding till the end hence limitless. As the name suggests, some adds-on exclusively useful in mining engineering have been segregated and compiled in this chapter to serve as a ready reckoner in a single place. We have not given an elaborate description/narration for each bulleted point as the written words are sufficient enough for understanding.

7.3   Concerned websites related to mining sector

a.   UNEP – United Nation Environment Programme (https://www.unep.org/)
b.   ILO – International Labour Organization (https://www.ilo.org/global/lang--en/index.htm)
c.   IIED – International Institute for Environment and Development (https://www.iied.org)
d.   ICMM – International Council on Mining and Metals
e.   IISD – International Institute for Sustainable Development
f.   Statuary authority of the respective country in which the mine is located
g.   Federal/State Govt. (Ministry of mines/coal/mineral resources) of the country concerned under which the administrative control of mine lies

Updated information Related to Mining Sector (*through the world wide web*)

1.   International mining websites, e.g. www.glacierrig.com/Infomine
2.   Mining magazine websites
3.   Global mining review websites
4.   Online electronic magazines and newsletters related to mining, geology, and exploration, e.g. Genonesis, India.
5.   Mineral Federations and their websites for the country concerned
     [*e.g. Federation of Indian Mineral Industry (FIMI), New Delhi, India*]
6.   Association/Chambers of commerce and industry (mineral sector) of the respective country in which the mine is located, e.g. ASSOCHAM, CII.

7.4   Mining companies/MDOs

All around the globe, every mine developer and operator called MDO, be it small, medium, or large, have to maintain their official website for the professional works with which they are engaged. It is now made mandatory by most countries and mining companies as well that the MDOs should be transparent in their approach to sharing and dissemination of information about their work and workforce. The mining companies' websites portray salient unrestricted information about the company for general use.

7.5   Mining software for technical solutions

(An exclusive list of computer software for mining engineering applications only)
The given list is a partial list of the most used software. The software field is a fast-changing field, and more and more new software keeps on adding to the mining field too.

i. Mine and civil survey:

- LISCAD software
- Terrain software
- Surfer software

ii. Geological modelling, mine planning, and designing:

- GEMCOM – Geologic modelling software that has drill hole data input as well as sectional and 3D modelling capabilities
- Minex – A software package for geological modelling and mine planning for coal mines Minex
- MineScape – Coal mine planning and geological assessment software.
- SURPAC software (GEOVIA)
- VULCAN 3D – Geologic modelling software of *Maptek Home*
- DATA MINE-3D – Geologic modelling software, stereonet plots, data log (drill hole) logs
- Geo Soft – A software for rock block construction, rockfall, cluster analysis, seismic amplification, and rock mass classification
- MICROMINE – A software that provides exploration field data analysis for geological resource estimation. Modules of this software are available for mine design, mine planning, mine production management systems, and data management systems
- HEXAGON – For digital mining solutions in surveying, mine design, fleet management, production optimization, etc. (https://hexagonmining.com/)

iii. Drill hole logging:

- ROCKWARE – A drill hole logging programme with the module name LOGPLOT
- GEOSYSTEM – A drill hole logging software
- DATA MINE – A module of this software can analyse the data of drill hole logs

iv. Slope stability for surface mines:

- GEOSLOPE – A 2D slope stability programme with probabilistic capabilities (SLOPE/W), rock wedge stability and slope wall stability, slope indicator data reduction, etc
- GEOSYSTEM software
- GALENA – A software for slope stability analysis
- FLAC Slope – A numerical modelling software for slope design, analysis, and safety management
- INTERSTUDIO SRL – For circular failure analysis (Macintosh-based software)
- TAGA SOFT – Slope stability analysis and seismic analysis (2D and 3D; irregular solid slope)

In addition, several other software, namely ROCSCIENCE, DIPS, UN-WEDGE, and GEOROK, are also used in slope design and analysis for mines. Depending on the technical requirement of the analysis, the software is selected and used.

v. Numerical modelling and rock mechanics analysis software for mining excavations:

- FLAC 3D, 2D 3DEC, UDEC Numerical codes (ITASCA consulting group).
- DYNAFLOW – FEM-based 2D–3D analysis (A finite element programme by Princeton University)
- SAGE CRISP – An advanced geotechnical finite element analysis programme that is Windows-based and developed by Cambridge University
- ROCSCIENCE Inc. – A geomechanics software with multiple modules (For research and practical application in mines and other underground excavations; or the finite element based)
- PLAXIS 2D, 3D – FEM-based geotechnical package for stress and strain analysis including the mining dynamics. The *PlaxFlow* and *Thermal* module of the package is a powerful and user-friendly finite element package intended for 2D and 3D analyses of deformation and stability in geotechnical engineering and rock mechanics. PLAXIS is used worldwide by top engineering companies and institutions in the civil, mining, and geotechnical engineering field in both industry and academia. Applications range from excavations, embankments, and foundations to tunnelling, mining, and reservoir geomechanics (https://www.plaxis.com)
- ARMPS: For analysis of retreat mining pillar stability (A software for pillar design)

vi. Tunnelling and underground excavations:

- ROCSCIENCE – A finite element code (software) for geomechanics analysis.
  Software named RS2 from RocScience, Canada, is the most commonly used tunnel design software which is user-friendly. It was called Phase2 in the past. With more features including 3D analysis, RS3 is also available. These are FEM-based software. There is other software from RocScience, e.g. Roc-Support, EX3, and RS Data. Dips (for stereonet analysis) and UnWedge (for wedge failure analysis) as mentioned in the software list of this section are also available for tunnelling and tunnel engineering applications (RocScience.com)
- GEOROK – A Windows-based geotechnical software package for tunnelling, underground excavations, and rock mechanics design, testing, monitoring, and support. The complete suite of 12 programmes/packages also includes rock–soil interaction, rock mass classification, stereonets, wedge analysis, and rock blasting analysis
- Tunnel design software: (*i*) *FLAC2D and FLAC3D (ITASCA)* – A finite difference method (FDM)-based complete software package useful for civil engineering problems. This ITASCA software is a better option for continuum modelling. FLAC or any continuum-based models are best suited when you don't have to explicitly model the joints. Few joints can be modelled in FLAC using interfaces, but FLAC does not allow slip along the joints. (*ii*) *UDEC and 3DEC (ITASCA)* – This FDM-based software simulate the tunnels in the discontinuous media. When the predominant behaviour is controlled by joints/

discontinuities, UDEC offers the best features for multiple joint modelling and allows slip and discontinuity in deformations. However, the solution scheme and modelling are relatively complicated in UDEC. (*iii*) *PLAXIS 3D:* The word PLAXIS stands for the 'plane strain and axial symmetry'. This software is based on FEM analysis and was developed in the Netherlands. In this software, you can model a tunnel in 3D space and there are many soil behaviour features in this software that have different versions. However, the PLAXIS3D tunnel software has some marked deficiencies, which makes it an incomplete package for tunnel modelling and other special cases such as a shaft or other underground excavations

- Particle Flow Code (PFC) – A software based on the DEM method (Discrete element method software package)
- ANSYS – A software based on carrying out finite element analysis
- ABACUS – A FEM-based software
- GTS software – A software of NX Midas, South Korea
- C-Tunnel software – A software of ITECH & CETU, France
- GEO5 (FEM Method) – The GEO5 FEM is used to compute the displacements, internal forces in structural elements, stresses and strains, and plastic zones in the soil and other quantities in every construction stage. With extension modules, the programme also performs the Tunnel excavation analysis, the steady-state or transient Water Flow analysis, or the coupled Consolidation analysis
- DeepEX – A finite element software for tunnels, USA
- TUNA - Linear Tunnel Analysis Programme. TUNA employs a static, two-dimensional, linear elastic finite element method. To get details, visit www.geotechpedia.com

vii. Ventilation Planning of underground mines:

- VENTSIM software
- VENTPC software

viii. Expert Systems for mining:

- SPONCOM software – Spontaneous Combustion assessments (Prediction software for potential assessment in an underground coal mine)

ix. Virtual reality in mining:

- Equipment simulators and virtual reality in mining engg. (https://5dt.com/)

x. Soil mechanics/Soil investigation related:

Many civil engineering firms/companies around the globe make soil mechanics-related software and programmes for soil bearing capacity, elastic stress, load analysis, etc., based on the number of available and proven methods namely the limit equilibrium method. Both 2D and 3D analyses of soil are performed for mining and tunnelling applications involving geotechnical engineering, soil mechanics, and rock mechanics.

- SOIL VISION softwar
- SLIDE software
- EJGE software
- GEOS software
- GGU software

7.6   Some unit conversions useful for mining engineering application

Unit conversion tables given here (Table 7.1 to 7.8) including Figure 7.3 are suitably tailored for mining-specific applications.

Table 7.1 Area

Area					
1 km^2	=	100 hectares (ha)	1 ha	=	$10,000 \text{ m}^2$
1 km^2	=	$1,000,000 \text{ m}^2$	1 acre	=	0.40469 ha
1 ha	=	2.47105 acres	1 mile	=	1.60934 km
1 ha	=	0.00386 miles^2	1 cm^2	=	100 mm^2
1 mile^2	=	258.99881 ha	1 m^2	=	10.76391 ft^2
1 m^2	=	$10,0000 \text{ mm}^2$	1 m^2	=	$10,000 \text{ cm}^2$

Table 7.2 Weight

Weight					
1 g	=	0.001 kg	1 g	=	100 centigram
1 g	=	1000 mg	1 centigram	=	0.01 g
1 kg	=	1000 g			

Table 7.3 Volume

Volume					
1 litre (L)	=	100 c L	1 K L	=	1000 L
1 mL	=	1 cc	1 gal	=	16 cc
1 mL	=	0.001 L			

Table 7.4 Length

Length					
1 centimetre (cm)	=	10 mm	1 metre (m)	=	100 cm
1 metre (m)	=	1,000 mm	1 metre (m)	=	3.3 ft.
1 kilometre (km)	=	1,000,000 mm	1 kilometre (km)	=	100,000 cm
1 inch (in)	=	25.4 mm		=	

Foot, yard, and inch conversions					
1 cubic foot	=	1,728 cubic inches	1 sq.yard	=	9 sq.ft
1 cubic yard	=	46,656 cubic inches	1 acre	=	4,840 sq. yards
1 cubic yard	=	27 cubic ft.	1 sq.mile	=	640 acres
1 cubic inch	=	1 6.38706 cubic cm	1 cubic metre	=	35.31467 cubic ft.
1 cubic ft.	=	0.02832 cubic metres	1 cubic cm	=	0.06102 cubic inches
1 cubic yard	=	0.76455 cubic metres	1 sq.inch	=	6.4516 sq.cm
1 sq.ft.	=	144 sq. inches	1 cu metre	=	1.30795 cubic yards
1 sq.yard	=	1,296 sq. inches	1 sq.ft.	=	0.09290 sq.m
1 sq.ft.	=	929.0304 sq.cm	1 sq.yard	=	0.83613 sq. m
1 inch (in)	=	2.54 cm	1 inch	=	2.54 cm

Table 7.5 Temperature

Temperature				
°C (Degree Celsius) [Celsius / Centigrade]	=	5/9 (°F-32)	K (Kelvin)	=	°C +32
°F (Degree Fahrenheit)	=	9/5(°C) +32	°K (Degree Kelvin)	=	5/9 (°F-32) + 273

Table 7.6 Mass

Mass				
I kg	=	2.2046 lb	I lb	=	16 oz
I lb	=	0.454 kg	I oz	=	28.35 g

Table 7.7 Energy

Energy				
I Joule (J)	=	2.39×10^{-4} kcal	I cal	=	4.184 J
I eV	=	1.602×10^{-19} K		=	

Table 7.8 Pressure

Pressure				
I atmosphere	=	101.3 k Pa	I Pa (Pascal)	=	$I N/m^2$
I atmosphere	=	760 mm Hg	I Pa	=	$I kg/m- s^2$
I bar	=	10 5 Pa	I psi	=	6.893 Pa
I MPa	=	10.1971 kgf /cm^2	I kgf /cm^2	=	0.09806 MPa

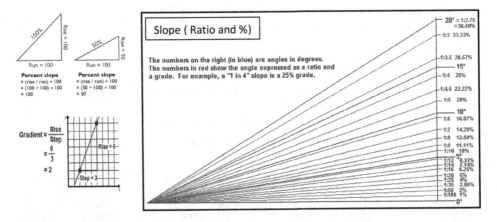

Figure 7.3 Slope/gradient.

Note: Some of the unit conversions given here for mining engineering applications can be extended by the reader easily depending on his/her requirement.

References

Appelo, C. A. J. and Postma, D. (2005), *Geochemistry, Groundwater and Pollution*, 2nd Edition, Rotterdam, the Netherlands: A.A. Balkema, p. 649.

Bandopadhyay, L. K., Chaulya, S. K., and Mishra, P. K. (2010), *Wireless Communication in Underground Mines – RFID Based Sensor Networking*, New York: Springer Publication, p. 477.

Bello, A. A. (2013), *Introductory Soil Mechanics*, 1st Edition, Lagos: Tony Terry Prints, p. 117.

Bieniawski, Z. T. (1989), *Engineering Rock Mass Classification*, New York: Wiley.

Brady, B. H. G. and Brown, E. T. (2005), *Rock Mechanics for Underground Mining*, 3rd Edition, Amsterdam, the Netherlands: Kluwer Academic, p. 614.

Bustillo Revuelta, M. (2018), Mineral Deposits: Types and Geology. In: *Mineral Resources*, Cham: Springer Textbooks in Earth Sciences, Geography and Environment, pp. 49–119. DOI: 10.1007/978-3-319-58760-8_2.

Canter, L. W. (1996), *Environment Impacts Assessment (Civil Engineering Series)*, New York: McGraw Hill Inc., p. 660.

CBIP. (2010), *Manual on Rock Mechanics*, Publication No 310, Edited by T. Ramamurthy, KG Sharma, AC Gupta and ML Baweja, New Delhi: Central Board of Irrigation and Power (CBIP), p. 585.

Deshmukh, D. J. (2001), *Elements of Mining Technology*, Vol. 1, Nagpur, India: Central Techno Publication, p. 423.

Dwivedi, R. D. and Soni, A. K. (2021), Eco-friendly Hill Mining by Tunnelling Methods, In: *Mining Technique -Past, Present and Future*, edited by Abhay Soni, London: IntechOpen. DOI: 10.5772/intechopen.95918.

Dwivedi, R. D., Soni, A. K., Goel, R. K., and Dube, A. K. (2000), Fracture Toughness of Rocks under Sub-Zero Temperature Condition, *International Journal of Rock Mechanics and Mining Sciences*, Vol. 37, pp. 1267–1275.

Goel, R. K., Jian Zhao, and Bhawani Singh. (2012), *Underground Infrastructures: Planning, Design, and Construction*, Amsterdam, the Netherlands: Elsevier Science & Technology, p. 352.

Hartman, H. L. and Mutmansky, J. M. (1987), *Introductory Mining Engineering*, New York: Wiley-Interscience, p. 570.

Hoek, E. (2007), *Practical Rock Engineering*, 2nd Edition, Toronto, Canada: Roc Science Inc.

Hoek, E. and Bray, J. W. (1981), *Rock Slope Engineering*, London: Institution of Minerals and Metals.

Hudson, J. A. and Harrison, J. P. (2005), *Engineering Rock Mechanics: An Introduction to the Principles*, 4th Edition, Oxford: Pergamon Press (Elsevier), p. 441.

Hustrulid, W. A. (2005), *Blasting Principles for Open Pit Mining: Theoretical Foundation* (Two Volumes), 1st Edition (Paperback), Boca Raton, FL: CRC Press.

Hustrulid, W. A. and Kuchta, M. (2006), *Open Pit Mine Planning and Design* (Two Volumes), 3rd Edition, Boca Raton, FL: Taylor & Francis Publications, CRC Press.

Jaeger, J. C., Cook, G. W., and Zimmerman, R. (2007), *Fundamentals of Rock Mechanics*, 4th Edition, Hoboken, NJ: John Wiley & Sons Ltd, p. 488.

Mathur, S. P. (1999), *Coal Mining in India*, Bilaspur, India: M.S Enterprises Publication, p. 612.

More, K. S., Wolkersdorfer, C., Kang, N., and Elmaghraby, A. S. (2020), Automated Measurement Systems in Mine Water Management and Mine Workings – A Review of Potential Methods, *Water Resources and Industry*, Vol 24, December, p. 12. DOI: 10.1016/j.wri.2020.100136.

National Mineral Policy (NMP). (2019), Ministry of Mines, Government of India, p. 12.

NTS. (2014), *Norwegian Tunnelling Technology*, Publication No. 23, Oslo, Norway: Norwegian Tunnelling Society (NTS/NFF). http://www.tunnel.no

Porathur, J. L., Roy, P. P., Shen, B., and Karekal, S. (2017), *Highwall Mining: Applicability, Design and Safety*, London UK: CRC Press (Taylor & Francis), p. 323.

Raina, A. K., Soni, A. K., Vajre, R. and Sangode, A. G. (2022), Impact of Topography on Noise Attenuation from a Cement Plant in a Hilly Terrain, *Science of the Total Environment*, Vol. 835, 155532. DOI: 10.1016/j.scitotenv.2022.155532.

Rai, S. S. (2020), Digital Transformation for Improving the Productivity of Mining – An Approach, *MGMI News Journal*, Vol. 46, No. 3, pp. 34–41.

Ramamurthy, T. (2007), *Engineering in Rocks for Slopes, Foundation and Tunnels*, New Delhi: Prentice -Hall of India Private Limited, p. 731.

Sawyer, C. N., Mccarty, P. L., and Parkin, G. F. (2003), *Chemistry for Environmental Engineering and Science*, 5th Edition, Noida, India: Tata McGraw-Hill Education Pvt. Limited, p. 752.

Sheorey, P. R. (1997), *Empirical Rock Failure Criteria*, 1st Edition, Rotterdam, the Netherlands: A.A. Balkema.

SME. (1998), *Mining Engineering Handbook of Society of Mining Engineers (SME)*, 2nd Edition, Available in CD-ROM (Multiple Volumes), Published by a Professional Society (Not a single author; SME as publisher), https://www.smenet.org/.

Soni, A. (2021), *Mining Technique – Past, Present and Future*, London: Intechopen, 130 pages.

Soni, A. and Nema, P. (2021), *Limestone Mining* in India, Singapore: Springer (Materials Horizons: From Nature to Nanomaterials series), p 180.

Soni, A. K. (2017), *Mining in the Himalayas – An Integrated Strategy*, Boca Raton, FL: CRC Press (Taylor & Francis); 225 pages. DOI: 10.1201/9781315367552.

Soni, A. K. (2020), History of Mining in India, *Indian Journal of History of Science (IJHS)*, Vol. 55, Issue, pp. 218–234. DOI: 10.16943/ijhs/2020/v55i3/156955.

Tatiya, R. R. (2005), *Surface and Underground Excavations –Methods, Techniques and Equipment*, Rotterdam, the Netherlands: A.A. Balkema Publishers, Taylor & Francis Group, p. 595.

Tatiya, R. R. (2005), *Surface & Underground Excavations – Methods, Techniques and Equipment*, Rotterdam, the Netherlands: A.A. Balkema, Taylor & Francis, p. 562.

Umathay, R. M. (2002), *Textbook of Mining Geology*, Nagpur, India: Dattsons Publishers Paperback Edition, p. 230.

WMD. (2021), *World Mining Data (WMD)-2021, A Brochure of World Data on Iron and Ferro-Alloy Metals, Non-Ferrous Metals, Precious Metals, Industrial Minerals and Mineral Fuels*, Volume 36, Vienna, A document of Austrian Federal Ministry of Agriculture, Regions and Tourism in Association with the International Organizing Committee of World Mmining Congress, Prepared by C. Reichl and M. Schatz. https://wmc.agh.edu.pl/world-mining-data-2021.

Wyllie, D. C. and Mah, C. W. (2005), *Rock Slope Engineering: Civil and Mining*, 4th Edition, London: Spon Press (Taylor & Francis), p. 431. (An improved version of *Rock Slope Engineering by* Evert Hoek & John Bray, 3rd Edition, 1981.)

Index

Note: **Bold** page numbers refer to tables and *italic* page numbers refer to figures.

Printed in the United States
by Baker & Taylor Publisher Services

Printed in the United States
by Baker & Taylor Publisher Services